INTERNATIONAL SERIES OF MONOGRAPHS IN
NATURAL PHILOSOPHY
GENERAL EDITOR: D. TER HAAR

VOLUME 56

THE LOGICAL ANALYSIS OF QUANTUM MECHANICS

OTHER TITLES IN THE SERIES IN NATURAL PHILOSOPHY

Vol. 1. DAVYDOV—Quantum Mechanics
Vol. 2. FOKKER—Time and Space, Weight and Inertia
Vol. 3. KAPLAN—Interstellar Gas Dynamics
Vol. 4. ABRIKOSOV, GOR'KOV and DZYALOSHINSKII—Quantum Field Theoretical Methods in Statistical Physics
Vol. 5. OKUN'—Weak Interaction of Elementary Particles
Vol. 6. SHKLOVSKII—Physics of the Solar Corona
Vol. 7. AKHIEZER *et al.*—Collective Oscillations in a Plasma
Vol. 8. KIRZHNITS—Field Theoretical Methods in Many-body Systems
Vol. 9. KLIMONTOVICH—The Statistical Theory of Non-equilibrium Processes in a Plasma
Vol. 10. KURTH—Introduction to Stellar Statistics
Vol. 11. CHALMERS—Atmospheric Electricity (2nd Edition)
Vol. 12. RENNER—Current Algebras and their Applications
Vol. 13. FAIN and KHANIN—Quantum Electronics, Volume 1—Basic Theory
Vol. 14. FAIN and KHANIN—Quantum Electronics, Volume 2—Maser Amplifiers and Oscillators
Vol. 15. MARCH—Liquid Metals
Vol. 16. HORI—Spectral Properties of Disordered Chains and Lattices
Vol. 17. SAINT JAMES, THOMAS and SARMA—Type II Superconductivity
Vol. 18. MARGENAU and KESTNER—Theory of Intermolecular Forces (2nd Edition)
Vol. 19. JANCEL—Foundations of Classical and Quantum Statistical Mechanics
Vol. 20. TAKAHASHI—An Introduction to Field Quantization
Vol. 21. YVON—Correlations and Entropy in Classical Statistical Mechanics
Vol. 22. PENROSE—Foundations of Statistical Mechanics
Vol. 23. VISCONTI—Quantum Field Theory, Volume 1
Vol. 24. FURTH—Fundamental Principles of Theoretical Physics
Vol. 25. ZHELEZNYAKOV—Radioemission of the Sun and Planets
Vol. 26. GRINDLAY—An Introduction to the Phenomenological Theory of Ferroelectricity
Vol. 27. UNGER—Introduction to Quantum Electronics
Vol. 28. KOGA—Introduction to Kinetic Theory: Stochastic Processes in Gaseous Systems
Vol. 29. GALASIEWICZ—Superconductivity and Quantum Fluids
Vol. 30. CONSTANTINESCU and MAGYARI—Problems in Quantum Mechanics
Vol. 31. KOTKIN and SERBO—Collection of Problems in Classical Mechanics
Vol. 32. PANCHEV—Random Functions and Turbulence
Vol. 33. TALPE—Theory of Experiments in Paramagnetic Resonance
Vol. 34. TER HAAR—Elements of Hamiltonian Mechanics (2nd Edition)
Vol. 35. CLARKE and GRAINGER—Polarized Light and Optical Measurement
Vol. 36. HAUG—Theoretical Solid State Physics, Volume 1
Vol. 37. JORDAN and BEER—The Expanding Earth
Vol. 38. TODOROV—Analytical Properties of Feynman Diagrams in Quantum Field Theory
Vol. 39. SITENKO—Lectures in Scattering Theory
Vol. 40. SOBEL'MAN—Introduction to the Theory of Atomic Spectra
Vol. 41. ARMSTRONG and NICHOLLS—Emission, Absorption and Transfer of Radiation in Heated Atmospheres
Vol. 42. BRUSH—Kinetic Theory, Volume 3
Vol. 43. BOGOLYUBOV—A Method for Studying Model Hamiltonians
Vol. 44. TSYTOVICH—An Introduction to the Theory of Plasma Turbulence
Vol. 45. PATHRIA—Statistical Mechanics
Vol. 46. HAUG—Theoretical Solid State Physics, Volume 2
Vol. 47. NIETO—The Titius–Bode Law of Planetary Distances: Its History and Theory
Vol. 48. WAGNER—Introduction to the Theory of Magnetism
Vol. 49. IRVINE—Nuclear Structure Theory
Vol. 50. STROHMEIER—Variable Stars
Vol. 51. BATTEN—Binary and Multiple Systems of Stars
Vol. 52. ROUSSEAU and MATHIEU—Problems in Optics
Vol. 53. BOWLER—Nuclear Physics
Vol. 54. POMRANING—Radiation Hydrodynamics
Vol. 55. BELINFANTE—A Survey of Hidden Variable Theories
Vol. 56. SCHEIBE—The Logical Analysis of Quantum Mechanics

THE LOGICAL ANALYSIS OF QUANTUM MECHANICS

by

ERHARD SCHEIBE
University of Göttingen, Germany

Translated by
J. B. SYKES

PERGAMON PRESS
OXFORD · NEW YORK
TORONTO · SYDNEY · BRAUNSCHWEIG

Pergamon Press Ltd., Headington Hill Hall, Oxford
Pergamon Press Inc., Maxwell House, Fairview Park, Elmsford, New York 10523
Pergamon of Canada Ltd., 207 Queen's Quay West, Toronto 1
Pergamon Press (Aust.) Pty. Ltd., 19a Boundary Street, Rushcutters Bay, N.S.W. 2011, Australia
Vieweg & Sohn GmbH, Burgplatz 1, Braunschweig

First edition 1973

Library of Congress Cataloging in Publication Data

Scheibe, Erhard.
The logical analysis of quantum mechanics.

(International series of monographs in natural philosophy, v. 56)
Bibliography: p. 196.
1. Quantum theory. I. Title.
QC174.1.S33 1973 530.1'2 73–652
ISBN 0–08–017158–3

Printed in Great Britain by Bell and Bain Ltd., Glasgow

TO MY WIFE

Contents

Contents

Introduction

When a scientific book is placed before the public, the author finds himself finally, after all the labour of composition, faced with the worrying question why the book had to be in that form and no other. There will certainly be cases where the question is easily or routinely answered, or does not arise at all. It may happen, for instance, that there is something essentially novel to be said, or that the discussion relates to a current research problem, or that a first general review is to be given on a topic previously accessible only through scattered articles in the literature, or that the book is simply a textbook. Such cases afford an easy justification before the public to whom the books are offered, and the fact that this is so does not even need explaining nowadays. The present book, unfortunately, does not come under any of the obvious types of contemporary scientific publication. Its aim was in the first place a fairly modest one; later, one which proved to be much too ambitious; and the outcome is a compromise, determined largely by extraneous circumstances, between these extremes, though one which seems all too dubious to me now that I have been a witness of its evolution. In this introduction I want to try to state what the reader ought to know and bear in mind, before he proceeds to the subsequent discussion with the hope of profiting therefrom.

The book deals with non-relativistic quantum mechanics. This physical theory arose in 1925–7 from earlier very fragmentary approaches to the quantum theory, and it underwent a remarkable development, in the following respects. Firstly, although it is an adequately complete theory as regards its field of application, as soon as it had been constructed it was thought to be in need of extension, since it did not take account of two important topics: the quantization of classical fields, and relativity. Thus, in addition to its application in atomic physics and especially in nuclear physics, it was soon the subject of a great variety of attempts to enlarge it in these two directions. This extension of non-relativistic quantum mechanics to form a relativistic quantum theory, using the field concept, is of course a purely physical development, which moreover cannot yet be regarded as completed. Secondly, non-relativistic quantum mechanics itself led to quite deep and continually revived debates of a more philosophical type regarding certain epistemological and ontological principles of the older classical physics, since it seemed that these principles were no longer tenable. These debates have a significance to physics which is by no means negligible, despite their philosophical flavour, since they essentially concern directly the question of the content of the new mechanics. Like the purely physical further development of quantum mechanics, they are today, more than forty years after the theory was established, still far from being settled. There is thus the remarkable situation that a vigorous discussion continues among those more inclined to the philosophical type of basic research, regarding the content of a theory that has long been left behind by the main stream of research in theoretical physics.

In historical retrospect, this argument can be roughly divided into two stages. First came the purely physical setting-up of quantum mechanics, together with an epistemological basis, the latter being usually referred to as the "Copenhagen interpretation of quantum mechanics". I shall prefer to use the somewhat more neutral and more comprehensive term "orthodoxy". As will be described later, this does not imply that there is a single doctrine. But the basic ideas set down in the crucial papers by Bohr (1925, 1927, 1929a, b, c), Heisenberg (1930), Born and Jordan (1930), Dirac (1930), von Neumann (1955; German original published in 1932) and Pauli (1933) can with some justice be together described as an "orthodoxy" and regarded as being of decisive importance to the whole subsequent development of the discussion of principles. Jammer (1966) has given an excellent and detailed account of both the physical and the epistemological development of quantum mechanics during the early period from 1925 to 1935, and his book should always be borne in mind as a reference.

The proceedings of the Fifth Solvay Congress in 1927 (Solvay 1928) already show that the orthodox view was disputed from the start, in particular by physicists such as Einstein, de Broglie and Schrödinger, whose purely physical investigations made a vital contribution to the origin of quantum mechanics. In addition, from the more philosophically oriented side there was early evidence of dissatisfaction with certain basic features of the orthodox view, e.g. from Margenau (1931, 1932) and Popper (1935). This initial criticism of orthodoxy was, however, unsuccessful. Although the debate between Bohr and Einstein (1949, pp. 199 ff., 666 ff.) acquired a certain celebrity, it must be admitted that in its first quarter of a century the orthodox view largely held the field, and in particular was accepted in the textbooks.

In the early fifties, however, the scene changed, the initial impulse probably being due mainly to a paper by Bohm (1952), which for the first time propounded a reasonable classical theory of hidden parameters for quantum mechanics. This opened up a line of research which made rapid progress and led to a variety of proposals for theories of hidden parameters of the classical and in particular the deterministic type. The significance of this new trend is very instructively shown by its effect on a man like de Broglie, who had originated the idea of the wave nature of matter, but who did not manage to enforce on orthodoxy his own further development of the idea as the pilot-wave theory (Solvay 1928); he had therefore abandoned it and accepted orthodoxy, but now, twenty-five years later, returned to it with renewed enthusiasm (de Broglie 1953). It is impossible to survey rapidly the various causal interpretations of quantum mechanics. The ground is roughly mapped out by de Broglie (1960), Takabayashi (1952, 1953), Vigier (1956), Jánossy (1952), Jánossy and Ziegler (1963), Fényes (1952), Weizel (1953a, b, 1954) and Wiener, Siegel, Rankin and Martin (1966), to mention only some of the more important publications. Freistadt (1957) has given a review of the theories of de Broglie, Bohm and Takabayashi. Bohm also has developed his original theory in a series of papers, and the most recent situation is described by Bohm and Bub (1966). There have been isolated orthodox replies by Pauli (1953) and Heisenberg (1955). But physicists are not generally aware of this literature, with the possible exception of Bohm's historic paper of 1952. This is not surprising, in that none has yet led to any purely physical results. It must be remembered, however, that the authors mentioned were usually not directly seeking any such results; they wished rather to show that the orthodox interpretation of quantum mechanics is not, as it for a while seemed to be and no doubt was by many believed to be, the only possible one.

In addition to the more or less direct critique of orthodoxy, involving the establishment of classical theories of hidden parameters, or at least causal interpretations of quantum mechanics, there is another critical approach which, together with the first, forms the second and much more lively stage in the development of the philosophical problems raised by quantum mechanics. The authors representing this approach include Margenau (1963) and Popper (1967), already mentioned in connection with the early period, and also Landé (1965), Bopp (1961), Bunge (1967), Feyerabend (1962) and others. They do not seek to return to the concepts of classical physics. They agree with the orthodox view to the extent that quantum mechanics is an essentially new type of physics, and are therefore closer to orthodoxy than are the "determinists". The criticism which they nevertheless put forward is difficult to reduce to a common denominator, since the positive counter-proposals which occur in each instance are quite noticeably different. It is perhaps not an over-simplification if the following brief description of their common features is used. The two most striking points in the orthodox formulation of quantum mechanics are the occurrence of probabilities for the description of states and the explicit reference to measurements. These two non-classical constituents of quantum mechanics are regarded by orthodoxy as simply irreducible. Whereas the "determinists" seek to eliminate both constituents, the group now under consideration directs its fire mainly at the assertion that measurements cannot be excluded, and at its philosophical implications, whereas in respect of the probabilities there is a tendency to give their independent occurrence even more emphasis than orthodoxy did. Thus there is a shift of stress from measurements to probabilities, in comparison with orthodoxy, which actually gives only a half-hearted treatment of the probabilistic nature of quantum mechanics, and emphasizes the independence of the quantum-mechanical measuring process. Another characteristic is firm rejection of the subjective interpretations which offer themselves for the probabilities as an expression of lack of knowledge, and for the measurements as the only road to the acquisition of knowledge. Whether or not orthodoxy has in fact followed this subjectivist direction, it has often been understood in that sense, and here it is opposed by a treatment having a strictly objectivist approach.

These two principal directions by no means exhaust the range of critical activity relating to the orthodox interpretation of quantum mechanics during the last twenty years. More extensive reviews are given by Heisenberg (1955), Bunge (1956), Polikarow (1962) and d'Espagnat (1965); see also the bibliography by Scheibe (1967) on the whole subject of fundamental problems in quantum mechanics. One must at least refer to the many papers, some of them full of technical details, regarding the quantum-mechanical process of measurement, which also for the most part involve some degree of criticism of the orthodox view; see the asterisked items in the References. But the reader's impression of the present importance of basic research in quantum mechnaics, gained from the explicit references in the two preceding paragraphs to publications based on almost two hundred papers, may suffice to clarify the aims of the present book. I do not seek to participate directly in the debate that has arisen, but only to elucidate some questions referring to the orthodox view of quantum mechanics, in order thereby to enable the reader himself to take part in that debate. That is, the immediate question is: *what* is being debated?

The first possibility here, of course, would be a prescription apparently unsurpassable: to study thoroughly the works of the orthodox school (above all, those of Bohr, Heisenberg and von Neumann); then, equally thoroughly, those of the critics. This will show what the debate is about. Each person would have to traverse such a path for himself, and it is the only one which affords complete certainty in assessing the situation, but it is a thorny path, where help is welcome. Orthodoxy is not without obscurity, ambiguity and incompleteness; the opposing theories, especially the second group mentioned above, are not without gaps, misunderstandings and prejudices. These circumstances make it difficult to investigate the present situation in research. There are several ways of seeking to alleviate the difficulties. The customary one at present is to make a comparison of the two sides' views while being both as objective as possible and yet critical. I considered using this approach, but my investigations have led me in a different direction and become so extensive that such a plan could not be carried out in the present book. Even the analysis of orthodox quantum mechanics becomes considerably expanded if an exact study is made and an introductory account, even at a high level, is to be given. Consequently, there is here no considered presentation of orthodoxy, giving equal attention to all the important problems. Instead, only a few groups of problems are treated, though these *are* fully discussed; and there is also some lack of balance as regards methods.

The very first chapter occupies an unusual position, though largely because of its subject, namely Bohr's views on quantum mechanics. It has long been acknowledged that these have a special position relative to the other basic treatments of quantum mechanics, and even among the orthodox treatments. It can be said of Bohr, if it can bc said of anyone, that his papers on complementarity are no more than leisure reading even for the physicist with a fairly deep interest in the fundamentals of the science. One reads them, and the next day one goes on with quantum theory. My own commitment to Bohr's ideas, too, is the result not of a spontaneous feeling but of a self-imposed decree that I should spend at least a certain period in reading and pondering on Bohr's publications, not only in the evening but the whole day through. This, of course, does not qualify me to act as Bohr's interpreter. I can therefore offer no more than a kind of philological study such as results from asking *precisely what* was intended by this and that passage; but this may be just what we need in contemplating Bohr's work today. At least I may presume that anyone who is also faced with the question just stated will be able to use the first chapter of this book as a guide through the mansion of Bohr's thought.

The approach used is described in the introduction to Chapter I and still more clearly in section 1 of the chapter. The starting point is not, as it usually is, the concept of complementarity. The emphasis is instead placed on the *argumentational* aspect of Bohr's discussion, which has not previously been considered in the secondary literature: his fairly clear intention to arrange his thoughts in a logical structure. I do not, however, go beyond a certain proposal for the ordering of the entire train of thought, emphasizing a number of central concepts, postulates and deductions. The next step, a critical assessment of the resulting ordering, is not taken here. It would involve very extensive studies, during which one would have to subject the Bohr system (if this can be said to exist) to a considerable enlargement. Here it must be recalled that there is *no single* formulation of quantum mechanics based entirely and consistently on the principles proposed by Bohr. A possible exception is the book by Hund (1954), in which the principle of field-particle duality is fairly consistently applied. But this principle, quite apart from the question of what it asserts, cannot be immediately equated with even some of Bohr's principles, and it certainly does not cover everything that Bohr had to say, in particular as regards the measurement process. Bohm's textbook (1951) should also be mentioned here, since it gives in unusual detail a weaving of thoughts which are at least Bohr-like into a presentation of the content of quantum mechanics. The ordinary textbooks, however, contribute almost nothing to the problem raised here. At best, they mention in passing some fundamental idea of Bohr's, almost as if to comply with the canons of good taste, and one is bound to gain the impression that such ideas are some part of the subject, but do not provide the whole basis of the content of quantum mechanics. So long as, for example, Bohr's concept of a phenomenon or his postulate of the classical description of measuring apparatus is treated merely as an independent subject of discussion in a presentation of quantum mechanics, and does not appear as an irreducible part of the formulation itself, like the concept of a straight line or the axiom of parallel lines in Euclidean geometry, these ideas are far from having been tested. From this situation, the critics have deduced that the ideas in question are not necessary for the understanding of quantum mechanics. I shall not try to determine here whether they are right in this conclusion; I wish only to assert that we have in fact no technically elaborated formulation of Bohr's quantum mechanics and that *consequently* it is not at present possible to make a really useful assessment of his contributions.

In Chapters II and III I turn to non-Bohr orthodoxy, in particular that of Heisenberg and von Neumann. Here, in contrast to Bohr's treatment, the general nature of the investigation, if not the detailed content, is determined by the inclination to start from certain basic ideas specific to quantum mechanics and to define these in accordance with the available mathematical possibilities. In this way there is compliance with certain requirements of methodology that are orthodox in a more general sense, as against which Bohr is bound to appear unorthodox. This refers, broadly, to the Aristotelian–Euclidean tradition of the axiomatic formulation of a science, but it must not be thought that following such a method will weld the whole content of orthodoxy (in the quantum-mechanical sense) into a single unified theory. The existing statements are individually too scanty and jointly too dissimilar, particularly with reference to the fundamental concepts of quantum mechanics, e.g. the concept of probability and hence the concept of state. This has often been a subject of complaint, especially by a group of authors mentioned previously (Margenau, Popper, Landé, Bopp, Bunge, Feyerabend). I would express the opinion that these circumstances impose a twofold task, of which one part is easier than the other. First, some quite primitive distinctions concerning the elementary concepts are made, and continued until there is a clear increase in difficulty. On the basis of these distinctions, we then try to combine the resulting fragments into various self-consistent formulations of the theory. This completes the easier part, which I do not regard as forming part of the critique of orthodoxy; it simply puts things straight or rounds things off. A deeper analysis can be made only by using what is thus produced, and this is the more difficult part.

Chapters II and III deal with the first part. There may be surprise that this elementary preparation of the subject is in fact necessary for orthodox quantum mechanics, but it is so. The statements in the relevant original papers, including those of von Neumann, relating to the concepts of probability, state, measurement, and reduction of states are *not* sufficient to compile a single formulation of quantum mechanics that might be regarded as *the* orthodox formulation. The primitive distinctions to be made here have long since been known to the critics, but these have mentioned them only very sporadically, and appear to have generally used them immediately (and, it seems, sometimes even misused them) in order to make life difficult for orthodoxy. I have preferred to ask first the simple question of which among the various formulations of the *entire* theory are obtained if the orthodox ideas are examined and reconstituted by means of these distinctions. The introduction to Chapter II indicates how this question can be answered. The reader will have to read the chapter itself in order to get the details; in particular, section 3, which deals with classical statistical mechanics, should not be omitted, since it includes some of the decisions relevant to quantum mechanics, for example the distinction between an epistemic and a statistical concept of probability, and a corresponding distinction for measurements. For quantum mechanics only, there is also the distinction between formulations with and without the concept of the state of an individual object. These and similar distinctions lead already to different treatments of the reduction of states, and a further distinction must also be made in respect of this reduction itself. Only the possibilities thus explicitly drawn up in Chapter II form a reasonably usable basis for a further discussion of the orthodox view that is not bound to fail for lack of clarity.

The subsequent chapters (IV to VII) deal with a small number of aspects of a further analysis of orthodox quantum mechanics, on the basis of the decisions made in Chapters II and III. In Chapter IV, the comparison of quantum mechanics with classical statistical mechanics plays a vital part also. This has some value with regard to the problems of indeterminism in quantum mechanics, which are also dealt with in that chapter. At the present time, there is a tendency to bring ordinary classical mechanics closer to quantum mechanics by asserting that the former is also to some extent indeterministic and that the latter has certain deterministic features. These attempts at rapprochement are most easily judged by taking ordinary classical mechanics in its indeterministic form as given in Chapters II and III, and comparing this with quantum mechanics. In section IV.1 it is shown that the two theories are still deterministic if this term is taken to refer simply to the possibility of some reduction of their contingent statements by means of their laws. Both theories are, on the other hand, indeterministic if we further require the possibility of reduction to a single time, which exists in and is characteristic of ordinary classical mechanics. In these respects, then, there is similarity.

In another respect, however, significant differences are immediately found. This is shown in section IV.2 with reference to the concepts of non-simultaneous measurability and various uncertainty relations representing an instantaneous indeterminism. Here the analysis is again concentrated rather on problems specific to the orthodox view of quantum mechanics. The ambiguity of this view, already demonstrated in Chapter II, is redoubled when these concepts are taken into account, since even within a single one of the interpretations distinguished in Chapter II one can introduce a wide range of concepts to represent an instantaneous indeterminism, without any clarification on the part of orthodoxy as to which it prefers to use. One fundamental distinction here is whether the concepts are correlated with measurements or with descriptions of state.

In section IV.3, indeterminism is examined from the aspect of reduction of states, the real problem child of quantum mechanics. Here again, there are considerable differences from the corresponding classical case. In order to make these clear, however, certain distinctions must again be made, which is not always done by orthodoxy with all desirable clarity. One must distinguish between reductions of states in which the result of the measurement is and is not taken into account, and where it is not, between strong and weak reductions, according as the components obtained are treated separately or recombined.

Chapter V is an elementary introduction to the basic ideas of the von Neumann programme. This has so far had little attention from the critics, which is regrettable because only the attempts at characterization which it involves allow a deeper insight into the peculiarities of the formal structure of quantum mechanics. The customary formulations of quantum mechanics do not allow this, just because they represent the physical concepts forthwith as a mathematical structure regarded as given in its entirety, and only the final empirical success is regarded as deciding on the correctness of this procedure. In the von Neumann programme, however, the relevant mathematical structure is obtained stepwise by a direct axiomatization of the physical concepts. This provides from the start a theoretical check on the whole procedure. Of course, one can seek to attain this objective in various particular ways, depending on which physical concepts are taken as the basis. The relevant investigations are exceedingly difficult. Although they have not been satisfactorily completed, a number of profound results have been obtained. In Chapter V I do not propose to reproduce these results, but only to clarify essentially what is happening. I also seek to distinguish the earlier von Neumann result regarding the characterization of the quantum-mechanical expectation-value functions from what is called the von Neumann proof that there exists no classical theory of hidden parameters for quantum mechanics. The latter is discussed in section VII.1.

Chapter VI is concerned with one of the central problems in the principles of quantum mechanics: whether the reduction of states can be eliminated by using a continuous Schrödinger process between the object and the measuring apparatus. There have been numerous papers on this subject in the last few years. Here again, I shall not deal with the recent discussion, but simply attempt to expound as clearly as possible the orthodox answer to this question in relation to the distinctions made in Chapter II. In section VI.1 the reproduction of the weak reduction of states of an object by an interaction with the measuring apparatus is analysed (such a reproduction is in fact possible), and a number of characteristic conditions are stated. Similarly in section VI.3 for the assumption (which is found to be a stronger one) that the interaction allows a measurement of the object to be replaced by a measurement on the apparatus. It is also shown that this possibility in fact exists, and therefore so does the one discussed in section VI.1. In section VI.2, on the other hand, a fresh presentation is given of von Neumann's earlier negative result that the strong reduction of states cannot, strictly speaking, be eliminated by a Schrödinger process.

Chapter VII deals with the last group of problems in this book, which relate to the completeness of quantum mechanics. Orthodoxy is here supported by the von Neumann proof that hidden parameters for quantum mechanics do not exist. This proof was later refined and also generalized by other writers. There have also been explicit proposals of classical theories of hidden parameters, as already mentioned. Thus the present position in research has, as its main problem, to show that these apparently contradictory groups of results are compatible. The proof must evidently depend on clarifying the concept of a classical theory of hidden parameters for quantum mechanics. In section VII.1 this concept is specified for several proofs, where it is relevant to them. It appears that von Neumann's concept was so narrowly defined that it would not even admit ordinary classical mechanics as a theory of hidden parameters for classical statistical mechanics. This point has since been dealt with, and is of course a purely academic one. However, the concepts and proofs given in section VII.1 for the problem of hidden parameters will enable the reader to go on to a comparison with the counter-proposals that are mentioned but not further discussed here.

The remainder of Chapter VII, finally, is devoted to an explicit criticism of orthodoxy, although one which originates from the "classical" period. It relates to the paradox of Einstein, Podolsky and Rosen, which has been much discussed recently. Here I have considered a question not previously treated: in what form can their argument be stated so as to lead to a logical contradiction? The answer is given in section VII.2. Since Einstein, Podolsky and Rosen regard quantum mechanics as non-contradictory though incomplete, the desired contradiction can be found only in an extension of quantum mechanics. In fact the authors use, though in a somewhat concealed manner, a theory of hidden parameters whose details are left largely unsettled, and which is connected with ordinary quantum mechanics (which they accept) by a reality condition. This theory of hidden parameters may also be related to ordinary quantum mechanics by completeness conditions. It can then be shown that the two conditions cannot be simultaneously satisfied in the extended theory without accepting what is strictly a logical contradiction. In section VII.3 it is shown that the orthodox quantum mechanics with states of the individual object (Chapter II) is essentially the only theory of "hidden" parameters which satisfies the completeness conditions. A trivial model is used to show that Einstein, Podolsky and Rosen's reality conditions are capable of being logically satisfied. Last comes a discussion of these conditions in terms of their orthodox counterparts, with Bohr's view once more brought forward.

Finally, I must say something concerning the methodological limits of these investigations. The title of the book shows that the analyses are predominantly logical ones. This is not intended to imply that logical analysis is pre-eminent, nor do I consider that it is the only way to examine a physical theory. It is of course profitless if not accompanied by an empirical analysis. Only their combination will provide a basis for a penetrating epistemological analysis. We have at present a more general (and therefore less binding) philosophy of science which makes us conscious of methodological purity and the separation of fundamental problems that are of different kinds. This philosophy of science suffers, however, from the fact that it is based on easily available and indicative but scientifically insignificant examples, not on advanced theories of any one branch of science. This is particularly clear from the many specialized works on the principles of quantum mechanics, which give the impression that all the general ideas of the philosophy of science have been forgotten and that the special problems concerned in each case are dictating the corresponding approaches. This is pardonable, to the extent that quantum mechanics, which raises so many epistemological problems, lacks any general guide to procedure. But in that case one might simply reverse the weapon and try to particularize to quantum mechanics the few available general ideas, correcting these if necessary. Far from attempting to do so here, I have simply made use of the fact that problems of the logical structure of a theory can be fairly definitely separated from other problems and treated by themselves. The profit is correspondingly small and the argument is highly abstract; but the gradual elucidation of detail and the creation of logical order are indispensable complements of the great inspirations of science.

CHAPTER I

A Chapter on its own:
Bohr's Interpretation of Quantum Mechanics

THE fundamental epistemological problems raised by quantum mechanics are still thought of by physicists in general as having been largely, if not entirely, discussed and resolved in the "Copenhagen interpretation" of quantum mechanics. On the other hand, as has been mentioned in the Introduction, in philosophical studies of the foundations of quantum mechanics the Copenhagen interpretation certainly does not occupy any such unique position. But it was also pointed out that attempts at a true understanding of the Copenhagen interpretation, or at least expressions of such attempts, have sometimes been inadequate, whether in terms of monographs or of textbooks of quantum mechanics. The first thing that must be realized if such attempts are to succeed is that there is no point in looking for *the* Copenhagen interpretation as a unified and consistent logical structure. Terms such as "Copenhagen interpretation" or "Copenhagen school" are based on the history of the development of quantum mechanics; they form a simplified and often convenient way of referring to the ideas of a number of physicists who played an important role in the establishment of quantum mechanics, and who were collaborators of Bohr's at his Institute or took part in the discussions during the crucial years. On closer inspection, one sees quite easily that these ideas are divergent in detail and that in particular the views of Bohr, the spiritual leader of the school, form a separate entity which can now be understood only by a thorough study of as many as possible of the relevant publications by Bohr himself.

This chapter is therefore concerned with *Bohr's* interpretation of quantum mechanics and with nothing else. The distinctiveness of this interpretation is generally appreciated insofar as Bohr is regarded as the creator of the *complementarity principle*, a key concept in understanding the new epistemological problems of physics. Bohr himself contributed a great deal to the propagation of the idea that the concept of "complementarity" was his main concern. Indeed, the application and analysis of this concept are discernible in all his work on the subject. There can also be no doubt that Bohr hoped to generalize this concept, as is seen from repeated attempts to introduce complementarity outside physics and make it a fundamental category of ideas. On the other hand, Bohr himself tells us that "the purpose of such a *technical term* is to avoid, so far as possible, a repetition of the *general argument*" (1929a, p. 19; italics E.S.). This contrasting of technical term and general argument is typical of Bohr's approach: he ultimately preferred a clear line of thought, based on generally understood concepts, to the use of technical terms as such. He was over-sensitive to the unavoidable introduction of a conceptual structure to describe a field of study, and such comments as the one just quoted were intended to prevent the reader from taking the terms introduced and making them into catchwords whose repeated and gradually more mechanical use will finally inhibit creative thought.

With this warning in mind, the principal objective of the following discussion will be to illuminate Bohr's "general argument" and allot to the concept of complementarity its due place within that argument. But even if the error mentioned earlier of trying to assign a single meaning to the "Copenhagen interpretation" of quantum mechanics is avoided, and only Bohr's own view is considered at first, a proper account of his approach is no easy task and involves considerable difficulties of understanding. Bohr's mode of expression and manner of argument are individualistic sometimes to the point of being repellent, so that the secondary literature includes few attempts to follow the detail of his arguments, and even Einstein describes Bohr's complementarity principle as a thing "the sharp formulation of which, moreover, I have been unable to achieve despite much effort which I have expended on it" (1949, p. 674). One must nevertheless guard against negatively interpreting this individualism as a refusal to be understandable. Bohr undoubtedly had considerable difficulty in making his statements sufficiently definite to be generally understood by physicists, but on the other hand we know that he took his writing very seriously. His son testifies that "those familiar with my father's way of working will know what great efforts he devoted to the preparation of all his publications. The text would always be re-written many times while the matter was being gradually elucidated, and until a proper balance was achieved in the presentation of its various aspects" (Aa. Bohr 1963). Thus we must assume that Bohr's obscurity does have something behind it and perhaps confuses only because he did not choose to display his thoughts by means of the techniques which were about that time being developed for the benefit of the analysis of science.

Since the following is an attempt to re-interpret Bohr's ideas on quantum mechanics by direct and detailed reference to his own publications, there will be no analysis of the secondary literature. This literature can be roughly divided into three classes: (1) that which seeks fuller interpretations, by authors who regard themselves as Bohr's disciples; (2) critical contributions which nevertheless promote understanding of Bohr's work; (3) more or less polemical and usually brief remarks which are hardly likely to assist understanding. Among the references given here, class 1 includes in particular the works of Meyer-Abich (1965), Petersen (1963, 1968), and Rosenfeld (1953, 1963), of which the first and third enter into considerable detail. In class 2 are, first of all, the papers by Feyerabend (1958, 1961, 1968, 1969), which are especially meritorious in their clarification and elimination of misunderstandings, and also those of Bunge (1955), Grünbaum (1957), Hanson (1959a, b), Shimony (1963), Bub (1968), Bohm (1971) and von Weizsäcker (1955, 1971). The widely dispersed comments in class 3 need not be identified here. Meyer-Abich (1965) and Jammer (1966) have given useful historical analyses.

I.1 Bohr's style of argument

First of all, a comment on Bohr's style of argument. He did not favour enveloping the content of physics, especially its fundamental aspects, in an extensive mathematical formalism. He expressed respect for von Neumann's axiomatic formulation of quantum theory (Bohr 1939, pp. 16 and 38), and mentioned the quantum-mechanical formalism in every relevant paper he wrote, but he himself did not incline to follow these lines of thought. Heisenberg writes regarding his early collaboration with Bohr: "I noticed that mathematical clarity had in itself no virtue for Bohr. He feared that the formal mathematical structure would obscure the physical core of the problem, and in any case, he was convinced that a complete physical explanation should absolutely precede the mathematical formulation" (Heisenberg 1967, p. 98); and from a much later period there is similar testimony, which even casts doubt on the respect for von Neumann mentioned above: "Bohr would on occasions be critical of von Neumann's approach which, according to him, did not *solve* problems, but created *imaginary difficulties* (such as the problem where the 'cut' between the observer and the things observed should be placed). During a seminar in Askov in 1951 I also had the impression that he definitely preferred his own qualitative remarks to the machinery of the famous 'von Neumann Proof'. Nor was he too fond of the rising axiomania" (Feyerabend 1969, p. 87, note 66). But, quite apart from this evidence, a glance at Bohr's own contributions to the interpretation of quantum mechanics is sufficient to show that it is not the work of someone who has great faith in the formalization of the basic ideas and arguments of physics. A person accustomed to the formal logical analyses of physical theories, or one who welcomes a tendency to provide such analyses, might gain from Bohr's papers the impression that they are no more than popular versions which spare an uninitiated public all technical details of the kind under discussion here. Such an impression would be quite wrong, however. The way in which Bohr expressed his thoughts is undoubtedly the one that he considered appropriate to the subject. Bohr's publications concerning the interpretation of quantum mechanics make no concessions, and it would be wrong (as regards this particular aspect of the method) to suspect anything behind them which they do not state explicitly. They are, as they stand, everything that Bohr wished to say on the matter.

These remarks about what Bohr's approach was *not* are necessary not only because from Chapter II onwards we shall be taking precisely the road of logical analysis of certain fundamentals of quantum mechanics (thus deviating from Bohr's keynote, as it were), but also because Bohr's arguments have a certain "logic" of their own which, to avoid serious misunderstandings, must be most carefully distinguished from the otherwise customary procedures of presenting a physical theory (both formalism and interpretation) in some logical order. It is difficult to specify the exact nature of the "Bohr logic", but anyone who reads his work must notice that he uses, at certain critical points, verbs and conjunctions of inferential argument, such as "entails", "implies", "follows"; "since", "because", "therefore". A typical instance which may be quoted here is the sentence "Indeed the finite interaction between object and measuring agencies *conditioned* by the very existence of the quantum of action *entails—because* of the impossibility of controlling the reaction of the object on the measuring instruments *if* these are to serve their purpose—the necessity of a final renunciation of the classical ideal of causality and a radical revision of our attitude towards the problem of physical reality" (Bohr 1935, p. 697; italics E.S.). This sentence, which explicitly includes four logical constants of the deductive type, will not be analysed here. Our purpose is only to comment that anyone who writes such a sentence and understands its meaning must have some idea of which things come first, and second, and third, i.e. must have in mind some kind of logical order. Bohr has expressly emphasized that this was indeed his concern. Dissatisfied with his previous arguments (1935), he referred to it in 1949 (p. 61) together with the hope "that the present account . . . may give a clearer impression of the necessity of a radical revision of basic principles for physical explanation *in order to restore logical order in this field of experience*". (Italics E.S.)

Thus, although Bohr himself testifies to his wish to present his view at least partly as a complex of causes or a hierarchy of concepts and statements having various degrees of priority, there is still good reason to question whether he succeeded in doing so intelligibly.

For, if we first inquire as to the elementary bricks from which Bohr proposes to construct his edifice of thought, there are already difficulties, especially of identification: Bohr's continually varying modes of expression make it sometimes almost impossible to decide whether passage A is intended to mean the same as passage B. However, the elementary themes can eventually be identified with some certainty because, in the course of time, a number of stereotyped formulations appear, and looking back from these it is possible to recognise their earlier forms. Here, then, there is a fairly clear convergence which assists interpretation; but the situation is much less satisfactory as regards the way in which Bohr juxtaposes his various statements and in particular gives them a "logical" connection. It is true that the elementary steps gradually become apparent, and on the large scale too some very rough canonical order can be seen. Between the two, however, the order is very difficult to grasp, on account of the many different combinations in which Bohr presents his thoughts. Yet there is always the impression that he *intends* there to be a well-defined order. This, of course, does not at all affect the difficulty of grasping the content of the individual ideas and the entire argument, even after a certain formulation and a certain order have been accepted.

Anyone who makes a serious study of Bohr's interpretation of quantum mechanics can easily be brought to the brink of despair by all this. There is an unaccustomed wilfulness and sometimes obscurity, which yet suggests precision and profundity of insight. Elucidation is laborious but still perhaps rewarding, since Bohr was, in somewhat paradoxical terms, a conservative revolutionary—holding to the well-tried classical physics until the last thread snapped, but thereafter able to accept a radical upheaval (cf. Heisenberg 1969, pp. 157 and 279f.). A person with this attitude will have a much more precise idea of the reasons for a revolution such as was implied by the change from classical physics to quantum theory than one whose response to such problems is a mobility that is more free but in the long run less far-reaching. It is for this reason that the analyses thus derived are worthy of close attention. Another important point is that logical analysis in the usual sense (that is, the kind which Bohr does *not* use) has its own laws which, if it alone is used, exclude certain other aspects of understanding. Since the logical method will be predominantly used from Chapter II onwards, it is desirable to demonstrate first a methodologically unorthodox approach such as that of Bohr.

I.2 The main lines of Bohr's thought†

It is evident from the discussion in section 1 that the following account of Bohr's thoughts concerning the new situation created by quantum mechanics cannot be a mere repetition of his writings: at least the arrangement of the arguments as a whole was not specified unambiguously by Bohr, and even in the formulation of the individual points covered by these arguments it is easy to go astray, because of the variations which occur in Bohr's works. Certain interpolations, standardizations and refinements are necessary in order to arrive at a fairly self-contained presentation which makes Bohr's views appear neither one-sidedly problematic nor one-sidedly dogmatic. As regard completeness, all the significant points mentioned by Bohr in connection with physics will be included, except one, namely the comparison which he frequently advanced between the new situation in quantum mechanics relative to classical mechanics (both non-relativistic) and the new situation of Einstein's theory of space-time relative to the older Galilean theory. This comparison is instructive (though not devoid of problems), but, being no more than a comparison, can be omitted

† In this section, citations are of works by Bohr unless otherwise stated.

from an account of the subject without risk of misunderstanding; the most detailed and connected account is that given in 1939 (§3). There will also be no discussion of the application of the concept of complementarity outside physics. Unity of presentation will be attempted, although Bohr's views did evolve to some extent. The clearest change is in the lesson that Bohr learnt from a paper by Einstein, Podolsky and Rosen (1935), to which reference will be made in subsection (d). A detailed account of this paper will be given in sections VII.2 and VII.3, in a wider discussion not relating to Bohr alone. Thus the present description, strictly speaking, represents the position after 1935, and includes earlier views where these are compatible with the later ones.

I.2(a) Some characteristics of classical physics

Bohr's analysis of quantum mechanics is founded on classical physics to an extent which is significant and indeed typical of his approach. This is true not only of his own contributions to the physics (in the narrow sense) of the older quantum theory, which were entirely governed by correspondence arguments "which give expression for the exigency of upholding the use of classical concepts to the largest possible extent compatible with the quantum postulates" (1939, p. 13). In addition, we shall see that his final interpretation of the new quantum mechanics depends essentially on the role which classical physics retains in that theory. Accordingly, Bohr's discussion often includes repeated reference to characteristics of classical physics that do not appear in the quantum theory, and it is desirable to begin by setting out which characteristics Bohr regards as falling into this category.

The most important such characteristic is probably the ideal of *causality* or *determinism* in classical physics. The following quotation will serve to introduce this ideal, which was fundamental to the conception of classical physics throughout the nineteenth century; it does not reflect any feature peculiar to Bohr's own viewpoint regarding classical physics. "In Newtonian mechanics, where the state of a system of material bodies is defined by their instantaneous positions and velocities, it proved possible, by the well-known simple principles, to derive, solely from the knowledge of the state of the system at a given time and of the forces acting upon the bodies, the state of the system at any other time. A description of this kind, which evidently represents an ideal form of causal relationships, expressed by the notion of *determinism*, was found to have still wider scope. Thus, in the account of electromagnetic phenomena, in which we have to consider a propagation of forces with finite velocities, a deterministic description could be upheld by including in the definition of the state not only the positions and velocities of the charged bodies, but also the direction and intensity of the electric and magnetic forces at every point of space at a given time" (1958, p. 1). Similar formulations are prominent, usually as the preamble to a statement of the problem, in the later works in particular (1948, p. 312; 1954, p. 69; 1955, p. 84; 1956, p. 85f.; 1957, p. 97), whereas in reality both Bohr and others had been teaching since 1927 that the new quantum mechanics was incompatible with the determinism of classical physics (see, e.g., 1927, 1929a, b, c). It is remarkable that Bohr did not so much state the concept of causality or of determinism (he regards the two terms as essentially synonymous) as being the classical ideal in contrast to the fundamentally statistical nature of quantum mechanics, but mainly compared it with the concept of complementarity which he himself introduced as a natural generalization of the concept of causality; cf. the titles of his papers (1937; 1948; 1958) and the typical phrase (1937, p. 291) that the new situation "forces us to replace the ideal of causality by a more general viewpoint usually termed 'complementarity' ".

Together with the deterministic nature of classical physics, and quite clearly as a parallel to it, Bohr repeatedly mentions as a second feature the *pictorial* nature of the description given by classical physics; he says, for example, "that the *pictorial description* of classical physical theories represents an idealization" (1958, p. 2; italics E.S.). The expressions *causal pictorial description* and *deterministic pictorial description* to refer to the classical approach occur in several places (1954, p. 71; 1955, p. 85; 1956, p. 87, col. 1; 1960, p. 11), and there are continual references, for which no citation need be given, to the *pictures* devised by classical physics for the description of objects. The most relevant to the discussion of the basis of the quantum theory are probably the two pictorial representations of an elementary object which stand in dualistic fashion at the start of classical physics: the object is regarded either as a rigorously localized particle or as a field extended in space, its changes with time being ascribed entirely either to the motion of the particle along a defined path or, more abstractly, to the time variation of the field strength at every point in space. We shall see later that Bohr always found a natural starting-point for his reasoning by illustrating the failure of classical physics by means of the duality of waves and particles, which shows the impossibility of maintaining the classical picture of a particle or a wave.

Bohr must have taken as a third feature of classical physics the concept of an object which is *independent* both in its instantaneous properties and in its behaviour with time. He most often states this, as will be seen from later citations, in the negative assertion that the concept is no longer applicable in quantum mechanics, but there are also such statements as the following: on the one hand, "the mechanical conception of nature, as characterized by attribution of *separate properties* to physical systems" (1954, p. 74) and "such *inherent attributes* as the idealizations of classical physics would ascribe to the object" (1937, p. 293); on the other hand, "all description of experiences has so far been based upon the assumption, already inherent in ordinary conventions of language, that it is possible to distinguish sharply between the *behaviour* of objects and the means of observation. This assumption is not only fully justified by all everyday experience but even constitutes the whole basis of classical physics" (1938, p. 25). (Italics in this paragraph E.S.) This clearly relates to the sharp distinction which classical physics always makes between a physically real situation (whether an instantaneous or permanent property of an object or the behaviour of an object at all times) and the conditions in which this can become a phenomenon perceivable by a conscious being. Classical physics certainly derived this distinction, at the same time making it a sharper one, from the world of everyday life governed by customary human observation and speech, as Bohr himself says in the passage we have just quoted. Everyone knows, and can see from countless instances, that we immediately express any particular contingent view of our environment, observed with the senses, in a language which is essentially a language of things; this language comprises statements that things *have* properties and ways of behaving, regardless of the conditions that permit us to be aware of them. Likewise, no reference to observation, experiment, measurement or anything else having the slightest connection with the subjective possibilities of awareness forms part of the meaning of the statements which describe the state of an object in classical physics. This even extends to our being unconscious of these possibilities of awareness in normal circumstances, i.e. when remaining within the orbit of classical physics. As we shall see, however, it is just this realization of the role of the measuring equipment (without involving the observer) for the description of physical phenomena that is regarded by Bohr as being the key to the understanding of all the phenomena which only quantum mechanics can explain.

The independence of an object, as a principle of classical physics, is not completely separate from the characteristics of determinism and pictoriality: Bohr seems to have held the view that a necessary condition for the causal description of an object is a description of it which is unrelated to any observational procedures. For example, he writes of "the assumption *underlying* the ideal of causality, that the behavior of a physical object . . . is uniquely determined, quite independently of whether it is observed or not" (1937, p. 290). And in the same paper (p. 293) he says with regard to quantum mechanics that "the renunciation of the ideal of causality . . . is *founded logically* only on our not being any longer in a position to speak of the autonomous behavior of a physical object". (Italics in this paragraph E.S.) Here we have for the first time a case of the kind indicated in section 1, where Bohr asserts certain apparently or allegedly logical relationships without explaining their exact nature. All that is clear is that he cannot be referring to a strict logical implication. The idea of the independence of an object is rather to be regarded, as Bohr himself says (1948, p. 317), as a conceptual basis without which it would not be possible even to formulate a deterministic theory.

Whereas it was not at all unusual to stress the foregoing three characteristics of classical physics, the fourth one mentioned by Bohr is, at least in his formulation of it, an instance of his typical wilfulness (see section 1). This characteristic is what Bohr terms the *combined use of space-time concepts and dynamical conservation laws*, which he again asserts in various places—though only as regards mechanics—to be typical of the classical conception of physical reality, for example in his reference to "the simultaneous use of space-time concepts and the laws of conservation of energy and momentum, which is characteristic of the mechanical mode of description" (1929a, p. 11) and similarly elsewhere (1935, p. 699, col. 1; 1949, p. 40f.; 1958, p. 5). This alleged characteristic has a variant formulation in which causality replaces the conservation laws, e.g. in "the space-time coordination and the claim of causality, the union of which characterizes the classical theories" (1927, p. 54; 1929b, p. 92). This phrasing might be simply an incompletely worked-out earlier form of the one given above, since it does not appear in the later publications; in these, as we shall see in a moment, a relationship is predicated between the causal description and just this same combination of space-time concepts and dynamical conservation laws, a procedure which is meaningful only if the later phrasing given previously is the basis. The earlier form might also, however, have been intended as a more general one, embracing electrodynamics as well as mechanics. Since Bohr subsequently abandoned it, there is probably no point in racking one's brains over the matter; the later and final form should be used in continuing the discussion.

The meaning of such references to the combined use of space-time (or, as Bohr sometimes calls them, kinematic) concepts and the dynamical conservation laws as one of the chief principles of classical mechanics is difficult to elucidate without considering at the same time the situation in quantum mechanics. This is evidently an aspect of the classical theory which was recognized as characteristic only *in retrospect*, i.e. after the realization that it did not occur in quantum mechanics. Consequently, Bohr himself emphasizes this difference much more often (see below) than simply the existence of this characteristic of classical mechanics. But, because of the connection already indicated between the combination in question and causality, one feature can be recognized, which Bohr mentions in his later work: like the independence of an object, the simultaneous use of space-time concepts and the dynamical conservation laws seems to have been regarded by him as a *necessary condition* for the causal character of classical mechanics. For example, he writes that "the deterministic description of classical physics *rests on* the assumption of an unrestricted compatibility of space-time coordination and the dynamical conservation laws" (1955, p. 89; italics E.S.), and entirely similar statements occur elsewhere (1954, p. 72; 1956, p. 87, col. 1; 1960, p. 11). Now it is known that the deterministic description of a system of particles in classical mechanics is possible because an initial state can be defined by giving the values of the momenta of the particles as well as of their spatial coordinates, i.e. because the concepts of position and momentum can be used simultaneously. Coordinates and momenta at a single instant, together with the forces acting, determine the whole course of events in time, and if conversely this course is specified in a purely space-time form, i.e. by the coordinates of each particle as functions of time, then the momenta can be *defined* by means of the velocities, and the conservation of total momentum follows from this description. In quantum mechanics, on the other hand, momentum can no longer be defined by means of the space-time description, but only by a conservation law. Bohr regarded this as important (1927, p. 60; 1937, p. 293; etc.), and it evidently dictated his terminology. He is ultimately concerned to show that, according to classical mechanics, position and momentum and also energy as the basis of a deterministic description of a process can be simultaneously measured, and this is expressed in a system of concepts recalling the *means* of such measurement, which are still available according to quantum mechanics. For momentum and energy these means are just the conservation laws; for position and time, the bringing about of coincidences. In this sense Bohr usually refers to space-time *coordination*, thus again indicating that the space-time localization of an object must be achieved by means of certain experimental adjuncts.

Bohr, however, seems to have regarded the pictorial, as well as the causal, description of an object in classical mechanics as being based on this ability to combine space-time coordination with dynamical conservation laws; for he speaks of "the unlimited combination of space-time coordination and the conservation laws of momentum and energy *on which the causal pictorial description of classical physics rests*" (1960, p. 11), and perhaps an even clearer statement is that of 1949 (p. 41): "the combination of these concepts *into a single picture* of a causal chain of events is the essence of classical mechanics". (Italics in this paragraph E.S.) In fact the idea of a particle moving in a defined trajectory implies the simultaneous use of the concepts of position and velocity; and if we take account of Bohr's way of describing the latter, as explained in the last paragraph, the two quotations just given express precisely this implication.

I.2(b) The failure of classical physics, and Bohr's basic idea as regards the interpretation of quantum mechanics

Bohr was like any other physicist of his time in being aware that classical physics is unable to explain a large number of phenomena in the atomic range of physical processes, discovered even before the turn of the century and increasingly in the early decades of the present century. But he was certainly among those who felt most deeply the transformation that was occurring and whose whole thinking most clearly shows the effects thereof. There is not space to enter into detail here, but attention must be drawn to the aspect which Bohr mentioned repeatedly in his work on the interpretation of quantum mechanics from 1927 onwards, as the typical instance of the failure of classical physics. This aspect is what is commonly referred to as the *duality of waves and particles.* As already indicated above, classical physics developed two basic (pictorial) views of the nature of an elementary physical object: that of a rigorously localized particle, and that of a field extended in space. The state of a particle (at any given time) is defined by its position in space and its velocity; the state of a field is defined by its components at every point in space. These ideas form the basis of the whole further theory of particles or fields and their interaction. Now it is obvious that the two views, of a particle and of a field, are mutually exclusive, in the sense that one and the same object cannot be imagined as a particle and a field at the same time. Such a conjunction would not only surpass our capacity for visualization; the corresponding theories and all their concepts are fundamentally different, since they are based on the particle and field concepts respectively. The new situation is nevertheless distinguished precisely by experiments which seem to force on us a simultaneous employment of these mutually exclusive pictures or theories. The classical experiments suggested that light should be described as a field, whereas the photoelectric effect and the Compton effect, for instance, seemed to be compatible only with a corpuscular picture of individual photons. The classical experiments on matter, contrariwise, indicated that it consisted of individual particles, whereas later experiments on reflection and diffraction gave interference effects which, classically, could be explained only by the wave picture.

The duality of waves and particles thus arises first as an internal contradiction affecting the classical nature of a pictorial description, and it is hardly surprising that Bohr regarded the experimental evidence relating most directly to this duality as being especially illustrative of the problem, and mentioned it in all his work on the subject. At first it seemed that he was concerned with the systematic development of a dualistic theory in which the classical concepts of fields and particles would come together to form a new "chimerical" object concept. In order to observe this tendency, it is sufficient to look at Bohr's first paper (1927, §2) on the interpretation of the new quantum mechanics, where in particular he showed for the first time that Heisenberg's uncertainty relations for the position and momentum of a *particle* can be derived from the *wave* picture of matter, using the de Broglie relation between the momentum and the wave vector. Heisenberg (1930 (German edition), p. 7) also referred to Bohr's work and further developed at least a dualistic viewpoint whereby the wave and particle pictures are simultaneously used, so as to maintain as far as possible the intuitiveness of physics: this duality of the pictorial representation, he says, has been shown by Bohr to form also an obvious starting-point for the critique of the pictures and concepts used in the theory. "For, obviously, uncritical application of both the particle picture and the wave picture will lead to contradictions. From the simultaneous existence of both pictures it can be concluded at once that nature limits the applicability of each picture with respect to the other. In this way one obtains e.g. the limitations of the concept of a particle by considering the concept of a wave. As N. Bohr has shown, this is the basis of a very simple derivation of the uncertainty relations between co-ordinate and momentum of a particle. In the same manner one may derive the limitations of the concept of a wave by comparison with the concept of a particle". Later, looking back, Heisenberg (1955, p. 15) tells us that Bohr "intended to work the new simple pictures, obtained by wave mechanics, into the interpretation of the theory".

Despite an initial approach of this kind, which may certainly be reckoned to Bohr's credit, the core of his contribution to the interpretation of quantum mechanics must be sought elsewhere. In this respect the reader will have to see for himself that the wave-particle duality, which Bohr indeed mentions over and over again, is only a part of the *negative* side of his arguments taken as a whole; he uses it to illustrate the difficulty of the then existing position, but it does not itself contain any positive method of resolving the problems. The duality is always presented as a puzzle, a dilemma, a paradox or even a contradiction; always as something to be overcome and done away with. Bohr may also have given support to a negative view of the fact that the application of classical concepts to atomic objects becomes *ambiguous* if (and this is where the positive attitude should be brought in) we follow classical physics in ignoring the *experimental conditions* which permit the processes in question to be perceived. It is true that the wave-particle duality can be (very crudely) expressed by saying that an object behaves as a particle or as a wave according to which experiment is being performed on it. From this basis, Bohr considered the chief aim of a consistent quantum theory to be *an unambiguous description of atomic phenomena, obtained by including in their description the experimental conditions in which the phenomena occur*.

This tendency and aim, repeatedly stated by Bohr with the use of the terms "ambiguous" and "unambiguous", is worth documenting in detail. Its origin in the wave-particle duality, for example, is clearly shown: "Under these circumstances an essential element of *ambiguity* is involved in ascribing conventional physical attributes to atomic objects, as is at once evident in the *dilemma* regarding the corpuscular and wave properties of electrons and photons, where we have to do with contrasting pictures, each referring to an essential aspect of empirical evidence" (1949, p. 40). Similarly: "Very striking illustrations are afforded by the well-known *dilemmas* regarding the properties of electromagnetic radiation as well as of material corpuscles, evidenced by the circumstances that in both cases contrasting pictures as waves and particles appear equally indispensable for the full account of experimental evidence. Here we are clearly in a situation where it is no longer possible to define *unambiguously* attributes of physical objects independently of the way in which the phenomena are observed" (1956, p. 87, col. 1; cf. also 1948, p. 313f.). Bohr continually repeats the argument that a return to an unambiguous description of atomic events is possible only by explicit inclusion of the experimental conditions under which such events can be perceived. A typical expression is the statement that atomic phenomena lie in "a domain of experience where *unambiguous* application of the concepts used in the description of phenomena depends essentially *on the conditions of observation*" (1957, p. 99; 1961, p. 60; similarly 1958, p. 5). Again, we learn that the consideration of what Bohr regards as the real reason for the failure of classical physics (to be discussed in more detail in the next subsection) "involves . . . the question of the scope of *unambiguous* application of classical physical concepts in accounting for atomic phenomena" and that this "*unambiguous* description demands a specification of all significant parts *of the experimental arrangement*" (1962b, pp. 91, 92). Finally, the interpretation of the quantum-mechanical formalism is viewed similarly: "there can be no question of any *unambiguous* interpretation of the symbols of quantum mechanics other than that embodied in the well-known rules which allow to predict the results to be obtained *by a given experimental arrangement*" (1935, p. 701, col. 2; cf. also 1939, p. 20). (Italics in this paragraph E.S.)

I.2(c) Further characteristics of classical physics

Thus we have arrived at the basic idea of Bohr's programme. In his opinion, the description of atomic objects and their behaviour by means of the concepts of classical physics involves certain ambiguities, and he found the wave-particle duality the most suitable way of illustrating this. His aim was to eliminate these ambiguities by explicitly considering the dependence of the phenomena on the experimental conditions in which they occur. Now this dependence must clearly be of a special kind. The relationship of an object to an experimental arrangement by means of which the conditions are produced that allow the behaviour of the object to be observed cannot, obviously, be regarded in the same manner as it was in classical physics. Otherwise there would be no point in expressly including the experimental arrangement in the description of a phenomenon that is in contradiction with classical physics. How classical physics in fact viewed this relationship, according to Bohr, is one of the most frequent statements in his works on the subject. In one place he writes: "within the scope of classical physics we are dealing with an idealization, according to which . . . *the interaction between the measuring instruments and the object under observation can be neglected, or at any rate compensated for*" (1962b, p. 91; italics E.S.; and similar statements in 1927, p. 53; 1929a, p. 11f.; 1935, p. 701, col. 1; 1937, p. 291; 1939, p. 19; 1954, p. 72; 1956, p. 87, col. 1; 1957, p. 98; 1961, p. 59; 1962a, p. 24).

The idealization by classical physics of the relationship between the object of measurement and the measuring apparatus, to which Bohr here refers, is indeed one of the implicit postulates of the classical approach, but it must be made explicit alongside the characteristics already listed in subsection (a), as soon as anyone (such as Bohr) feels obliged to include in a consistent and unambiguous formulation of quantum mechanics the experimental conditions under which the phenomena occur. Bohr evidently held the view that the independent description of an object (see subsection (a)), regardless of any observational process, is possible precisely because any interaction between the object and the means of observation during the measurement can be ignored. Accordingly he considered the non-existence of the independent description for atomic objects as a consequence of the fact that there the interaction between the object and the means of observation can be neither neglected nor determined. The contrary situation thus becomes a relevant and distinctive feature of classical physics.

In the present context, Bohr sought to describe further the position in classical physics by the statement that, "within the frame of classical physics, *there is no difference in principle between the description of the measuring instruments and the objects under investigation*" (1955, p. 89; italics E.S.; see also 1935, p. 701, col. 1). Whereas the emphasis on the ignorability of the interaction in measurement as a characteristic of classical physics may seem natural to every physicist, at least at first sight, a statement such as the one just quoted, which afterwards appears in a reversed form as a characteristic of quantum mechanics, may need some explanation to be even provisionally understandable. In classical physics too, the object must be distinguished from the apparatus of measurement if an experiment is to be described; and this distinction cannot be made in just the wholly trivial sense that the object and the apparatus simply are two distinguishable things. In order to understand the experiment, one must see clearly that they play different parts in it, one *being* the object of measurement and the other *being* the apparatus. Nevertheless, this distinction is not a fundamental one, in that the theoretical analysis of the experiment treats both the apparatus *as such* and the object *as such* with the methods of classical physics. We shall see that one of the main points of Bohr's interpretation of quantum mechanics is that there the theoretical treatment depends decisively on whether the thing concerned is functioning as an object of measurement or as an apparatus of measurement.

For completeness, one further characteristic of classical physics must be mentioned; it is entirely dependent on the terminology introduced by Bohr himself, which was based on his view of the quantum theory, but, for just that reason, is not to be overlooked. We shall see that Bohr regarded a certain individuality or wholeness as essential to truly quantum phenomena, and sought thus to find a new use for a genuine ontological category. In contrast to the situation in quantum theory, he writes (and here we continue a previous quotation which was shown incompletely) that "within the scope of classical physics we are dealing with an idealization, according to which *all phenomena can be arbitrarily subdivided*, and the interaction . . .", etc. (1962b, p. 91; italics E.S.). This formulation, which also occurs elsewhere (1957, p. 99; 1958, p. 4; 1961, p. 59; 1962a, p. 24) seems to depend entirely on the alleged individuality of a quantum phenomenon, which has just been mentioned and will shortly be further examined. But Bohr is not afraid to say that "the demand of unrestricted divisibility on which classical physical description rests is also clearly incompatible with that feature of wholeness in typical quantum phenomena" (1957, p. 99). If this statement is not meant in the sense of philosophical definitions, it must be possible even without reference to the quantum theory to attach a meaning to the statement that in classical physics phenomena can be unrestrictedly subdivided. Bohr, however, did not explain this meaning further.

I.2(d) The concept of a phenomenon: introductory remarks

Having stated in subsection (b) the Bohr programme for the interpretation of quantum mechanics by means of the inclusion of the experimental conditions in which phenomena occur, which according to subsection (a) is a procedure unknown in classical physics, and having then in subsection (c) indicated the features of classical physics which allow it to ignore those conditions, we can now begin to describe the positive execution of the programme. For this purpose, some introductory comments must first be made concerning the concepts on which this execution is based. These include, firstly, the concept of a *phenomenon*, introduced by Bohr himself, which occupies a key position; secondly, concepts which Bohr borrowed from classical physics and modified, such as those of *object*, *measuring apparatus* and *interaction* between an object and the apparatus that is used to measure it.

Bohr did not initially use the concept of a phenomenon with the necessary clarity and consistency. His aim was evidently, from the start, to express by means of this concept the fact that an object, an apparatus and their mutual interaction form a whole in a manner which is unknown to classical physics but fundamental to quantum theory, and which no longer allows the object to be said to have any behaviour of its own, thus vitally affecting the meaning of the concept of an object, in particular. This aspect, however, does not appear clearly at first, e.g. in 1927 (p. 53) he writes that in classical physics "the phenomena concerned may be observed without disturbing them appreciably", whereas according to the quantum theory "any observation of atomic phenomena will involve an interaction with the agency of observation not to be neglected". According to Bohr's later statements, the word "phenomena" in this passage should be replaced by the word "objects"; the concept of a phenomenon includes not only the object but also the agency of observation and their interaction too. Bohr probably did not fully appreciate this point before his study of the "EPR paradox" (Einstein, Podolsky and Rosen 1935; Bohr 1935); as he later put it, "The unaccustomed features of the situation with which we are confronted in quantum theory necessitate the greatest caution as regards all questions of terminology. Speaking, as is often done, of disturbing a phenomenon by observation, . . . is, in fact, liable to be confusing, since all such sentences imply a departure from basic conventions of language which . . . can never be unambiguous. It is certainly far more in accordance with the structure and interpretation of the quantum mechanical symbolism, as well as with elementary epistemological principles, to reserve the word 'phenomenon' for the comprehension of the effects observed under given experimental conditions" (1939, p. 24). Similarly, "Phrases often found in the physical literature, as 'disturbance of phenomena by observation' . . . represent a use of words like 'phenomena' and 'observation' . . . hardly compatible with common usage . . . As a more appropriate way of expression, one may strongly advocate limitation of the use of the word *phenomenon* to refer exclusively to observation obtained under specified circumstances, including an account of the whole experiment" (1948, p. 317; cf. also 1949, p. 63f.; 1954, p. 73; 1958, p. 5). But, even apart from the establishment of a consistent use of the word "phenomenon" and the rejection of unsuitable phraseology, Bohr did not expressly emphasize the decisive point in his interpretation of quantum mechanics until after 1935, although he then began to repeat it very frequently. From the beginning, he had referred to the limitations of classical concepts when applied to atomic objects, and the corresponding interaction between object and apparatus of measurement, but only after 1935 are there unmistakable expressions of his realization that an unambiguous interpretation of the quantum-mechanical formalism can be achieved only by an equally unambiguous description of the phenomena, with the experimental arrangements consistently included in this description: "The essential lesson of the analysis of measurements in quantum theory is thus the emphasis on the necessity, in the account of the phenomena, *of taking the whole experimental arrangement into consideration*, in complete conformity with the fact that all unambiguous interpretation of the quantum mechanical formalism involves the fixation of the external conditions, defining the initial state of the atomic system concerned and the character of the possible predictions as regards subsequent observable properties of that system. Any measurement in quantum theory can in fact only refer either to a fixation of the initial state or to the test of such predictions, *and it is first the combination of measurements of both kinds which constitutes a well-defined phenomenon*" (1939, p. 20; italics E.S.).

Thus we see here that the concept of a phenomenon has been made part of just the programme laid down in subsection (b) for taking into consideration the experimental conditions which alone allow atomic processes to be perceived. The question remains how far this treatment should tend towards the actual *observer* as a conscious being who is *aware* of the experimental results. On this question it must be stated quite clearly that Bohr never regarded the role of the conscious observer as peculiar to quantum mechanics, so that the observer as *subject* remains entirely outside a phenomenon. The only assertion made by Bohr in this respect is that there is some *analogy* between the consequences of a certain conception of the object-apparatus relationship in physics and the consequences of the corresponding conception of the object-subject relation in psychology. For example, using the earlier terminology as regards the concept of a phenomenon, he writes "The impossibility of distinguishing in our customary way between physical phenomena and their observation places us, indeed, in a position quite similar to that which is so familiar in psychology where we are continually reminded of the *difficulty of distinguishing between subject and object*" (1929a, p. 15). Analogies of this kind were first put forward in 1929b (p. 92 ff.), and recur in the same form in almost all later publications on the subject of complementarity. But in respect of the problem in quantum physics, which is always discussed first, reference is made only to apparatus and to measurements or observations, never to the observer as subject. It was certainly in accordance with this view of Bohr's that Heisenberg as early as 1930 expressly pointed out that in the Copenhagen interpretation there need be no observer-subject: "The observing system need not always be a human being; it may also be an inanimate apparatus" (Heisenberg 1930, p. 58), and Bohr himself expressly opposed von Neumann's view that the observer cannot be eliminated from quantum theory. In the discussion at a conference in Warsaw in 1938, at which Bohr and von Neumann were present, the former's view on this matter appears as follows: "As for Professor von Neumann's remark about the separation of the phenomenon and the observer, he [Bohr] thought that the problems we were dealing with in quantum theory were naturally defined as soon as the phenomenon was fixed into the descriptive framework of daily life, whether it were a case of finding the place where an electron or a photon was absorbed into a photographic plate rigidly attached to the reference system, or if it concerned direct visual impressions, for which the retina acted as a photographic plate. Any question of observation beyond those limits was, in his opinion, a philosophical problem, common to all domains of knowledge, and for which atomic theory was in no way distinct from classical physics" (1939, p. 45). Explicit statements along the same lines occur later (1954, p. 74; 1955, p. 90f.; 1958, p. 3; 1961, p. 60; 1962a, p. 24); notice especially the following: "The description of atomic phenomena has . . . a perfectly objective character, in the sense that *no explicit reference is made to any individual observer*' (1958, p. 3; italics E.S.).

Bohr also makes clear how a phenomenon is completed, namely by the generation of records from irreversible processes: "The observational problem in atomic physics is free of any special intricacy, since in actual experiments all evidence pertains to observations obtained under reproducible conditions and is expressed by unambiguous statements referring to the registration of the point at which an atomic particle arrives on a photographic plate or to a corresponding record of some other amplification device. Moreover, the circumstance that all such observations involve processes of essentially irreversible character lends to each phenomenon just that inherent feature of completion which is demanded for its well-defined interpretation within the framework of quantum mechanics" (1948, p. 317; and similarly in 1949, p. 51; 1954, p. 73; 1955, p. 88f.; 1957, p. 98; 1958, p. 3; 1961, p. 61; 1962b, p. 92). Thus it is irreversible amplification effects in macroscopic objects acting as measuring apparatus which complete an atomic process and make it a phenomenon. The subsequent observation of these effects by the experimental physicist has no relevance to the actual quantum-theory problem of the concept of observation. This lies rather between the atomic object and the apparatus which records an effect. The final observation of a phenomenon therefore offers no problem from the standpoint of quantum theory. The only question is how to *define* any (atomic) phenomenon, since this is presumed to convey some information about an atom or other object, yet such information cannot be meaningful without a proper inclusion of the measuring apparatus and the recorded results.

I.2(e) The essential conditions for a quantum phenomenon

The foregoing discussion of the concept of a phenomenon is of course purely preliminary: the inclusion of the experimental arrangement in the description of the object, when discussed merely as a possibility, is not a specific feature of quantum mechanics; it could equally well be proposed in classical physics. Such a situation can become important only when arguments are given to show that in classical physics this inclusion is not necessary. Bohr (1927) attempted to do this for the actual *quantum phenomena* in what seems at first to be a very crude manner, by establishing a postulate which he provisionally called the "quantum postulate". This name, which will also be used here, was originally intended to refer to the much more specialized quantum postulates in the older quantum theory of the hydrogen atom (cf. 1939, p. 13). In its general form, using phraseology typical of Bohr, it can be stated as follows:

Quantum postulate. Every quantum phenomenon has a feature of *wholeness* or *individuality* which never occurs in classical physics and which is symbolized by the Planck quantum of action (1927, p. 53; 1929a, p. 4; 1935, p. 697, col. 2; 1937, p. 291; 1939, p. 13; 1948, p. 313; 1949, passim; 1954, p. 71; 1956, p. 86, col. 2; 1957, p. 98; 1958, p. 2).

The physicist's initial reaction to the statement of such a postulate is likely to be somewhat reluctant, with the demand for a further explanation of what is meant here by "wholeness" and "individuality", since otherwise it is not clear what use the postulate will be. But it must be remembered that the only choice is either to omit completely any such description as is given by Bohr, or to appeal to a certain appreciation of fundamental concepts like wholeness and individuality, such as everyone possesses, before they appear in the new context of the quantum theory. In taking the second option, Bohr continues from the resolution of a phenomenon into object, apparatus and their interaction, as described in subsection (d), and asserts that *these* form an inseparable whole in a quantum phenomenon, but not in a classical phenomenon. This can be stated, even though we have already distinguished the three components in a quantum phenomenon (object, apparatus and interaction), and it does not exclude the possibility of giving further details about each of these. It implies only that the treatment of any one of the components must pay significant regard to the others.

If we first consider the relevant object in this way, there is no doubt that Bohr would refer to the *quantum-mechanical formalism* for the treatment of the object, to the extent that such treatment can take place in isolation, though warning us that this formalism will not furnish any new physical concepts. Although, as has been described in section 1, Bohr had no enthusiasm for formalisms, he nevertheless regarded the establishment of the mathematical apparatus of quantum mechanics as a convincing performance, which in particular guarantees the self-consistency of the entire theory. In most of his publications he refers to this formalism and selects from it the parts which he considers particularly important, such as the non-commutativity of the algebra, the canonical commutation formulae, the Heisenberg uncertainty relations (both the latter involving the Planck quantum of action), and perhaps also the Schrödinger wave function (see in particular 1935, passim; 1939, p. 14ff.; 1948, p. 314; 1949, p. 38f.; 1956, p. 87, col. 2; 1958, p. 5). He also uses the phrase "the interpretation of the quantum-mechanical formalism", although here we must begin to impose the warnings against any tendency to infer independence, which have already been mentioned and will appear again later. Furthermore, Bohr (see especially 1935, passim) uses the term "quantum-mechanical description" of a phenomenon, process or even object, meaning in every case that the *object* is treated by the quantum-mechanical formalism and especially that the uncertainty relations apply to it.

The formal quantum-mechanical treatment of an object is, however, not something that by itself could have any physical significance. Here the wholeness of a phenomenon comes into play, and demands that the corresponding apparatus should also be taken into account, for only by means of the apparatus can the properties of the object be meaningfully defined: "a measurement can mean nothing else than the unambiguous comparison of some property of the object under investigation with a corresponding property of another system, serving as a measuring instrument" (1939, p. 19; see also 1935, p. 699f.; 1937, p. 291). Here the reference is made classically to a property of the object, which is determined by being correlated with a property of the apparatus. In a quantum phenomenon, however, measurement does not *determine* a property of the object so much as essentially *define* it in the first place, as far as is possible. If the procedure is to be meaningful, it is certainly necessary that the properties of the apparatus should be well established, unlike those of the object; otherwise, there would be no benefit in using the apparatus. Bohr expressed this point in a further requirement, to which he gave no name; for clarity, we shall call it (for want of a better name) the "buffer postulate". The function of the postulate is to use classical physics as a buffer against the quantum-mechanical treatment of a phenomenon. Here again, Bohr's work provides no standard formulation; like the quantum postulate, it was embodied in his discussions with many varieties of wording and shades of meaning. Our tentative formulation is as follows:

Buffer postulate. The description of the apparatus and of the results of observation, which forms part of the description of a quantum phenomenon, must be expressed *in the concepts of classical physics* (including those of "everyday life"), eliminating consistently the Planck quantum of action. (See 1927, p. 53; 1929a, pp. 5 and 15f.; 1929b, p. 94f.; 1935, passim; 1937, p. 293; 1938, p. 25f.; 1939, p. 19; 1948, p. 313; 1949, p. 39; 1954, p. 72; 1955, p. 88f.; 1956, p. 87, col. 1; 1958, p. 3; 1962b, p. 91.)

The buffer postulate is to be regarded as one of the essential features of Bohr's interpretation of quantum mechanics. It is true that the aspect in which quantum mechanics deviates from classical physics is expressed by the quantum postulate, and the description of a quantum phenomenon as regards the object contained in it calls for the use of the quantum-mechanical formalism. But the non-classicality of a quantum phenomenon has an impassable limit in the requirement that it can be experimentally perceived. The buffer postulate is designed to take account of this fact. It restricts the quantum postulate by demanding that experimental perception must be possible; it begins, so to speak, with a "But". The individuality of a quantum phenomenon is classically irrational. *But* (and now follows a verbatim quotation) "*however far the phenomena transcend the scope of classical physical explanation, the account of all evidence must be expressed in classical terms.* The argument is simply that by the word 'experiment' we refer to a situation where we can tell others what we have done and what we have learned and that, therefore, the account of the experimental arrangement and of the results of the observations must be expressed in unambiguous language with suitable application of the terminology of classical physics" (1949, p. 39).

It is necessary to avoid any misunderstanding of the buffer postulate, and in particular to emphasize that the requirement of a classical description of the apparatus is not designed to set up a special class of objects differing fundamentally from those which occur in a quantum phenomenon as the things examined rather than measuring apparatus. This requirement is essentially epistemological, and affects an object only *in its role as apparatus.* A physical object which may act as apparatus may in principle also be the thing examined, and is then likewise subject to the quantum postulate: "the construction and the functioning of all apparatus like diaphragms and shutters, serving to define geometry and timing of the experimental arrangements, or photographic plates used for recording the localization of atomic objects, will depend on properties of materials *which are themselves essentially determined by the quantum of action*" (1948, p. 315; see also 1937, p. 294; 1949, p. 51; 1958, p. 2; 1962b, p. 91). As well as this positive formulation, we find also the negative formulation that "an independent reality in the ordinary physical sense can neither be ascribed to the phenomena *nor to the agencies of observations*" (1927, p. 54; see also p. 66). Concerning the limit of applicability of classical concepts, in accordance with the quantum postulate, Bohr again uses a negative formulation, but with epistemological instead of ontological concepts: "the limitation in question applies to any use of mechanical concepts and, hence, *applies to the agencies of observation as well* as to the phenomena under investigation" (1929a, p. 11; italics in this paragraph E.S.). All this shows very clearly that the buffer postulate demands the classical description of a measuring apparatus only if a physical object is being used *as such apparatus*.

Since therefore every physical object is basically governed by quantum theory, it must in turn be explicitly emphasized that certain objects are *capable* of classical description as measuring apparatus, and hence that the buffer postulate is capable of being satisfied. It is evident from the above that the conditions to be stated here must be approximate ones, as Bohr himself stressed in various places (1948, p. 315f.; 1957, p. 98; 1958, p. 3; 1962b, p. 91). The objects that are actually used as measuring apparatus are sufficiently heavy for the limitations imposed by the quantum of action on their localizability in space and time to be negligible. Thus there is in fact (though not in principle) an absolute distinction in this approximation between objects which can be described classically and therefore used as apparatus, and those which cannot.

The result of the quantum postulate and buffer postulate together was also sometimes summarized by Bohr in the following terms: "the essentially new feature in the analysis of quantum phenomena is . . . the introduction of a *fundamental distinction between the measuring apparatus and the objects under investigation*" (1958, p. 3; see also 1935, p. 701, col. 1; 1948, p. 315; 1949, pp. 50 and 55f.; 1955, p. 89). The comprehension of this has been prepared above, by making it clear that the difference between the measuring apparatus and the object under investigation affects both of these only in their mutual relationship. For the theoretical treatment of a quantum phenomenon, a decision has to be made as regards which is the apparatus and which is the object, and this distinction is a fundamental one, since, in contrast to the situation in classical physics, the two are described in totally different ways. The apparatus is governed by classical physics, the object by the quantum-mechanical formalism: "While . . . in classical physics the distinction between object and measuring agencies does not entail any difference in the character of the description of the phenomena concerned, its fundamental importance in quantum theory . . . has its roots in the indispensable use of classical concepts in the interpretation of all proper measurements, even though the classical theories do not suffice in accounting for the new types of regularities with which we are concerned in atomic physics" (1935, p. 701, col. 1).

The relative and (at the same time) fundamental nature of the difference between apparatus and object is also emphasized by what is sometimes referred to as the movable division or "cut" between the two (1927, pp. 54 and 67; 1935, p. 701; 1939, p. 23f.; 1948, p. 315). The point here is that, firstly, there is some arbitrariness as regards the choice of what is apparatus and what is object in a quantum phenomenon, i.e. where the boundary is placed between the classical and quantum-mechanical descriptions. The existence of such freedom is clear evidence that the difference between apparatus and object is relative. On the other hand, this freedom is restricted not only by the fact that both components with their different descriptions must be present, but also by the fact that the division between them has to be placed "within a region where the quantum-mechanical description of the process concerned is effectively equivalent with the classical description" (1935, p. 701, col. 2). Any movement of the division must therefore leave unchanged the sharp distinction in the treatment of apparatus and object.

I.2(f) Consequences as regards the interaction in measurement

In subsection (e) the features of a quantum phenomenon have been exhibited mainly in terms of the object and the apparatus. But they are not fully elucidated until the *interaction* between the two has also been considered. It is indeed typical of Bohr that, in order to characterize the situation in quantum theory, he made statements about the interaction in measurement and gave these a central position in the argument of all his relevant work. This interaction is stated to be *finite*, *inevitable*, *not negligible*, and *not separately accountable*, and it seems as if Bohr desired to present these results as a *consequence* of the two postulates described in subsection (e), leading in turn to further consequences, in particular the non-independence of the object (1927, pp. 53–68 passim; 1929a, p. 11f.; 1935, p. 697, col. 1; 1937, p. 293; 1938, pp. 26, 39; 1939, p. 19; 1948, p. 313; 1955, p. 89f.; 1956, p. 87, col. 1; 1961, p. 60; 1962a, p. 24; 1962b, p. 91f.).

Careful citation is again necessary to make clear Bohr's opinions on this subject. Let us take first a fairly well-considered formulation from a late work (1961, p. 60): "The element of wholeness, symbolized by the quantum of action and completely foreign to classical physical principles has, however, the *consequence* that in the study of quantum processes any experimental inquiry implies an interaction between the atomic object and the measuring tools which, although essential for the characterization of the phenomena, evades a separate account if the experiment is to serve its purpose of yielding unambiguous answers to our questions". This assertion follows immediately after the one mentioned in subsection (c), that in classical physics it is assumed that the interaction in measurement can be either neglected or compensated. The text now quoted might suggest that the first possibility, neglecting the interaction, is excluded by the individuality of a quantum phenomenon, whereas the second possibility, explicit allowance for the interaction, after neglecting it has been shown to be impossible, is likewise impossible, because of the classical description of the apparatus. There are other passages also which indicate Bohr's desire to follow this line of argument, for example: "The quantum postulate *implies* that any observation of atomic phenomena will involve an interaction with the agency of observation not to be neglected" (1927, p. 54); "the finite interaction between object and measuring agencies *conditioned* by the very existence of the quantum of action" (1935, p. 697, col. 1); then, more briefly, the "interaction *implied* by the quantum" (1955, p. 89f.). Conversely, the buffer postulate is stressed in relation to the precise surveillance of the interaction: "*Just* the necessity of accounting for the functioning of the measuring agencies on classical lines *excludes* in principle in proper quantum phenomena an accurate control of the reaction of the measuring instruments on the atomic objects" (1956, p. 87, col. 1; italics in this paragraph E.S.).

There is some plausibility in assigning the two components concerned here (the impossibility of neglecting the interaction and the impossibility of precisely surveying it) to the quantum postulate and to a *combination* of this with the buffer postulate respectively. The buffer postulate *alone* cannot have any consequences which contradict classical physics: for its essential function is to establish classical physics as an indispensable part of quantum theory. It therefore cannot alone imply the impossibility of neglecting or precisely surveying the interaction, since both these are classically permissible (cf. subsection (c)). Conversely, Bohr seems to imply that the quantum postulate *by itself* implies the occurrence of a finite interaction, which cannot be made arbitrarily small. The idea of the individuality of a quantum phenomenon is here directly applied to the interaction between two objects. In the sense that there is a lower bound to the interaction, represented in fact by the quantum of action, it is *in principle* impossible to neglect the interaction, although this does not of course exclude its being neglected *in practice* in the classical limit. In certain critical cases, however, the interaction cannot be neglected; then the only other possibility would be that of surveillance of the interaction. This path, however, is blocked by the buffer postulate, and the reason is as follows. "The necessity of basing the description of the properties and manipulation of the measuring instruments on purely classical ideas implies the neglect of all quantum effects in that description, and in particular the renunciation of a control of the reaction of the object on the instruments more accurate than is compatible with the relation (5)" (1939, p. 19), the latter being the Heisenberg uncertainty relation. From this it appears that (1) precise surveillance of the interaction is impossible because the quantum of action is neglected as regards the apparatus, in accordance with the buffer postulate, (2) the quantum of action in fact specifies the degree of impossibility. This shows once again that the apparatus too is basically subject to the quantum-mechanical description, but nevertheless the buffer postulate implies that "in each case some ultimate measuring instruments . . . must always be described entirely on classical lines, and consequently kept outside the system subject to quantum mechanical treatment" (1939, p. 24). The error thus committed in respect of the apparatus is then held responsible for the impossibility of precise surveillance of the interaction.

I.2(g) The non-independence of an object, and the limited applicability of classical concepts to it

Bohr proceeded to draw further conclusions from the impossibility of neglecting or surveying the interaction. Of these, we shall first of all describe those which involve the non-independence of the object, and the corresponding limitation in the applicability of classical concepts to it. The arguments are difficult to analyse, but it will probably be in accord with Bohr's intention to begin by saying that, as a consequence of the impossibility of neglecting or surveying the interaction in a quantum phenomenon, the interaction must be an *integral constituent* of the phenomenon. For example: "The fundamental difference with respect to the analysis of phenomena in classical and in quantum physics is that in the former the interaction between the objects and the measuring instruments may be neglected or compensated for, while in the latter *this interaction forms an integral part of the phenomena*" (1954, p. 72; the same contrast occurs in 1937, p. 291; 1939, p. 19; 1962b, p. 92). The meaning is clearly that in quantum theory, unlike classical physics, the assumption of neglect or surveillance of the interaction in measurement is prohibited, and therefore that this interaction is an essential part of the phenomenon itself. This shows once more the individuality or wholeness of a quantum phenomenon. The passage just quoted continues as follows: "The essential wholeness of a proper quantum phenomenon finds indeed logical expression in the circumstance that any attempt at its well-defined subdivision would require a change in the experimental arrangement incompatible with the appearance of the phenomenon itself" (similarly 1937, p. 291; 1939, p. 19f.; 1948, p. 313; 1949, p. 40; 1955, p. 90; 1956, p. 87, col. 1; 1958, p. 4). This "attempt at subdivision" of the phenomenon is intended to mean an attempt to survey the interaction by changing the experimental arrangement, which is a part of the phenomenon: one variation of the last quotation reads "The individuality of the typical quantum effects finds its proper expression in the circumstance that any attempt of subdividing the phenomena will demand a change in the experimental arrangement introducing *new possibilities of interaction* between objects and measuring instruments which in principle cannot be controlled" (1949, p. 40; italics in this paragraph E.S.; similarly in 1948, p. 313). The interaction is therefore an integral constituent of the phenomenon, in the sense that the phenomenon is destroyed by attempting a complete surveillance of the interaction. To understand this point correctly, one must remember that the interaction concerned is not itself the thing examined; in that case there would be nothing surprising in the destruction of the phenomenon by influencing the interaction. The interaction is rather between measuring equipment and an object whose behaviour is to be investigated, and such an interaction is normally thought of as lying outside the phenomenon itself.

Next, we return to a consideration of the object; this can no longer be regarded as having any behaviour distinct from the interaction in measurement. Bohr (1937, p. 293) refers to "our not being any longer in a position to speak of the autonomous behavior of a physical object, *due to* the unavoidable interaction between the object and the measuring instruments which in principle cannot be taken into account, if these instruments according to their purpose shall allow the unambiguous use of the concepts necessary for the description of experience"; similarly (1938, p. 25; italics in the rest of this paragraph E.S.) "the unavoidable interaction between the objects and the measuring instruments *sets an absolute limit* to the possibility of speaking of a behaviour of atomic objects which is independent of the means of observation"; (1939, p. 19) "above all, we must realize that this interaction cannot be sharply separated from an undisturbed behavior of the object, *since* the necessity of basing the description of the properties and manipulation of the measuring instruments on purely classical ideas implies . . . the renunciation of a control of the reaction of the object on the instruments" (cf. also 1927, p. 54; 1929a, p. 11f.; 1949, p. 39f.). The understanding of Bohr's views is vitally dependent on close attention to the consequences of this non-independence of the object as regards both its properties and its behaviour within a quantum phenomenon: "From the above considerations it should be clear that the whole situation in atomic physics *deprives of all meaning* such inherent attributes as the idealizations of classical physics would ascribe to the object" (1937, p. 293). In this connection, there are also stern warnings against such improper phraseology as "the creation of physical properties by measurement", which might suggest itself if things do not comprise their own properties and if the measurements are regarded as essentially included in the physical interpretation of the phenomena (1939, p. 24; 1948, p. 317; 1954, p. 73; 1958, p. 5). On the other hand, Bohr still does not deviate in the slightest from his conviction, already embodied in the buffer postulate, that the quintessence of all that can be experienced can be expressed only by the concepts of classical physics and everyday life. The non-independence of a quantum-mechanical object can accordingly be expressed only in negative terms, as a *limited applicability of the concepts of classical physics* in its description. Bohr expressed in these terms too the consequences of the non-independence of the object: "In this situation, an inherent element of *ambiguity* is involved in assigning conventional physical attributes to atomic objects" (1948, p. 313f.; similarly in 1949, pp. 40, 51). There belong here also all the statements, quoted in subsection (b) in connection with the programme as a whole, which relate to the ambiguity of classical concepts as applied to atomic objects. As we have now seen, this ambiguity can be eliminated by returning to the measuring apparatus and simply making it subject to classical physics.

The resulting unambiguous position relates, however, only to the apparatus, and the subsequent transfer of some of its properties to objects remains essentially restricted, in a manner and to an extent which are still to be considered.

I.2(h) COMPLEMENTARITY

The restriction mentioned a moment ago must, according to the preceding discussion, be that the measurements used to define a quantum phenomenon do not lead to a description of the object which is *classically* complete. The completeness demanded by classical physics can be imposed only at the cost of destroying the phenomenon concerned, since the different experimental arrangements that are necessary belong to different phenomena. This situation was repeatedly depicted by Bohr on the basis of two fundamental types of measurement: "Any phenomenon in which we are concerned with tracing a displacement of some atomic object in space and time necessitates the establishment of several coincidences between the object and the rigidly connected bodies and movable devices which, in serving as scales and clocks respectively, define the space-time frame of reference to which the phenomenon in question is referred. Just this situation implies, however, a renunciation of any sharp control of the amount of momentum or energy exchanged during each coincidence between the object and the separate bodies entering into the experimental arrangement. Inversely, every phenomenon in which we are essentially concerned with momentum and energy exchanges—and which therefore necessitates an experimental arrangement allowing at least two successive determinations of momentum and energy quantities—will, in principle, imply a renunciation of the control of any precise space-time co-ordination of the objects in the time intervals between these measurements" (1939, p. 22, and similarly in *all* his other relevant publications).

Thus we have here a failure of the combination of space-time coordination and dynamical conservation laws, which has been quoted in subsection (a) as characteristic of classical mechanics, and which is the basis of a deterministic and pictorial description of the behaviour of an object. Bohr evidently regarded this failure as a direct *consequence* of the impossibility of complete surveillance of the interaction in measurement. This is expressly stated: "the renunciation in each experimental arrangement of the one or the other of two aspects of the description of physical phenomena,—the combination of which characterizes the method of classical physics, . . .—depends essentially on the impossibility, in the field of quantum theory, of accurately controlling the reaction of the object on the measuring instruments, i.e., the transfer of momentum in case of position measurements, and the displacement in case of momentum measurements" (1935, p. 699, col. 1). Furthermore, the limitation of classical concepts for describing the behaviour of the object becomes qualitatively clear: in one quantum phenomenon, only a part of the classically possible characterization of an object can occur, while the remaining part can arise only through a new experimental arrangement incompatible with the old, and hence only in a different quantum phenomenon. For example, surveillance of the interaction suffices *either* only to determine the position of a particle within a specified region of space *or* only to determine its momentum in a corresponding manner.

Bohr devised the idea of *complementarity* to describe such situations. This is generally regarded as a decisive step in his interpretation of quantum mechanics, probably largely because Bohr himself sought to apply the concept of complementarity to other situations having no relevance to physics but analogous to those which exist in the quantum theory.

The Logical Analysis of Quantum Mechanics

Without seeking to deny the considerable significance of complementarity in Bohr's interpretation of quantum mechanics, one must point out that Bohr's contribution to our understanding of the new situation in physics is not confined to the introduction of this concept; the entire sequence of arguments enumerated here constitutes Bohr's true achievement. The concept of complementarity does appear in every one of the many ways in which Bohr put forward his arguments, but it is never the whole thing, though it may be the crowning part.

This is clear, if only because a correct understanding of the concept of complementarity demands a correct drawing together of the many threads that run through Bohr's arguments and bear the mark of this concept. The task is not eased by the extremely vague and continually changing phraseology used. The most obvious fact is that complementarity is a binary relationship: some A is complementary to some B. This insight tends to be nullified by the occasional use of the expression "complementary description" as applied to one thing, e.g. as contrasted with the "causal description"; but this is evidently only a compact figure of speech. If then we take complementarity as a binary relation, it is much less clear (quite apart from the content of the relation) what things it purports to relate. On examining all the various passages in which Bohr mentions complementarity in the context of physics, one finds a great variety of things said to be related.

For example, there is said to be *complementarity between space-time coordination* (or description) *and the requirement of causality*: "The very nature of the quantum theory thus forces us to regard the space-time coordination and the claim of causality, the union of which characterizes the classical theories, as complementary but exclusive features of the description" (1927, p. 54; also pp. 60, 68, 84, 87). This way of expressing the matter does not appear in later works (although the matter itself does), whereas in the work cited it is given a very prominent place. At the same period, Bohr *seems* also to regard the *classical wave picture as complementary to the classical particle picture*, although with a certain reservation whose nature is not made clear. In an explanation of the wave-particle duality of radiation and matter, he writes: "In fact, here . . . we are not dealing with contradictory but with complementary pictures of the phenomena, which only together offer a natural generalization of the classical mode of description" (1927, p. 56). Since the wave picture and the particle picture in fact have some contradictory consequences, this comment must include an implied limitation. Somewhat later, a new complementarity appears, between Schrödinger's wave mechanics and Heisenberg's matrix mechanics, as regards the treatment of the exchange of energy between atoms. Not only is this a further instance of complementarity, but that of the wave and particle pictures is asserted outright: "Indeed, the two formulations of the interaction problem might be said to be *complementary in the same sense as the wave and particle idea* in the description of the free individuals" (1927, p. 75; italics E.S.). In 1929a (p. 10), the discussion from 1927 is (allegedly) repeated in terms of *complementarity between classical concepts*, with allusion to "a new mode of description designated as *complementary* in the sense that any given application of classical concepts precludes the simultaneous use of other classical concepts which in a different connection are equally necessary for the elucidation of the phenomena". Which classical concepts are meant here is immediately explained by a reference to the wave-particle dilemma, presented in terms of light as "a dilemma as regards the choice between the *wave description* of the electromagnetic theory and the *corpuscular conception* of the propagation of light in the theory of light quanta" (italics E.S.), which in turn indicates the complementarity of the wave picture and the particle picture, i.e. of classically *incompatible* ideas. But shortly afterwards there is the following remark concerning the Heisenberg uncertainty relations: "This indeterminacy exhibits, indeed, a peculiar complementary character which prevents the simultaneous use of space-time concepts and the laws of conservation of energy and momentum, which is characteristic of the mechanical mode of description" (*ibid.*, p. 11). Thus this now refers to the *complementarity between space-time coordination and the dynamical conservation laws* and, consequently, to the complementarity between two aspects which are classically *compatible.* Somewhat further on, there is again a general insistence that the complementarity refers to features "which are *united* in the classical mode of description but appear separated in the quantum theory" (*ibid.*, p. 19; italics E.S.). In agreement with this statement and in a more precise manner, reference is made in a subsequent paper to complementary classical concepts (1935, p. 699, col. 2), but also to *complementary physical quantities* (*ibid.*, p. 700, col. 2, perhaps Bohr's only explicit mention of these), the coordinates and momenta of a particle being quoted as examples. The same paper speaks of "*two aspects of the description of complementary physical phenomena* [italics E.S.]—the combination of which characterizes the method of classical physics, and which therefore in this sense may be considered as *complementary* to one another" (*ibid.*, p. 699, col. 1). The context shows that the reference here is again to the two aspects of space-time coordination and the dynamical conservation laws.

Later, however, a new definition once again appears: "The apparently incompatible sorts of information about the behavior of the object under examination which we get by different experimental arrangements can clearly not be brought into connection with each other in the usual way, but may, as equally essential for an exhaustive account of all experience, be regarded as 'complementary' to each other" (1937, p. 291). Since the information got by different experimental arrangements, which is here said to be "complementary", is also said to be "apparently incompatible", the reference can scarcely be to those classical concepts, quantities or aspects whose *combination* was previously asserted to be characteristic of the classical theories. For "apparently incompatible" surely means incompatible on classical considerations alone. Nevertheless, this terminology has to be taken quite seriously, since from that time onwards it occurs repeatedly, with the word "information" sometimes replaced by "evidence" or "experience" (1938, p. 26; 1948, p. 314; 1949, p. 40; 1958, p. 4; 1962b, p. 92). Another passage (1938, p. 26) makes it particularly clear that the incompatibility is to be taken in the ordinary sense: "Information regarding the behaviour of an atomic object obtained under definite experimental conditions may, however, according to a terminology often used in atomic physics, be adequately characterized as *complementary* to any information about the same object obtained by some other experimental arrangement *excluding* [italics E.S.] the fulfilment of the first conditions. Although such kinds of information *cannot be combined into a single picture by means of ordinary concepts* [italics E.S.] they represent indeed equally essential aspects of any knowledge of the object in question which can be obtained in this domain".

The terminology of complementary information or experience forms the penultimate stage of development, very close to Bohr's final formulation of the topic, in which he refers to *mutually complementary phenomena*; this is noteworthy as representing another of Bohr's technical terms (if such can be said to exist in his work). The formulation in question was first used in 1939 (p. 24), following a passage in which the complementary relationship of the wave and particle pictures was discussed with unusual clarity: "It is evident however that any concrete *wave picture* is as unable to account for basic experience regarding the individuality of the electron as a *corpuscular picture* for the superposition properties of radiation fields. In the two cases we are dealing, in fact, with two *complementary aspects* of experience"

(1939, p. 15; italics E.S.). Then follows a mention of *complementarity between the principle of superposition and the dynamical conservation laws* (*ibid.*, p. 23); this seems to combine crosswise the wave picture and the mechanical conservation laws as relata of the complementarity relation, but will be seen later to have an explanation. Finally, Bohr lays stress on the experimental conditions as the only basis for the definition of the concepts by which a phenomenon is described: "It is just in this sense that *phenomena* defined by different concepts, corresponding to mutually exclusive experimental arrangements, can unambiguously be regarded as *complementary aspects* of the whole obtainable evidence concerning the objects under investigation" (*ibid.*, p. 24; italics E.S.). This becomes a standard formulation which differs from the formulation in terms of "information", "evidence" or "experience" (these being normally non-specific words in Bohr's usage) only in that they are replaced by the word "phenomenon": "the impossibility of combining phenomena observed under different experimental arrangements into a single classical picture implies that such apparently contradictory phenomena must be regarded as complementary in the sense that, taken together, they exhaust all well-defined knowledge about the atomic objects" (1962a, p. 25); compare the variable use of "information", etc. and of "phenomenon" (1948, pp. 314, 316, 317; 1949, pp. 40, 41, 45, 47), followed by more consistent use (1954, p. 74; 1955, p. 90; 1956, p. 87, col. 1; 1957, p. 99; 1961, p. 60).

What can we make of this, at first sight, somewhat confusing picture which results if we examine closely Bohr's *actual words* used in describing the concept of complementarity? First of all, if the various publications are traversed in chronological order, we do find a clear convergence towards the preferred expression of a *complementarity between phenomena.* One may therefore reasonably regard this as the final form of the intended assertion and treat all other wordings as derivative. Then two phenomena would be called complementary if they refer to the same object, and if the two corresponding experimental arrangements and the information which they yield are mutually exclusive while at the same time being equally important for a complete description of the object concerned. Bohr sees this exclusiveness as a sure consequence of the individuality of each quantum phenomenon, and still more directly as a consequence of the fact that there exists an interaction in measurements (see above): it would be impossible to maintain one phenomenon by using the experimental arrangement pertaining to the other phenomenon. On the other hand, both phenomena comprise experience of the *same* object, and this is the basis of the mutual completion of complementary phenomena by each other. The two aspects of complementarity thus explained in a somewhat *ontological* manner, namely *incompatibility* and *mutual completion*, can be further elucidated by reference to the possible *description* of a phenomenon. According to Bohr's fundamental view, classical physics with its pictures and concepts will here predominate, and in this special context it will be necessary in particular to test whether the argument involves the two fundamental types of classical theory which use the particle picture and the wave picture, or certain contrasted pairs of aspects of one only of these two theories.

We start from two kinds of complementary phenomena, as specified in the quotation at the beginning of this subsection. If we have, firstly, an experimental arrangement which uses diaphragms and shutters in a screen in order to obtain a *space-time coordination* between the object and the apparatus, then the experience which is obtained by means of a second screen is that of an interference pattern. The classical non-statistical interpretation of this experience is based on the *wave picture*. Secondly, from an experimental arrangement which allows a measurable exchange of momentum and energy of the object and hence a test of the *dynamical conservation laws*, these are found to be valid in each case, e.g. for the Compton effect. The classical interpretation of this experience is based on the *particle picture*.

Thus there is, first of all, an assignment of the wave picture to one of the two mutually complementary phenomena, and of the particle picture to the other. On the basis of this assignment Bohr feels able to say, as we have seen, that the *wave picture and the particle picture are mutually complementary*. With only an alternative between these two pictures, the assignment is in fact unambiguous, since the Compton effect contradicts the classical wave theory and the interference phenomenon contradicts classical particle mechanics, whereas each can be explained by the other theory. In this way the complementarity between the wave picture and the particle picture is associated with the complementarity of phenomena, and makes evident the incompatibility of the two phenomena in one classical picture, their contrastedness, and other aspects mentioned by Bohr in this connection: the corresponding classical pictures are obviously *incompatible*, and a decision in favour of the one or the other is not in question, since experience of the *same* object has to be interpreted. Although this makes clear the *incompatibility* of two quantum phenomena from the classical standpoint, the complementarity of the particle and wave pictures cannot by itself throw light on the *mutual completion* of two quantum phenomena; this is shown only by the fact that the other picture has to be invoked in order to explain certain experiences relating to the object concerned.

In order to reflect in classical physics this second feature of the complementarity of two quantum phenomena, another approach is needed, this time *within one classical picture only*. Taking the particle picture, for example, and the corresponding classical mechanics of particles, let us again consider the above two types of phenomena. The first involved a space-time coordination, and the second the dynamical conservation laws. Here there is a further correlation, that of the space-time coordination with the first phenomenon and of the conservation laws with the second, and here too Bohr states that the *space-time coordination and the dynamical conservation laws are complementary*. The assignment is again unambiguous, since the converse is impossible because the interaction in measurement cannot be surveyed, whereas the assignment stated in fact allows a prediction of the position or momentum immediately after the interaction with the diffracting screen or the auxiliary object under investigation. But now, in clear contrast to the case discussed previously, there is an associated complementarity between aspects which are *combined* in a classical theory, and whose compatibility in such a theory Bohr indeed stresses as characteristic of the latter; see subsection (a). Thus this complementarity certainly cannot give a classical account of the incompatibility of two complementary quantum phenomena. Instead, following Bohr's effort to give classical physics the maximum possible significance in the quantum theory, it must apparently serve to emphasize the *mutual completion* of complementary phenomena. The fact that this emphasis was indeed intended by Bohr is shown by his resumption of the word "complementarity", which he had temporarily abandoned in favour of "reciprocity": "The term 'complementarity', which is already coming into use, may perhaps be more suited also to remind us of the fact that it is the *combination* of features which are *united* in the classical mode of description but appear separated in the quantum theory that ultimately allows us to consider the latter as a natural generalization of the classical physical theories" (1929a, p. 19; italics E.S.).

There is no need to give here a similar discussion with the wave picture in place of the particle picture, since we are concerned only with quantum mechanics, for which particle mechanics acts as the classical limit. To summarize again what has been said so far: the basis is taken to be a complementarity between phenomena; complementarity between information, evidence or experience can be regarded as essentially equivalent to this, since the information (etc.) is in one-to-one correspondence with the phenomena, and the varying phraseology is no doubt due simply to Bohr's wish to avoid any fixed expressions. Since he shows a definite preference for the concept of a phenomenon, this should be made the basis of the complementarity relation. Complementarity in this primary sense, i.e. viewed as relating phenomena, is an extensional relation. Next, a derived complementarity between the wave picture and the particle picture has been demonstrated. This is not, strictly speaking, an extensional relation which can exist between some things and not between others: the basis is simply the wave and particle pictures, and if these are said to be complementary, this has no extensional significance unless one says what is, in the same sense, *not* complementary. But the expression can be regarded as asserting that the antithesis of the wave and particle pictures as two mutually exclusive representations expresses a feature common to all pairs of complementary phenomena, namely their incompatibility. This is true to the extent that, of two complementary phenomena, one is better interpreted by the wave picture and the other by the particle picture. Lastly, there is a further derived complementarity between space-time coordination and dynamical conservation laws. Formally, the same is true here as for the complementarity of the wave and particle pictures: it is not a question of an extensional relation. But the point now is to clarify the other basic feature of any pair of complementary phenomena: their mutual completion is expressed by the fact that space-time coordination and the conservation laws occur in combination in classical particle mechanics (the classical limit of quantum mechanics). Such clarification is possible to the extent that, of two complementary phenomena, one involves a space-time coordination and the other a momentum-energy balance.

On this view, only one of the two basic features of the original complementarity between phenomena is expressed in each of the two derived complementarities: for the complementarity between the wave picture and the particle picture, their incompatibility for describing the same classical object; for the complementarity between space-time coordination and the dynamical conservation laws, their joint appearance to complete each other within the same classical theory. But the opposite correlations can and should be considered also: the mutual completion of the wave picture and the particle picture, and the incompatibility of space-time coordination and dynamical conservation laws. The previous correlations still reflect the situation in *classical physics*, whereas the new correlations are classically unintelligible, and give a first indication of the peculiar features of *quantum mechanics*. If the above two complementary types of phenomena are again considered, we must say, to begin with, that the wave picture and the particle picture *complete each other* to the extent that these phenomena to be represented by the two pictures relate to the same object, so that both pictures must be used for a full description of the object. The pictures being mutually exclusive from the classical standpoint, however, this situation can be understood only in quantum-mechanical terms. Correspondingly, space-time coordination and dynamical conservation laws are *mutually exclusive* to the extent that two phenomena cannot occur simultaneously if one involves a space-time coordination and the other an application of the dynamical conservation laws. Since space-time coordination and the application of the dynamical conservation laws occur together in classical mechanics, this again can be understood only in terms of quantum mechanics.

The interpretation proposed here may be said to cover all aspects of Bohr's use of the term "complementarity", with a certain reservation regarding a few obscure passages. The foregoing survey, though not complete, may assist the reader in checking that all aspects are in fact covered. The terminology employed does not exactly agree with the standardization proposed here, for example, in referring to complementarity between the principle of superposition and the dynamical conservation laws (1939, p. 23). By this, however, Bohr evidently means the complementarity between the wave picture and the particle picture, since the phenomena mentioned are interference and the Compton effect. The former can be explained classically by the wave picture, using the superposition principle in wave theory; the Compton effect can be explained classically by the particle picture, using the conservation laws. Thus the wording used by Bohr in this passage arises because he is taking from each picture the features which are ultimately decisive in explaining the corresponding phenomenon. The situation is somewhat different as regards accommodating the complementarity between space-time coordination and causality, on which Bohr laid such stress in 1927. In subsection (a) the combination of these two features in the classical theories has been described as an earlier form of the confrontation of space-time coordination and dynamical conservation laws, which alone appears in later papers. To be consistent, the same should now be asserted for the corresponding complementarity also: causality does not rank as being alongside and complementary to space-time coordination, but rather as embracing the combination of space-time coordination and dynamical conservation laws, whose complementarity implies the absence of causality from quantum theory. One might at most say that quantum theory contains, in the form of the conservation laws, whatever fraction of causality remains valid.

I.2(i) Uncertainty relations

It was Bohr's custom to mention and discuss the Heisenberg uncertainty relations

$$\Delta q \cdot \Delta p \geqslant \tfrac{1}{2}\hbar \tag{1}$$

in close connection with the concept of complementarity. Here q denotes the coordinate and p the momentum of a particle. There are passages in almost all Bohr's papers on the subject which show that as regards (1) he was considering the complementarity between space-time localization and dynamical conservation laws rather than the other types of complementarity described in subsection (h). In the non-relativistic case, time and energy can be ignored. Then the complementarity in question, for a free particle, comprises firstly a spatial localization at some instant, ultimately based on measuring instruments which create a fixed frame of reference in space and have a purely classical description, and secondly the determination of the momentum by applying the appropriate conservation law. Formally, the Δq and Δp in the uncertainty relations would be respectively allocated to these constituents, but there is a difficulty as regards the *significance* of the allocation: Bohr regarded the complementarity between space-time coordination and the dynamical conservation laws (which we have now converted into a complementarity between the determination of position and that of momentum) as being presented by *two* complementary phenomena, with one phenomenon involving a position determination and the other a momentum determination, depending on the experimental arrangements used, and the two determinations are mutually exclusive. Yet there is no doubt that the uncertainty relations must be interpretable in terms of *one and the same* phenomenon only.

How then did Bohr overcome this difficulty? If we successively examine his comments about the interpretation of the uncertainty relations, we first of all find the warning that "it is important to recognize that no unambiguous interpretation of such relations can be given in words suited to describe a situation *in which physical attributes are objectified in a classical way*" (1948, p. 315; italics in this paragraph E.S.; see also 1954, p. 72f.). More precisely, with reference to the two constituents (position and momentum) of the uncertainty relations: "It would in particular not be out of place in this connection to warn against a misunderstanding likely to arise when one tries to express the content of Heisenberg's well known indeterminacy relations . . . by such a statement as: 'the position and momentum of a particle *cannot simultaneously be measured with arbitrary accuracy*'. According to such a formulation it would appear as though we had to do with some arbitrary renunciation of the measurement of either the one or the other of the two well-defined attributes of the object, which would not preclude the possibility of a future theory taking both attributes into account on the lines of the classical physics" (1937, p. 292). This warns against the phrase "cannot simultaneously be *measured*", but there is also the statement that "it must here be remembered that even in the indeterminacy relation . . . we are dealing with an implication of the formalism which defies unambiguous expression in words suited to describe classical physical pictures. Thus, a sentence like '*we cannot know* both the momentum and the position of an atomic object' raises at once questions as to the physical reality of two such attributes of the object, which can be answered only by referring to the conditions for the unambiguous use of space-time concepts, on the one hand, and dynamical conservation laws, on the other hand" (1949, p. 40). Here it is the phrase "we cannot *know* both" which is rejected as misleading. The reason in both cases is that, according to the customary view, measurement ascertains the value of a quantity, i.e. the value which that quantity *has*, and knowledge is always knowledge of something that already exists. If now the possibility of measurement or knowledge is denied, this might first of all be taken in the corresponding sense, that something exists which could in principle be measured or known, but that this is prevented by some more accidental factors. But Bohr, it seems, means that the customary *concepts* of measurement and knowledge have to be modified in order to deal correctly with the new situation in quantum mechanics. For this purpose he makes, in the above quotations, the somewhat uncompromising proposal that we should *in certain circumstances* not even *speak* of measurement or knowledge. It is not always easy to decide whether Bohr himself has complied with this proposal, since his words do not always make clear whether he is considering those circumstances which he himself would regard as excluding reference to measurement or knowledge in the ordinary sense. There is a clear error in this respect, for example, in the following passage relating to the Compton effect: "Even if the positional co-ordinates of the particle were accurately known in the beginning, *our knowledge of the position* after observation nevertheless will be affected by an uncertainty" (1927, p. 65). This instance makes the situation especially clear, since knowledge is mentioned twice in rapid succession: once legitimately, since Bohr here envisages that the electron, *before* interacting with the photon, has been prepared in such a way that its position could be predicted in a manner capable of confirmation provided that the collision with the photon has not taken place. But the second reference to knowledge is illegitimate according to the prohibition described above, since, *after* the collision which is used in order to measure the momentum of the electron, we have exactly the situation which led Bohr to frame this prohibition: the electron then *has no* definite position. There is a similar inconsistency as regards position measurement in the statement that "the measurement of the positional co-ordinates of a particle is accompanied . . . *by a finite change* in the dynamical variables" (1927, p. 68; similarly in 1939, p. 18). Although Bohr also points out in the same passage the special nature of this "finite change", he nevertheless uses the expression, although his own approach would disallow it: a change in the momentum, in the usual sense of the term, signifies that one definite value of the momentum is replaced by another. Thus, even if Bohr is here thinking that the momentum was definite *before* the position measurement, in the sense that its value could have been predicted (cf. the previous case), the situation *after* the measurement is again one in which, according to Bohr, "*the* momentum" of the particle is a meaningless expression.

Despite these occasional inconsistencies, it is nevertheless sufficiently clear which interpretation of the uncertainty relations Bohr had in mind for the relevant problem. First of all, they represent in a purely qualitative manner his idea of complementarity, in the sense that the position and momentum of a particle cannot both be unambiguously and simultaneously defined by experiment in the same phenomenon, and hence by means of classical concepts according to the quantum postulate. Thus, in contrast to the statement that they cannot be simultaneously *measured*: "These circumstances find quantitative expression in Heisenberg's indeterminacy relations which specify the reciprocal latitude for the *fixation*, in quantum mechanics, of kinematical and dynamical variables required for the *definition* of the state of a system in classical mechanics. In fact, the limited commutability of the symbols by which such variables are represented in the quantal formalism corresponds to the mutual exclusion of the experimental arrangements required for their *unambiguous definition*. In this context, we are of course *not* concerned with a restriction as to the *accuracy of measurements*, but with a limitation of the *well-defined application* of space-time concepts and dynamical conservation laws, entailed by the necessary distinction between measuring instruments and atomic objects" (1958, p. 5; italics in this paragraph E.S.). And similarly, in contrast to the statement that they cannot be simultaneously *known*: "Indeed we have in each experimental arrangement suited for the study of proper quantum phenomena *not* merely to do with an *ignorance* of the value of certain physical quantities, but with the impossibility of *defining* these quantities in an unambiguous way" (1935, p. 699, col. 1; similarly in 1927, p. 63ff.; 1929a, p. 18). The complementarity of position and momentum, as discussed in subsection (h) and at the beginning of this subsection, could be interpreted as meaning that, whenever a phenomenon involves the determination of position, there can be *no* reference to momentum, and *vice versa*. It would then be obviously impossible to arrive at a positive interpretation of the uncertainty relations with respect to one phenomenon. But, contrary to this view, Bohr holds that *in addition* their content "may be summarized in the statement that according to the quantum theory a general reciprocal relation exists between the maximum sharpness of definition of the space-time and energy-momentum vectors associated with the individuals. This circumstance *may* be regarded as a simple symbolical expression for the *complementary nature* of the space-time description and the claims of causality. *At the same time, however*, the general character of this relation makes it possible to a certain extent to *reconcile* the conservation laws with the space-time coordination of observations, the idea of a coincidence of well-defined events in a space-time point being replaced by that of unsharply defined individuals within finite space-time regions" (1927, p. 60). Similarly, "*the proper rôle* of the indeterminacy relations consists *in assuring quantitatively the logical compatibility* of apparently contradictory laws which appear when we use two different experimental arrangements, of which only one permits an unambiguous use of the concept of position, while only the other permits the application of the concept of momentum defined as it is, solely by the law of conservation" (1937, p. 293). Here an apparent incompatibility is to be partly resolved by putting forward the *quantitative* aspect, while at the same time it becomes possible to interpret the uncertainty relations in respect of a single phenomenon.

To do so, Δp and Δq must evidently be defined *as quantities*. It has already been mentioned in subsection (b) that Bohr himself was the first to derive the uncertainty relations by combining the wave picture with the de Broglie relation between wave vector and momentum. Bohr of course knew also, and often referred to, the fact that shortly afterwards a derivation of the uncertainty relations from the commutation relations

$$pq-qp = (\hbar/i)\,\mathbf{1} \tag{2}$$

was given, using the quantum-mechanical formalism. Such derivations can be valid only if they are based on precise definitions of Δp and Δq. In both cases the definition (formally the same for each) represents Δp or Δq as a certain functional of the Schrödinger ψ-function: the full notation would therefore be $\Delta_\psi p$ and $\Delta_\psi q$. Since ψ represents something resembling the "state" of the object concerned, this shows once again that the two uncertainties belong to the same phenomenon, but it now seems that their interpretation depends on that of the ψ-function. If ψ is interpreted as a classical wave packet, then $\Delta_\psi q$ is a measure of the spatial extent of the packet. No direct interpretation of $\Delta_\psi p$ is possible, since the momentum belongs to the particle picture. Only a consideration of the corresponding wave vector gives an idea that the spatial concentration of the wave packet ψ increases with increasing number of partial waves forming it. A statistical interpretation, on the other hand, shows that the two uncertainties are the ordinary ranges of dispersion pertaining to the position and momentum distribution of ψ (as stated explicitly by Bohr in at least one place (1939, p. 18)). Then the treatment of position and momentum would be wholly symmetrical.

It is difficult to decide which interpretation of ψ Bohr preferred, but in either case it seems that the significance of $\Delta_\psi p$ and $\Delta_\psi q$ is detached from phenomena whose purpose is the determination of position or momentum: *any* phenomenon in which the particle concerned is described by a ψ-function has also the corresponding uncertainties $\Delta_\psi p$ and $\Delta_\psi q$.

Perhaps it is untrue to say that Bohr rejected this use of uncertainty relations detached from the determination of position or momentum. For example, the comments quoted previously about the uncertainty relations as the origin of compatibility *might* mean that these relations concern just the situations where *neither* the classical concept of position *nor* that of momentum can be unambiguously applied: "the *general* character of this relation makes it possible to a certain extent to reconcile the conservation laws with the space-time co-ordination of observations, the idea of a coincidence of *well-defined* events in a space-time point being *replaced* by that of *unsharply defined* individuals within finite space-time regions" (1927, p. 60; italics in this paragraph E.S.). And in the second quotation there was reference to assuring the compatibility "of apparently contradictory laws *which appear when* we use two different experimental arrangements, of which only one permits an *unambiguous* use of the concept of position, while only the other permits the application of the concept of momentum *defined as it is*, solely by the law of conservation" (1937, p. 293; see also 1939, p. 19). If these passages are interpreted to mean that the uncertainty relations should serve to form a bridge *also* between situations in which classical concepts are unambiguously applied, we should in fact consider something like the above statistical interpretation of both $\Delta_\psi p$ and $\Delta_\psi q$, and at the same time it would definitely be possible to use $\Delta_\psi p$ and $\Delta_\psi q$ within a single phenomenon.

On the other hand, it must not be forgotten that Bohr was mainly concerned to *explain* the classically unintelligible uncertainty relations, and this could not easily have been done *without explicitly connecting them with an experiment to determine position or momentum.* The explanation is based especially on the arguments given in subsections (d)–(g) above, and, as we have seen, the interaction in measurement is of vital importance here. In this respect there is a very concise account of Heisenberg's uncertainty principle as "expressing the reciprocal limitation of the fixation of canonically conjugate variables. This limitation appears not only as an immediate consequence of the commutation relations between such variables, but also *directly reflects the interaction* between the system under observation and the tools of measurement" (1962b, p. 91; italics in this paragraph E.S.; similarly in 1929a, p. 11; 1955, p. 89). It must also be remembered that Bohr warned against an over-schematic application of the Schrödinger ψ-function, which is the *only* basis of the above-mentioned symmetrical treatment of position and momentum, for example: "In the treatment of atomic problems, actual calculations are most conveniently carried out with the help of a Schrödinger state function, from which the statistical laws governing observations obtainable under specified conditions can be deduced by definite mathematical operations. It must be recognized, however, that we are here dealing with a *purely symbolic procedure*, the unambiguous physical interpretation of which in the last resort requires a reference to a complete experimental arrangement" (1958, p. 5).

Bearing this in mind, then, let us imagine a phenomenon which in particular allows a position coordinate q, say, to be determined. What is then the significance of Δq, and still more of Δp, and how do these uncertainties arise? The difficulty mentioned initially is now very evident, since position and momentum are by definition unsymmetrically concerned in this phenomenon. The claim of compatibility, cited above, might again be applied here also. For the particular phenomenon under consideration, the indications are that Bohr would make an unqualified statement that the concept of position can be unambiguously employed here, since it is imposed by the classical description of the instrument used to measure position. The only question then is what is meant by "unambiguous". The term might include the implication that q can be *exactly* determined to have a certain value. Then $\Delta q = 0$, and (improper) application of the uncertainty relations would give $\Delta p = +\infty$. No definite statement could be made about the momentum. The claim of compatibility might be that this case is excluded. For, to repeat once more one of the above quotations with yet a third point of emphasis, reference is made to the attainment of this compatibility, "the idea of a *coincidence* of well-defined events in a space-time *point* being *replaced* by that of unsharply defined individuals within *finite* space-time regions" (1927, p. 61). On this view, one would impose the condition $\Delta q > 0$ in order to make possible $\Delta p < +\infty$. At the same time, however, the position of the particle is no longer said to be unambiguously defined. This would be compatible with a classical and unambiguous description of the corresponding *measuring instrument* (which of course Bohr regards as essential) if Δq were interpreted as the length of the shortest interval (such as a finite slit-width) within which q would certainly be found if it were measured immediately after another determination of q (by virtue of passage through the slit). And indeed, in the discussion of a diffraction experiment with one slit, Bohr wrote "Now the width of the slit, at any rate if it is still large compared with the wave-length, *may be taken as the uncertainty* Δq of the position of the particle relative to the diaphragm, in a direction perpendicular to the slit" (1935, p. 697, col. 1; similarly in 1949, p. 43). This, however, would be a *re*definition of the position uncertainty.

What of Δp in the *same* phenomenon? Here we must first note that the interpretation of

Δp cannot in any way correspond to that of Δq, i.e. be based on the classical description of the measuring apparatus, since the phenomenon concerned does not (and, in Bohr's view, cannot) possess any experimental arrangement to determine the momentum. The limited applicability of the classical momentum concept to the phenomenon concerned therefore cannot now consist simply in the need to allow some finite range in the measurement of momentum (as in the case of position discussed here). In agreement with this conclusion, sections IV.2 and IV.4 will give, at least formally, a definition of Δp (for the *same* phenomenon) which corresponds to the definition of Δq just indicated and which leads to $\Delta p = +\infty$ even if $\Delta q > 0$. This approach, therefore, would not attain the objective stated above. The only escape seems to be to take the statistical interpretation of Δp, which is obtained from the ψ-function valid immediately after the measurement and which *can* be finite for $\Delta q > 0$. Bohr does not clearly state his position on this, but it would agree with his treatment in that he regards the impossibility of making Δp less than the value given by the uncertainty relations as being due to the impossibility of surveying the momentum exchange between particle and measuring instrument, and the resulting empirical statistical variation of p as being likewise a direct consequence of this latter impossibility. These aspects are respectively discussed in the following passages: with reference to the same diffraction experiment, "Obviously the uncertainty Δp is inseparably connected with the possibility of an exchange of momentum between the particle and the diaphragm" (1935, p. 697, col. 2); and immediately following the emphasis on the impossibility of surveying the interaction in measurement, "This last circumstance clearly shows that the statistical character of the uncertainty relations in no way originates from any failure of measurements to discriminate within a certain latitude between classically describable states of the object, but rather expresses an essential limitation of the applicability of classical ideas to the analysis of quantum phenomena. The significance of the uncertainty relations is just to secure the absence, in such an analysis, of any contradiction between different imaginable measurements" (1939, p. 19). The last sentence shows that, even in the present context where the statistical nature of the uncertainty relations (for p in our case) is regarded as a direct consequence of the impossibility of surveying the interaction in measurement, they retain their role of avoiding contradictions.

There is, however, no intrinsic inconsistency either. Firstly, the whole problem under consideration involves position and momentum unsymmetrically. It is, therefore, not surprising if different definitions emerge for Δp and Δq. Next, as will be further explained in sections IV.2 and IV.4, the usual dispersion $\Delta_\psi q$ from the ψ-function immediately after the measurement can be examined for q also, and is found to be *at most* equal to the Δq above. Although it has no direct application to the phenomenon concerned, this analysis shows that the uncertainty relations are at least theoretically valid even for the differently defined Δq and $\Delta_\psi p$.

I.2(j) Reduction of States

In the orthodox formulation of quantum mechanics, which will be described in subsequent chapters, using the Hilbert space formalism, the *reduction of states* or (as it is also called with regard to the coordinate representation of the ψ-function) the *reduction of wave packets* plays a very important part. Bohr seldom referred to it, possibly because its exact formulation is very much dependent on the use of the quantum-mechanical formalism, which, as we know, was little used by Bohr. In particular, it involves a peculiar change in the ψ-function, and subsection (i) has already shown that Bohr enjoined caution in the use of this function. It may therefore be expected that, when Bohr did refer to the matter, it was in his own manner, and there are indeed some early statements relevant to it, denying the causal behaviour of an atomic object and regarding the particular type of failure of causality that is concerned here as being a further consequence of the impossibility of surveying the interaction in measurement. A somewhat more detailed comment appears in 1927 (p. 68), where Bohr opposes a remark by Heisenberg that, strictly speaking, phenomena are created only by observation, even in the classical case: "It must not be forgotten, however, that in the classical theories any succeeding observation permits a prediction of future events with ever-increasing accuracy, because it improves our knowledge of the initial state of the system. According to the quantum theory, just the impossibility of neglecting the interaction with the agency of measurement means that every observation introduces a new uncontrollable element. Indeed, it follows from the above considerations that the measurement of the positional co-ordinates of a particle is accompanied not only by a finite change in the dynamical variables, but also the fixation of its position means *a complete rupture in the causal description of its dynamical behaviour*, while the determination of its momentum always implies *a gap in the knowledge of its spatial propagation*" (italics in this paragraph E.S.). This is an explicit epistemological formulation of the opposite of the classical case. Even in classical physics, our information concerning a particular phenomenon may be initially incomplete, but in classical physics it can be successively completed while *retaining* the information previously obtained. This is impossible for quantum-mechanical phenomena, as a consequence of the impossibility of surveying the interaction in measurement. This *loss of information* on making a further measurement is made particularly clear by another passage: "In fact, the indivisibility of the quantum of action demands that, when any individual result of measurement is interpreted in terms of classical conceptions, a certain amount of latitude be allowed in our account of the mutual action between the object and the means of observation. This *implies* that a subsequent measurement to a certain degree *deprives* the information given by a previous measurement of its significance for predicting the future course of the phenomena" (1929a, p. 18). The same appears in an ontological formulation: "To understand why a causal description is impracticable, however, it is essential to remember, as shown in the article, that the magnitude of the disturbance caused by a measurement is always unknown, since the limitation in question applies to any use of mechanical concepts and, hence, applies to the agencies of observation as well as to the phenomena under investigation. This very circumstance *carries with it* the fact that any observation takes place *at the cost* of the connection between the past and the future course of phenomena" (*ibid.*, p. 11).

I.2(k) Statistical description

Bohr made relatively brief statements (except perhaps in 1939, §1) on the statistical description in quantum mechanics in each of his papers. This of course does not mean that he underestimated or even overlooked the significance of the fact that the quantum-mechanical description of states is a statistical one. In particular, he repeatedly remarked (and it is noteworthy that he always added a comment to the effect that the remark was scarcely necessary) that in quantum mechanics the need to make use of a statistical description has a totally different origin from the corresponding need in classical statistical mechanics. A typical passage is: "In conformity with the circumstance that several individual quantum processes may take place in a given experimental arrangement, the predictions of the formalism concerning observations are of an essentially statistical character. It must be realized, however, that in this respect we are presented, *not* with an analog to the use of probability considerations in the account of the behavior of complicated mechanical systems, but with the impossibility of defining any directive for the course of individual processes beyond those afforded by the self-consistent generalization of deterministic mechanics" (1956, p. 87, col. 2; italics E.S.). On the other hand, Bohr does not regard the mere fact that quantum mechanics is statistical as being a suitable basis for illustrating the difference from the classical case. As he evidently sees no difference in the *meaning* of the statistical statements which occur in quantum mechanics and in classical statistical mechanics, there remains only the possibility of seeking different *reasons for the occurrence* of such statements in the two cases. These reasons are summarized, according to Bohr, in the argument given here and in his idca of complementarity: "The view-point of complementarity allows us indeed to avoid any futile discussion about an ultimate determinism or indeterminism of physical events, by offering a straightforward generalization of the very ideal of causality" (1939, p. 25).

I.3 Illustrations

In advancing the arguments described in section 2, Bohr sometimes referred to various experiments, either imaginary or already performed (see especially 1927, 1935, 1939, 1948, 1949, and Heisenberg 1930, passim). Here again, as in the general discussions, little use is made of the quantum-mechanical formalism, and the treatment is therefore more qualitative and heuristic. In the secondary literature there is considerable uncertainty as to the function of Bohr's analysis of experiments (in particular, of *Gedankenexperimente*) in relation to his interpretation of quantum mechanics, and what can be the possible value of such a treatment. In the present section, the view will be taken that they are intended simply to serve as examples for the illustration of the general arguments. In describing these arguments, an attempt has been made to show that they have a certain logical order peculiar to Bohr's thought. The aim was to demonstrate at the same time that Bohr sought to prove something to his reader. Although this procedure may raise problems of detail, our view is that the occasional discussion of experiments is not intended to prove anything involving consequences beyond those of the general discussion. For illustration, we choose the two diffraction experiments, with one slit and with two slits. These are discussed by Bohr mainly in two places (1935, 1949). These experiments have actually been performed, long ago for light and more recently (Jönsson 1961) for particles such as electrons.

Let us first consider an experimental arrangement in which, as indicated in Fig. 1, a beam of electrons or light coming from the left passes through a slit in the screen S and then strikes a photographic plate S_1 at a distance x_1 from S (see also Heisenberg 1930, p. 23f.; Shpol'skii 1969, p. 521ff.; Blokhintsev 1964, p. 49ff.; Messiah 1961, p. 142f.). To simplify subsequent calculations (section IV.4), the system will be regarded as having only two dimensions in space. Let a plane wave with a well-defined wavelength be present to the left of S and travelling in the x direction. Then S_1 will show an interference pattern in which the width of the principal peak is considerably greater than that of the slit in S, and which contains very narrow interference fringes around this peak. The phenomenon can be at least qualitatively explained in terms of the classical wave picture, and accordingly (cf. subsection 2(h)) any attempt to explain it by the complementary classical particle picture must show that the latter has only limited applicability in the present case. And in fact, to correspond to a harmonic plane wave in the x direction, the particle picture would call for particles to the left of S, having a well-defined velocity in the x direction. These would either be absorbed by S or else pass through the slit in S and arrive *en masse* at points on S_1 exactly opposite the slit. Neither the widening nor the interference fringes could occur. This lack of success cannot be explained by the participation of *large numbers* of particles in the phenomenon. The experiment can be conducted with very low intensities, but no change is found in the interference pattern on S_1. The electrons or photons then strike S_1 at intervals of time which are large in comparison with the time taken to traverse the entire apparatus, and are seen as individuals from the isolated points at which they blacken the plate (see Shpol'skii 1969, p. 523 for electrons; Jánossy and Náray 1957 for photons in the Michelson interferometer). The interference pattern is built up gradually, and in this way a kind of residual particle nature is seen to exist even in this phenomenon.

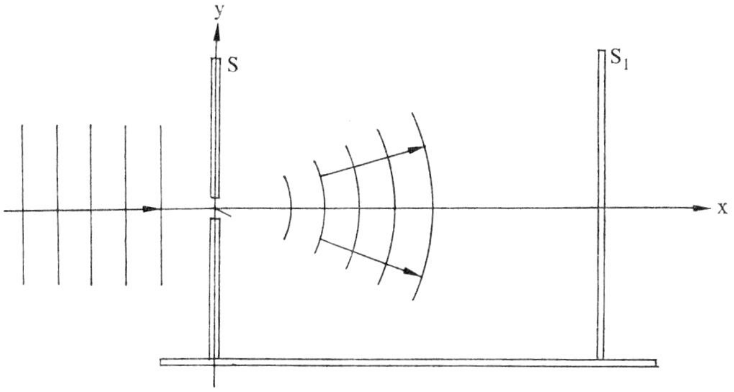

Fig. 1

We must now consider, following the general lines of discussion in section 2, precisely to what extent the particle picture fails, with the complementarity of space-time coordination and dynamical conservation laws appearing in the form of their (quantum-mechanical) incompatibility. The starting point is that the experiment in question comprises an arrangement (the screen S and the slit in it) which fixes in space the y component of position for a particle passing through the slit: if S is taken as a spatial frame of reference, this component at the instant of passage is known to be in the range $[-y_0, +y_0]$, where $-y_0$ and $+y_0$ are respectively the y coordinates of the lower and upper edges of the slit. But, if S is rigidly fixed to S_1, the experiment shows that a particle can appear on S_1 at a point which classically it could not reach. If, as assumed, the particle has momentum p^0 before passing through the slit, and this momentum is exactly (or with sufficient accuracy) in the positive x direction, then classically it will retain this momentum until it strikes S_1, and no deviation should occur. Whenever such a deviation does occur, therefore, there must have been an *interaction* between the particle and the screen S, which makes doubtful any kind of conservation of momentum. This is the interaction of which Bohr says (cf. subsections 2(e)–(g)) that it necessarily occurs in quantum phenomena, that it cannot be surveyed, that it is an essential constituent of the phenomenon, and that all these properties are to be regarded as a consequence of the individuality of a quantum phenomenon and the classical description of the measuring instruments. In the present case, however, it must be clearly understood that the occurrence of such an interaction is incomprehensible only in terms of the classical *particle picture*. In the wave picture it is obvious. The further assertion that this interaction cannot be surveyed would be false or unproved for all constituents concerned in the wave picture, and it would be pointless to emphasize that the interaction is an essential constituent of the phenomenon and that the measuring instruments are to be described in classical terms. These comments can be true or significant (as regards the phenomenon in question) only in relation to the particle picture. In particular, only this picture makes the screen S and its slit an instrument to determine a quantity describing the state of the object; in the wave picture it has a quite different function, and one which alters the state. Thus the need to complete the particle picture by means of the wave picture appears already in a general manner.

All the above-mentioned aspects which are critical to the particle picture combine into the question of what is the momentum of the particle after it has passed through the slit in S. According to Bohr, the initially well-defined momentum has to be regarded as becoming undefined, or defined only within certain limits, as a result of the occurrence and unsurveyability of the interaction of the particle with S (subsections 2(f), (g) and (i)). The occurrence of the interaction is in turn a consequence of the individuality of the quantum phenomenon, while its unsurveyability is due to the classical description of the experimental arrangement (subsection 2(f)). Now, from the particle picture, this interaction, insofar as it can be understood at all, has to be regarded as a *momentum exchange* between the particle and S (with negligible energy exchange). Then the dispersion visible on S_1 is empirical evidence for the occurrence of the interaction as a momentum exchange. On this basis we can explain the unsurveyability of the exchange as a consequence of the classical description of the whole arrangement: "Let us first assume that, corresponding to usual experiments on the remarkable phenomena of electron diffraction, the diaphragm, like the other parts of the apparatus, . . . is rigidly fixed to a support which defines the space frame of reference. Then the momentum exchanged between the particle and the diaphragm will, together with the reaction of the particle on the other bodies, pass into this common support, and we have thus voluntarily cut ourselves off from any possibility of taking these reactions separately into account in predictions regarding the final result of the experiment,—say the position of the spot produced by the particle on the photographic plate" (Bohr 1935, p. 697, col. 2).

The amount Δp_y of the resulting uncertainty of the momentum in the y direction is usually taken as

$$\Delta p_y = p^0 \sin \alpha \tag{1}$$

in heuristic analyses of the experiment under discussion; here p^0 is the (definite) momentum before the interaction, and α the angle between the x axis and the first diffraction minimum. If the slit width d in the y direction is taken as the uncertainty of position Δy after the passage through the slit:

$$\Delta y = d, \tag{2}$$

then the de Broglie equation,

$$p^0 = h/\lambda, \tag{3}$$

and the elementary equation in diffraction theory

$$\sin \alpha = \lambda/2d \tag{4}$$

give immediately

$$\Delta y \cdot \Delta p_y = \tfrac{1}{2}h \geqslant \tfrac{1}{2}\hbar, \tag{5}$$

and therefore the uncertainty relations. Equations (3) and (4) bring in the wave picture, and thus express the inadequacy of the particle picture. Concerning the assumption (1) which is fundamental in this derivation, we must first note that together with (5) it leads to the condition

$$\Delta y \geqslant \hbar/2p^0 \tag{6}$$

for Δy. Thus it cannot be used for a very narrow slit (cf. Blokhintsev 1964, p. 50; it may be mentioned in passing that, in the experiments of Jönsson (1961), $2p^0 \cdot \Delta y = 10^6\hbar$). Another question, of course, is the justification of the assumption (1). It may appear a plausible assumption if one has only a very rough idea of the appropriate uncertainty of momentum, and if the momentum in the x direction is assumed to be still p^0. This amounts only to making use of the freedom to specify the uncertainty, using completely macroscopic experimental results (in this case, the angle α of the first diffraction minimum). The only defect of such a specification is that it is largely arbitrary. It almost seems that in this way the uncertainty relations can always be satisfied. In contrast, Δy is specified as the length of the interval $[-y_0, +y_0]$ in a way which is not tied to the particular conditions of the experiment. The idea of determining a range for a quantity, in the sense that it can be expected with certainty to lie in that range, is a quite general one (cf. sections IV.2 and IV.4). In the present case, however, Δp_y cannot be treated in this way: there is no experimental arrangement for measuring p_y, and we know from experience that p_y may sometimes, though not often, lie outside the range $[-\Delta p_y, +\Delta p_y]$ (length 2 Δp_y), with Δp_y given by (1). The other idea that could be generalized is to regard Δp_y as the ordinary range of dispersion, as already mentioned in section 2(i). Whether this agrees *de facto* with the value given by (1) (apart from a constant factor) has never been tested.

Let us assume that, despite the difficulty of quantitatively determining the uncertainty Δp_y, the occurrence of the uncertainty is taken as established; there would then still be the possibility of actual surveillance of the uncertainty, provided that this could be shown to be compatible with the uncertainty relations for the screen S, which would then also have to be taken into account. Bohr's arguments as reproduced above indicate that this surveillance is rendered impossible by the assumption that S is strictly at rest during the entire passage of a particle from the source through the slit to S_1, which in particular involves neglecting the uncertainty relations for S. To this it might be objected that, by making S movable in the y direction, one could observe the recoil, apply the law of conservation of momentum to the system of screen and particle, and thus observe the final momentum of the particle also. In evaluating this proposal, it is vital to take note of the changed role of S which it implies: "The principal difference between the two experimental arrangements under consideration is, however, that in the arrangement suited for the control of the momentum of the first diaphragm, this body can no longer be used as a measuring instrument for the same purpose as in the previous case, but must, as regards its position relative to the rest of the apparatus, be treated, like the particle traversing the slit, as an object of investigation, in the sense that the quantum-mechanical uncertainty relations regarding its position and momentum must be taken explicitly into account" (Bohr 1935, p. 698). When we do so, the result is as follows. The accuracy δp_y with which the momentum of S in the y direction is measured after the interaction determines the accuracy Δp_y with which the momentum of the particle in the y direction is known after the interaction: since the law of conservation of momentum is to be applied, we must have

$$\delta p_y \approx \Delta p_y. \tag{7}$$

On the other hand, the purpose of the experiment demands that

$$\Delta p_y \ll p^0 \sin \alpha \tag{8}$$

in order that the particle momentum should be known much more accurately than is ensured in any case by formula (1). From (3), (4), (7) and (8) we have

$$\delta p_y \ll h/2d. \tag{9}$$

Together with the uncertainty relations *for S*,

$$\delta y \, \delta p_y \geqslant \tfrac{1}{2}\hbar, \tag{10}$$

this gives

$$\delta y \gg d. \tag{11}$$

Hence we see not only that S can no longer be used to determine the position of the particle with the accuracy given by (2), but also that there will no longer be an interference pattern on S_1. The additional surveillance of the momentum would destroy the very phenomenon that is being investigated.

Let us now turn to the two-hole experiment. This differs in its arrangement from the one sketched in Fig. 1 for the one-hole experiment only in that the screen S contains two slits instead of one. In this arrangement, the interference fringes which appear on S_1 are the essential feature of the phenomenon (cf. the experiments described in Jönsson 1961). The failure of the particle picture to explain the phenomenon is accordingly even more obvious than in the one-hole experiment. The following analysis will show this in relation to a typical Bohr enunciation, already suggested at the end of the discussion of the one-hole experiment: the attempt to bring the particle picture into full validity in a quantum phenomenon destroys the phenomenon itself—or, in somewhat more positive terms, enforces its replacement by a complementary phenomenon.

The two-hole experiment is a particularly good starting point for illustrating the complementarity of two phenomena. On the one hand, we have the interference phenomenon, which occurs when the experimental arrangement in Fig. 2 is used. If the screen S is here regarded as an apparatus for measuring the position of each incoming particle, we can say that the position of a particle which passes through the screen is determined, at a time immediately after the interaction of the particle with S, as lying in a region consisting of two quite separate parts just to the right of the two holes. The arrangement does not decide in which of these two parts the particle is then located or, equivalently, through which of the two holes it has passed. From the classical standpoint, with the particle picture fully utilized, one would say that the particle has passed through one of the holes, and the arrangement does not tell us which hole, but this also can be ascertained by a modification which would not disturb the interference phenomenon. This would be precisely what Bohr calls an attempt at further subdivision of the phenomenon, namely the attempt to analyse the phenomenon more precisely both theoretically (by saying that the particle must have passed through one hole or the other) and practically (by experimentally testing which path it has taken).

What are the practical possibilities for such a test? The most forceful way of determining the path of a particle in this respect is to block one of the holes. The experiment would then undergo a subdivision, in the sense that only the one hole is open for a certain time, and then only the other hole. According to classical particle mechanics, the structure of the pattern thus produced on the photographic plate S_1 should be the same as that of the pattern produced by leaving both holes open for a certain time. If not, we should have to conclude that what happens to a particle after it passes through hole 1 depends on whether hole 2 is open, and *vice versa.* The two experiments have not been conducted and compared in precisely this form. But from all the experiments that *have* been done it follows that the two patterns produced have quite different structures. In the first two-hole experiment, interference occurs; in the modified experiment, to the first order and when the slits are not too far apart, we get a distribution of intensity which has a peak in the centre and decreases

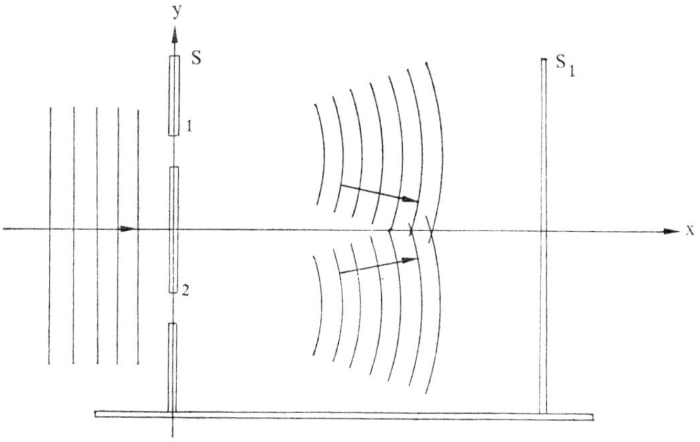

Fig. 2

steadily on either side. Thus the change in the experimental arrangement destroys the original phenomenon. Accordingly, a theoretical explanation of the different structures of the two patterns on S_1 is impossible on the basis of classical particle mechanics. The use of the wave picture to describe the process between S and S_1 does, however, give at least a qualitative explanation. In the two-hole experiment, two wave trains originate from holes 1 and 2, and interfere between S and S_1. In the modified experiment, they propagate independently. Accordingly, interference fringes appear on S_1 in the former case, while in the latter the intensities are simply added, and there is no interference (see e.g. Ludwig 1954, §I.6; Heisenberg 1959, p. 51f.; Feynman and Hibbs 1965, §1–1).

If this modified two-hole experiment is regarded as a statistical one, the individual particles are treated in different ways. Some particles encounter S when only hole 1 is open, others when only hole 2 is open. Other modifications can be devised in which all particles are treated in the same way. The literature contains a number of such *Gedankenexperimente*: illumination of the two holes from the right of S (Bohm 1951, §6.2; Feynman and Hibbs 1965, §1–1; Büchel 1965, ch. 4); path determination in a cloud chamber replacing the photographic plate (Bohm 1951, §6.2); observation of movable slits immediately to the left of the two holes in S (Tomonaga 1966, p. 273); measurement of the momentum of a one-hole screen placed to the left of S (Bohr 1935, p. 698, col. 2; 1949, p. 45); measurement of the momentum of S itself (Messiah 1961, ch. IV, §18; Feynman and Hibbs 1965, §1–1). All these experiments are designed to ascertain the hole through which an individual particle passes. In every case it is shown, usually by means of the uncertainty relations, that this additional determination of the path of the particle is incompatible with the occurrence of the interference phenomenon.

Bohr's argument is especially clearly illustrated by the experiment last mentioned, since here the modification of the two-hole experiment is directly aimed at an additional surveillance of the interaction between the particles and the screen S. There is partial surveillance of this interaction in the two-hole experiment itself, in that the particular form of S allows the position of an individual particle passing through it to be determined, to the extent stated above. The idea then is to determine *also* the momentum of the particle, to the extent that, together with the knowledge of its position after striking S_1, this will allow a decision as to which hole it has passed through. The momentum is to be measured by determining the momentum exchange in the interaction with S. The modification must therefore consist in detaching S from its support and making it movable in the y direction. If the momentum of S is then determined before and after its interaction with an individual particle, and if the momentum of the latter before the interaction is determined, the law of conservation of momentum can be used to calculate the momentum of the particle after the interaction. This therefore presents a typical case in which a part of the experimental arrangement that is initially described in classical terms in accordance with Bohr's buffer postulate (cf. subsection 2(e)) is converted into the object and thus must be described in quantum-mechanical terms, whereas the requirement of classical description is applied to different parts of the experimental arrangement, in this case the parts used in measuring the momentum of S. The quantum-mechanical description of S is expressed in the following arguments precisely in the fact that the uncertainty relation in the y direction for S is taken into account.

Let us assume, for simplicity, that before the interaction the particle has exactly the momentum p^0 in the x direction and that the screen S is precisely at rest. Let the result of measuring the momentum of S after the interaction be (in accordance with the conditions in position measurement by a one-hole screen) that the momentum of S in the y direction certainly lies between p_y and p'_y. Then the corresponding uncertainty is

$$\delta p_y = p'_y - p_y. \tag{12}$$

Figure 3 shows the effect of this on the particle momentum after the interaction. If the particle passes through hole 1, it continues in the upper shaded sector, and if it passes through hole 2 it continues in the lower shaded sector. The angles are plotted by using the conservation of momentum:

$$\mathbf{p}^0 = \mathbf{p}_y + \mathbf{p}_p, \quad \mathbf{p}^0 = \mathbf{p}'_y + \mathbf{p}'_p \tag{13}$$

for the two limiting cases given by p_y and p'_y. To decide for certain the hole through which the particle has passed, from the information just mentioned together with the position where the particle strikes S_1, it is necessary that the two shaded sectors should not overlap between S and S_1. In fact, the position uncertainty δy of S imposes an even stronger condition, but since the argument is valid without it, we shall neglect it; this is legitimate if δy is small compared with the distance between the two holes. Our condition is therefore

$$-y_1 > -y'_1. \tag{14}$$

Figure 3 gives

$$p'_y/p^0 = y'_0/x'_1, \quad (x_1 - x'_1)/x'_1 = -y'_1/y'_0,$$

$$p_y/p^0 = -(y_1 + y'_0)/x_1.$$

Hence, from (12) and (14),

$$\delta p_y < p^0 \cdot 2y'_0/x_1.$$

We now substitute for the momentum p^0 of the incident particle in terms of its de Broglie wavelength by means of equation (3):

$$\delta p_y < (h/\lambda)\,(2y'_0/x_1).$$

With the uncertainty relations (10) for S in the y direction, this gives

$$\delta y > (x_1/4\pi)\,(\lambda/2y'_0). \tag{15}$$

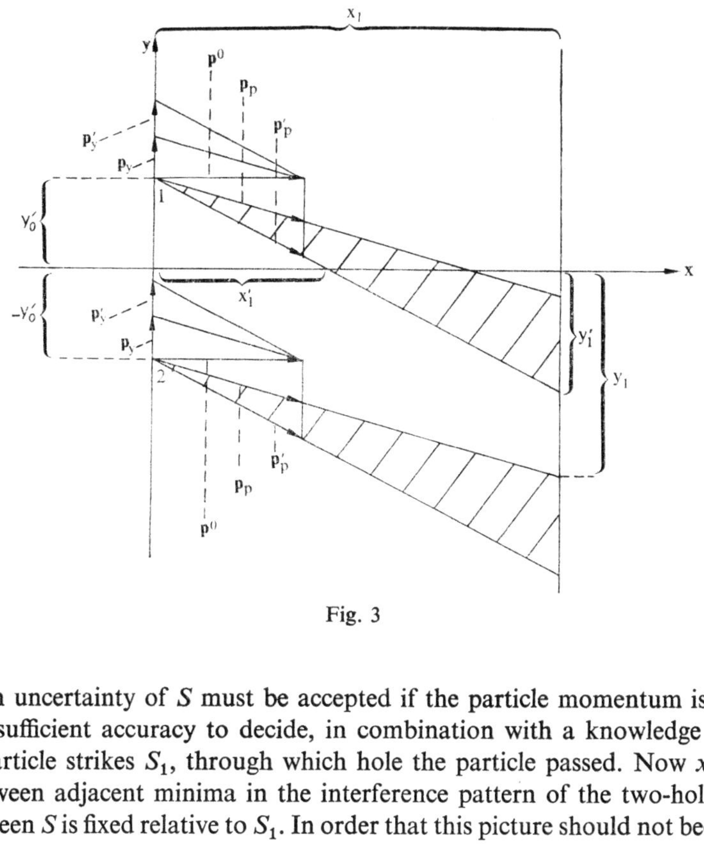

Fig. 3

This position uncertainty of S must be accepted if the particle momentum is to be determined with sufficient accuracy to decide, in combination with a knowledge of the point where the particle strikes S_1, through which hole the particle passed. Now $x_1\lambda/2y'_0$ is the distance between adjacent minima in the interference pattern of the two-hole experiment where the screen S is fixed relative to S_1. In order that this picture should not become blurred when S is movable, we must evidently have

$$\delta y \ll x_1\lambda/2y'_0. \tag{16}$$

This condition contradicts the conclusion (15), and the interference phenomenon will therefore not occur in the modified experiment.

CHAPTER II

Orthodox Quantum Mechanics in Hilbert Space: Formulation in Terms of Quantities

In Bohr's interpretation of quantum mechanics, as described in Chapter I, the quantum-mechanical formalism is remarkably little in evidence. We have seen that Bohr treated this formalism with respect but did not include it explicitly in his account of the theory. This does not necessarily signify that his treatment could not have been made more explicit in this regard, but only that Bohr was not especially concerned with it. As a counterpart to Bohr's treatment of the fundamental problems of quantum mechanics, this chapter will describe attempts to construct a quantum mechanics which make essential and fully explicit use of the formalism of this theory.

The most important such attempt is that of von Neumann, the first version of which (von Neumann 1927) appeared at the same time as Bohr's first paper on the new quantum mechanics; the final version appeared in 1932 (von Neumann 1955). The account below is based on his treatment, but is even less of a mere reproduction than the presentation of Bohr's ideas in Chapter I. This is mainly because we are here interested primarily in problems of physical interpretation, and von Neumann's formalism, though presented in a notably unambiguous manner, allows some freedom in matters of interpretation, even if we try to consider only what may be regarded as orthodox accounts. In fact, the very attempt to use a specified formalism of quantum mechanics to support a definite physical formulation of the complete theory shows most impressively that nothing is achieved which could be put forward as "the" orthodox quantum mechanics. Any treatment aiming at being complete and unambiguous will encounter authors who must be ranked as orthodox in the historical sense but who oppose one or other of the proposed assertions. Indeed, it is not even possible to succeed by restricting oneself to a single author. The position is in every instance the same as has already been shown in regard to Bohr: if a sufficiently sharp distinction is made, questions arise which have not been definitely resolved by the author concerned. If nevertheless a name is demanded whose owner initiated and defended as a physicist (and not as a mathematician like von Neumann) ideas which will be treated in the following account as typically orthodox, then the first choice is Heisenberg; another crucial presentation is that by Schrödinger (1935a) of the official view in the middle 1930s, although he himself opposed this view.

After expounding the historically significant case of Bohr in Chapter I, we shall now more systematically proceed, in accordance with the above-mentioned situation, by describing several alternative treatments. These will be taken as orthodox inasmuch as they all regard quantum mechanics as (a) including indispensable propositions regarding *probabilities*, (b) including likewise indispensable propositions regarding *measurements*, (c) including as a fundamental constituent some form of *reduction of states*. In contrast to this agreement, the main point of difference is the distinction between the formulations which do not include the concept of the *state of an individual object* (section 4) and those which do (section 5). In the first category there is a further distinction between the theories in which the propositions regarding probabilities are taken to express the *state of knowledge* concerning an individual object (epistemic interpretation) and those in which they are taken to express, essentially, relative frequencies in *statistical ensembles* of individual objects (statistical interpretation). This distinction may be correlated with an ambiguity of the propositions regarding measurements, which can be either propositions of newly acquired knowledge about an individual object or propositions about the creation of a new statistical ensemble.

In the theories which include the concept of the state of an individual object (ontic formulations), we have initially a purely quantum-mechanical interpretation of the probability propositions as propositions about the state of an individual object. This interpretation can be coupled with an objective interpretation of the measurement propositions, which also refer to an individual object. But a distinction is then again necessary between theories which express the idea that the state of the object concerned is incompletely known, and theories of statistical ensembles of individual objects in definite states (epistemic or statistical extension of the ontic formulation). This procedure gives a total of four treatments, each of which will be further examined, without implying that they embrace all orthodox treatments.

As a preliminary, it will be found useful to include a glance at classical mechanics, whose original form is a theory involving neither probability propositions nor measurement propositions (section 2). It can also, however, be formulated in a manner based on these propositions, and is thereby brought as close as possible to quantum mechanics; this form is especially suitable for purposes of comparison (section 3). All the presentations of physical theories in this chapter are based on a method which is briefly sketched in section 1. In Chapter V this will be contrasted with another method. Moreover, all the formulations are based on concepts which include that of a physical *quantity*; in Chapter III this will be replaced by the concept of a physical *property*.

II.1 A method for presenting a physical theory

The statement of the content of a physical theory is always somewhat arbitrary as regards the formal means of presentation used, even if the content itself is completely definite. For a concise presentation of the content of quantum mechanics, and also (by way of comparison) of ordinary and statistical classical mechanics, the following method of formulating a physical theory will be used. It consists of

(1) selecting certain *physical concepts* which are essential to the theory concerned;

(2) selecting a class of *mathematical structures*;

(3) *representing* the selected physical concepts in terms of concepts in the theory of the selected class of mathematical structures.

This does not define the way in which the mathematical theory involved is specified. It is usually constructive, i.e. a part of concrete mathematics based on the concept of number, which is assumed to be understood. Where axioms are introduced, they are intrinsically mathematical, and not aimed at the physical application. The application of the axioms in a physical theory retains traces of masterly insight and of historical accident, and the resulting logical interconnection of physical concepts is at first obscured. As these are gradually revealed by logical analysis, the mathematical theory takes a new form, in which it can be employed for a different method of presentation of a physical theory. The physical concepts are *directly* axiomatized as far as possible, and the resulting axioms are, apart from their physical significance, a part of abstract mathematics, forming together at least a part of the entire mathematical theory that is to be used. This other method of formulating a physical theory will not be brought into play until Chapter V.

II.2 Classical mechanics

In preparation for the discussion of quantum mechanics, the structure of ordinary classical mechanics will first be very briefly recapitulated. There are various ways of describing it, even within the scope of the method of presentation outlined in section 1. Of these, the Hamiltonian form is closest both to its probabilistic extension and to quantum mechanics; this formulation will therefore be taken as thc basis.

II.2(a) Objects

The mathematical description of the physical concepts relating to an object Σ in classical mechanics is based on a real manifold $\mathfrak{M}$ of dimension $2n$, usually called the *phase space of* Σ, in which a class of *canonical coordinate systems* is selected and which also possesses certain topological and analytical properties which need not be stated here. For example, if Σ is a system of N particles which can move in space without significant boundary conditions, then $\mathfrak{M}$ is a certain manifold of dimension $6N$, in which the canonical coordinates are formed by the position and momentum coordinates of all N particles, referred to a Cartesian spatial coordinate system.

II.2(b) Composite objects

One specification of the concept of an object deals with a relation between objects $\Sigma^1, \ldots, \Sigma^N$ and Σ, to the effect that the object Σ consists, or is composed, of the objects $\Sigma^1, \ldots, \Sigma^N$. As regards the mathematical description of the objects, the question arises in connection with this relation: what will be the phase space $\mathfrak{M}$ of the object Σ if the objects $\Sigma^1, \ldots, \Sigma^N$ have the phase spaces $\mathfrak{M}^1, \ldots, \mathfrak{M}^N$?

In classical mechanics, the answer is as follows. *If* $\mathfrak{M}^1, \ldots, \mathfrak{M}^N$ *and* $\mathfrak{M}$ *are the phase spaces of the objects* $\Sigma^1, \ldots, \Sigma^N$ *and* Σ, *and if* Σ *is composed of* $\Sigma^1, \ldots, \Sigma^N$ ($\Sigma = \Sigma^1 \times \ldots \times \Sigma^N$), *then* $\mathfrak{M}$ *is the Cartesian product of* $\mathfrak{M}^1, \ldots, \mathfrak{M}^N$ ($\mathfrak{M} = \mathfrak{M}^1 \times \ldots \times \mathfrak{M}^N$). This requirement is familiar in the case of a system of N particles, where the coordinates for the phase space of the whole object are obtained by simply taking all combinations of the position and momentum coordinates of the individual particles. It must be noted, however, that no two particles can be at the same point, and therefore certain singular surfaces must be omitted from $\mathfrak{M}$.

II.2(c) Quantities

In ordinary classical mechanics, the *quantities pertaining to an object* Σ are well known to be described by real-valued functions in the phase space of Σ. The fundamental problem which arises in this description is that of the precise delimitation of the domain of the functions which represent the quantities. The finite series of individual mechanical quantities which are usually the subject of principal interest in theoretical studies do not lead unambiguously to any conceptual features whereby conditions on the functions which describe them might be formulated as a general definition of the concept of a mechanical quantity. The considerable latitude which remains can therefore be reduced only on the basis of appropriateness for a particular type of investigation. In the investigations to be described here, the following definition, which embraces a very wide class of functions, has proved appropriate: *the quantities of* Σ *are described by a one-to-one correspondence with the set of Borel functions in the phase space of* Σ. (The Borel functions are described, for example, by Halmos (1950, Chapter X).) This is the first of a number of instances of one-to-one correspondence between the elements of a set of physical entities and those of a set of mathematical entities, in which it is an agreeable simplification, if not a strictly correct one, to use the same names for corresponding entities. This *identification convention* will be used here and in all similar cases, provided that no misunderstanding is likely.

II.2(d) Propositions regarding states

The concept of a quantity defined in subsection (c) becomes conclusive only when the propositions involving it are explained. For a mechanical object Σ actually present at a time t, in the simplest case, these propositions have the form that the quantity f of Σ has the value α at t. The obvious time-dependence shows that the propositions are contingent ones. The temporal aspect thus added does not, however, immediately assume its full significance. At first, all that is needed is to use a certain technique to demonstrate the available possibilities in the description of an object at a single instant. The permissible time variations of the object will be specified in subsection (e).

In classical mechanics, the description is based on the concept of the *state of an object* Σ. *The possible states of* Σ *are represented in a one-to-one manner by the set of points in the phase space of* Σ. Their relationship to the quantities of Σ is specified by the *universal propositions regarding states*, $\mathbf{S}(M, f, \alpha)$, which signify that *the quantity* f *of* Σ *in the state* M *has the value* α. Unlike the previous contingent propositions which involve a particular time, these universal propositions regarding states are determined only by the structure of the object Σ. Accordingly, the mathematical representation of the propositional function $\mathbf{S}$ is determined only by the phase space of Σ: *the universal propositions regarding states,* $\mathbf{S}(M, f, \alpha)$, *are mathematically described by the propositions* $f(M) = \alpha$. (Here the identification convention in subsection (c) should be noted. It would have been more accurate, as also in other similar cases, to write the representative propositions as $f'(M') = \alpha'$, where the function f' in $\mathfrak{M}$, the phase point $M' \in \mathfrak{M}$ and the number $\alpha' \in \mathbf{R}$ respectively represent the quantity f of Σ, the state M of Σ and the value α of the quantity f.) This mathematical description again emphasizes the universality of the propositions regarding states. In particular, a proposition having the form $\mathbf{S}(M, f, \alpha)$ says nothing at all as to whether the object Σ *is* in the state M, but only as to which quantities have which values *if* Σ is in the state M.

For a fixed state M, the propositions

$$on_M(f, \alpha) \rightharpoonup \mathbf{S}(M, f, \alpha) \tag{1}$$

form a collated class of propositions (in this case, belonging to one state). To each state M,

(1) assigns a propositional function which specifies, for each quantity f and each value α, whether f has the value α or not. The mathematical representation of S easily shows that the states are characterized by these propositional functions, i.e. that

$$\bigwedge_f \bigwedge_\alpha . \, on_M(f, \alpha) \leftrightarrow on_{M_1}(f, \alpha). \rightarrow M = M_1. \tag{2}$$

That is to say, different states of Σ correspond to different assignments of values to the quantities of Σ. The propositions $on_M(f, \alpha)$ still say nothing about the actual existence of the state M, but the different placing of the argument M in comparison with f and α indicates for a given M the propositional function which characterizes the state M.

II.2(e) Time dependence

The axiomatic treatment in subsections (a)–(d) has not referred to the possible changes with time of an object in classical mechanics. It was indeed not possible to do so, since only subsections (c) and (d) have introduced the concepts needed to formulate time-dependent processes.

The concept of *time* which is now brought into action will not be further analysed, but it should be explicitly noted, in relation to the programme set out in section 1, that the formulation of possible mechanical processes and the laws of these processes given below is based on the usual conception of time as comprising a sequence of *instants* and the usual mathematical description of these by *real numbers*. This procedure enables us to make the mathematical description of time-dependent processes a continuous extension of the correlations already established between mathematical and physical systems of concepts. These processes can now be regarded simply as time variations of the state of the relevant object Σ, i.e. as functions which assign a state of Σ to each instant. They are therefore described mathematically by functions which assign to every real number a point in the phase space $\mathfrak{M}$ of Σ. For any time dependence of the state, M_t, we then obtain in accordance with the definition (1) in subsection (d) the time-dependent family of *contingent propositions*

$$on_t(f, \alpha) \rightharpoondown S(M_t, f, \alpha), \tag{1}$$

which assert that *the quantity f of Σ has the value α at time t*. These functions again represent only possibilities (which will immediately be further restricted) for the behaviour of an object in the course of time, but any one of them can be identified as that which represents the actual behaviour of a specific object. The single proposition $on_t(f, \alpha)$ is then contingent in the sense that (except in trivial cases) it can be neither proved nor disproved by the theory of the object concerned, which is usually determined by its phase space and its Hamiltonian function (see next paragraph).

The further restriction of the possible variations of the state of Σ depends on whether Σ has its own dynamics. If it does, a function H, the *Hamiltonian of* Σ, must be defined in the phase space $\mathfrak{M}$ of Σ (in addition to $\mathfrak{M}$, which hitherto was sufficient for the mathematical representation), to describe this dynamics. The function H determines the Hamiltonian equations, which in a canonical coordinate system of $\mathfrak{M}$ have the form

$$\dot{p}_k = -\partial H/\partial q_k, \quad \dot{q}_k = \partial H/\partial p_k. \tag{2}$$

These equations must be satisfied by any change in state of Σ if Σ has a dynamics described by H. The more general case where Σ interacts with other mechanical objects and accordingly its changes with time depend on the states of the other objects can be reduced to the case of an object having its own dynamics, by combining all the objects concerned into a single system (cf. subsection (b)) and then assuming that this system has its own dynamics.

II.3 Classical statistical mechanics

Classical statistical mechanics is usually regarded as a physical theory devised as a basis for the phenomenological thermodynamics of systems having a finite but very large number of degrees of freedom. The formulation of this theory given below is based on the traditional one due to Gibbs, but is not bound to it. Classical statistical mechanics will be considered with a different motivation from that of providing a basis for thermodynamics. Firstly, it will act in many respects as a useful standard of comparison for quantum mechanics, occupying a kind of intermediate position between the latter and ordinary classical mechanics. Secondly, in this intermediate position it can be regarded as a modification of ordinary classical mechanics resulting from an inherent objection to the latter and thus already having certain indeterministic features without attaining the full indeterminism of quantum mechanics.

To say that the objection is inherent means that it is not enforced by the failure of classical mechanics as a physical theory but is the result of voluntary philosophical reflection. The basis of the objection emerged as soon as classical mechanics was employed to explain thermodynamic phenomena with an atomistic view of matter, that is, from the time of Maxwell, Boltzmann and Gibbs onward. For there the theory, previously tested chiefly (and with remarkable success) in astronomy, for the first time seriously found application in a different field which at the same time should have evoked deep philosophical questions. No such criticism really occurred, however, partly because of the fascinating possibility that was perceived of reducing thermodynamics to mechanics, partly because there soon arose the familiar difficulties which were to mark the end of the classical era in physics and whose purely physical resolution now occupied the full attention of the physicists.

The content of the criticism is to state as a problem the assumption made implicitly (or at least rendered very plausible) by ordinary classical mechanics, that it must be possible in principle to know the state of a classical-mechanical object with the same accuracy as that of its mathematical description in classical mechanics. Here, then, the question is explicitly raised of our possible knowledge of a mechanical object, accompanied by the suspicion that the mathematical description allotted by exact classical mechanics to the states of objects is fundamentally in excess of our possible knowledge of these objects. This leads to the question of a reformulation of classical mechanics incorporating explicitly the concept of knowledge of an object and at the same time expressing any fundamental limitation on our knowledge.

This topic will be further analysed mainly in the subsections (d) (i) and (d) (ii) below, which include the replacement of the exact description of state as in subsection 2 (d). The historical side of the criticism is almost lost, in retrospect, in the stability studies relating to celestial mechanics and to hydrodynamics, conducted at the end of the nineteenth century by Poincaré and by Lyapunov, but it is clearly shown, for example, by Duhem (1906, Chapters II and III), and was continued by Exner (1919, Chapter IV, especially p. 654) in his wide-ranging critique of the exact laws of nature. Von Mises found in it one of the principal fields of application of his probability theory (1930, §3). Von Neumann takes it as one of the bases of his classical-mechanical ergodic theory, deriving from it a mathematical formulation of mechanics, which is the main source of the formulation given later (1932, section I; see also Birkhoff and von Neumann 1936, §§4 and 5). From 1955 onwards Born made it the starting point of his discussion in a series of papers concerning indeterminism in classical mechanics (1955a, c, 1958, 1959, 1961). Reichenbach too based on it an indeterministic formulation of classical mechanics (1956, §11). It forms the background to much of an entire book by Brillouin (1964), and is already a standard item in modern textbooks on the philosophy of science (e.g. Frank 1957, §11.6; Nagel 1961, §10.1).

II.3(a) Objects and statistical ensembles

The *individual object* of classical statistical mechanics may be, firstly, any object Σ which is also able to be considered by ordinary classical mechanics and which in particular is unrestricted as regards its number of degrees of freedom. Then the mathematical description of Σ requires its association with a phase space $\mathfrak{M}$, as in subsection 2(a). The difference from ordinary classical mechanics in this case is that Σ is viewed in a different manner, and (as already indicated) a vital role is played by the concepts of a measurement of Σ having limited accuracy and a knowledge of Σ which is in some way restricted; see subsections (d) (i), (ii).

Secondly, we have also to include *statistical ensembles* of individual objects as being able to be considered by classical statistical mechanics. Such an ensemble is a series (in the ideal case an infinite series) of individual objects $\Sigma_1, \Sigma_2, \Sigma_3, \ldots$ identical in structure but possibly differing in their contingent behaviour. The structural identity is expressed in the first place by the fact that all the objects Σ_k in a statistical ensemble have the same phase space $\mathfrak{M}$. The consideration of statistical ensembles is to be seen as an alternative to the consideration of individual objects for which our knowledge is incomplete.

II.3(b) Composite objects and statistical ensembles thereof

Here again the situation as regards the individual object is the same as in subsection 2(b). The description of $\Sigma = \Sigma^1 \times \ldots \times \Sigma^N$ is effected in exactly the same way as regards the phase spaces. In considering a statistical ensemble $\{\Sigma_k\}_k$, the structural identity is again to be noted, in this case with reference to the possible composition $\Sigma_k = \Sigma_k^1 \times \ldots \times \Sigma_k^N$ of the Σ_k. For fixed n, all the Σ_k^n are structurally identical, and the resolution $\mathfrak{M} = \mathfrak{M}^1 \times \ldots \times \mathfrak{M}^N$ of the common phase space $\mathfrak{M}$ of the Σ_k is therefore the same for the entire ensemble.

II.3(c) Quantities

The concept of a quantity in statistical mechanics differs from that in exact mechanics in the following way. Since the determination of a state in exact mechanics always amounts to determining the values of certain quantities, whereas in statistical mechanics it is not possible to determine a state, it follows that certain quantities occurring in ordinary mechanics are not distinguishable in statistical mechanics. The quantities in the statistical theory therefore correspond to certain *classes* of quantities in the exact theory. The following specific definition of this new concept of a quantity is found to be suitable: *the quantities of an object Σ are in one-to-one correspondence with those classes of real Borel functions in the phase space* *of Σ which contain, together with any one such function f, all such functions g which differ from f only on a set having Lebesgue measure zero.* The resulting classes are, in one sense, fairly small: two continuous functions which are different (in the customary sense of the word) always belong to different classes. But the characteristic functions of phase space, which are always discontinuous, are more markedly affected by this class assignment: for example, the characteristic functions which belong to two exact energy values always belong to the same class. In considering a statistical ensemble $\{\Sigma_k\}_k$, we shall follow the assumption in subsection (a) that all the Σ_k have the same phase space $\mathfrak{M}$ by now assuming that all the Σ_k have the same range of quantities to be represented in the specified manner in $\mathfrak{M}$.

II.3(d)(i) Propositions regarding measurements

As has already been mentioned at the beginning of this section, it is the description of state in ordinary classical mechanics which undergoes a vital modification in statistical mechanics. This change from a description of things as they *are* to a description of things as they *are known*, in a physical theory, will of course bring into prominence the *measurements* that can be made on an object. If the change also involves, as here, a test of the achievable accuracy of our knowledge of states, the first question to be discussed must concern the accuracy of measurement that is in principle attainable.

With the establishment of modern analysis in the nineteenth century, mathematics furnished a notable ideal of quantitative exactitude, most persuasively expressed in a certain view of the continuum. This view consists in some form of reconstruction of the (one-dimensional) continuum as the domain of all real numbers, and was regarded as the keystone of the edifice, already long under construction, of the analysis of functions, their derivatives and integrals, differential equations and so on. The indispensable role of this analysis in the recent development of physics is familiar. But the consequences are inescapable. If the continuous quantities of physics can have values which range over the domain of the real numbers, it becomes possible, and in fact obligatory, to raise some very awkward questions regarding the physical significance of certain distinctions in the continuum. What is meant by saying that a mechanical quantity can have a value represented by a certain real number, for example an algebraic or a transcendental number? What is meant by saying that the meaning of an assertion about the value of a quantity is no longer to be understood purely objectively, but must be justified by stating the way in which such an assertion can in principle be supported by a measurement? Do there exist, even in principle, measurements of quantities which establish their exact values as real numbers? No physicist would have believed such things. The next best approach to this difficulty (or the next worst, according as one takes the mathematical or the physical reality as one's model) would then be to permit measurements of mechanical quantities which are not exact but can be made as nearly exact as is desired. The following paragraphs outline a possibility of incorporating measurements in classical mechanics (later to be supplemented by propositions regarding probabilities), whereby the idea of a measurement having arbitrary but not absolute accuracy can be made specific, though without at the same time removing the possible root of the trouble, namely the contemporary mathematical treatment of the continuum.

In the same way as we distinguished in exact classical mechanics between the states of an object Σ and the corresponding propositions regarding states, so here we shall distinguish between *measurements of* Σ and the corresponding propositions regarding measurements, and first consider the measurements themselves. A mathematical representation of these may be obtained as follows. In the phase space $\mathfrak{M}$ of Σ there is an infinity of systems $\{\chi_i\}_i$ of characteristic Borel functions χ_i such that the corresponding Borel sets $\mathfrak{M}_i$ defined by

$$M \in \mathfrak{M}_i \rightharpoonup \chi_i(M) = 1 \tag{1}$$

form a complete disjunctive division of $\mathfrak{M}$, i.e.

$$\mathfrak{M}_i \neq \emptyset, \tag{α}$$

$$\mathfrak{M}_i \cap \mathfrak{M}_j = \emptyset \quad \text{for} \quad i \neq j, \tag{β}$$

$$\cup_i \mathfrak{M}_i = \mathfrak{M}. \tag{γ}$$

In terms of the χ_i themselves, these formulae are equivalent to

$$\chi_i \neq \mathbf{0}, \tag{α_1}$$

$$\chi_i \cdot \chi_j = \mathbf{0} \quad \text{for} \quad i \neq j, \tag{β_1}$$

$$\Sigma_i \chi_i = \mathbf{1}, \tag{γ_1}$$

where **0** and **1** are the trivial characteristic functions; the product in (β_1) is the usual product of functions

$$(\chi\chi_1)(M) = \chi(M)\,\chi_1(M) \tag{2}$$

and the sum in (γ_1) is the "logical sum"

$$\left.\begin{aligned}(\chi+\chi_1)(M) &= 1 \quad \text{for} \quad \chi(M) = 1 \quad \text{or} \quad \chi_1(M) = 1,\\ &= 0 \quad \text{otherwise.}\end{aligned}\right\} \tag{3}$$

Such systems $\{\chi_i\}_i$ *could* be used to describe measurements if these are assumed to be *absolutely accurate*. Then $\{\chi_i\}_i$ would serve to describe the measurement which decides the set $\mathfrak{M}_i$ in which the phase point describing the state of Σ is situated. An additionally selected χ_{i_0} would then represent the result of such a measurement. But it is clear that measurements described in this way would have an accuracy unattainable according to the criticism outlined above: any quantity in exact mechanics could be determined with absolute accuracy by such a measurement.

In order to avoid this difficulty, we take the characteristic Borel functions only to within the equivalence relation defined in subsection (c) above, which will be denoted by $\sim$. The operations (2) and (3) are easily shown to be compatible with $\sim$. If we also take the system $\{\chi_i\}_i$ only to within the same equivalence relation of its elements, then the conditions corresponding to $(\alpha_1)-(\gamma_1)$ are

$$\chi_i \nsim \mathbf{0}, \tag{α_2}$$

$$\chi_i \cdot \chi_j \sim \mathbf{0} \quad \text{for} \quad i \neq j, \tag{β_2}$$

$$\Sigma_i \chi_i \sim \mathbf{1}, \tag{γ_2}$$

independently of the representatives selected for the formulation. Changing from (α_1)–(γ_1) to (α_2)–(γ_2) ensures that sets having Lebesgue measure zero play no part, and this is just the transition to measurements having *limited* and not absolute accuracy. We say that *the measurements of Σ are to be described in one-to-one correspondence by pairs* $(\{\chi_i\}_i, \chi_{i_0})$, *where the χ_i are characteristic Borel functions* (*to be taken to within the equivalence relation* $\sim$) *with* (α_2)–(γ_2) *in $\mathfrak{M}$, and χ_{i_0} is one of the χ_i*. The variables in the measurements, or their representatives, will be denoted by *me*, me_1, . . . when it is not necessary to state explicitly the $(\{\chi_i\}_i, \chi_{i_0})$.

Let us now consider the form of the propositions regarding measurements of Σ which correspond to the measurements thus defined. These propositions involve both the measurements and the quantities of Σ, and in order to obtain the mathematical relation between the representatives we must first elucidate the concept of the spectral resolution of a Borel function, taken to within the relation $\sim$, in the phase space $\mathfrak{M}$ of Σ. Let f be a Borel function in $\mathfrak{M}$. Then the characteristic Borel function

$$\left.\begin{aligned} \chi_a^f(M) &= 1 \quad \text{for} \quad f(M) \in a \quad (M \in \mathfrak{M}), \\ &= 0 \quad \text{otherwise,} \end{aligned}\right\} \tag{4}$$

is assigned to each Borel set $a \subseteq \mathbf{R}$ (the set of all real numbers), and χ_a^f as a function of a has the properties

$$\chi_a^f \cdot \chi_b^f = \mathbf{0} \quad \text{for} \quad a \cap b = \emptyset, \tag{α_3}$$

$$\chi_{\cup_i a_i}^f = \Sigma_i \chi_{a_i}^f \quad \text{for any countable set } \{a_i\}_i \text{ with } a_i \cap a_j = \emptyset \text{ for } i \neq j, \tag{β_3}$$

$$\chi_\emptyset^f = \mathbf{0} \quad \text{and} \quad \chi_{\mathbf{R}}^f = \mathbf{1}. \tag{γ_3}$$

Thus, for any given f, (4) defines a kind of measure χ^f on the real axis, whose values for the arguments a are characteristic Borel functions in $\mathfrak{M}$. With the assignment $f \rightarrowtail \chi^f$ we can now easily show that, if f and g are equivalent in the sense defined in subsection (c), then χ_a^f and χ_a^g are equivalent for all a. Hence (4) assigns to each equivalence class of Borel functions a mapping which in turn assigns to each Borel set $a \subseteq \mathbf{R}$ an equivalence class of characteristic Borel functions. This assignment $\chi \rightarrowtail \chi^f$ (to within the relation $\sim$) now in fact gives a characterization of the Borel functions (taken similarly) in $\mathfrak{M}$. If a *measure* denotes any mapping which uniquely (to within the relation $\sim$) assigns a characteristic Borel function χ_a to each real Borel set a in such a way that

$$\chi_a \cdot \chi_b \sim \mathbf{0} \quad \text{for} \quad a \cap b = \emptyset, \tag{α_4}$$

$$\chi_{\cup_i a_i} \sim \Sigma_i \, \chi_{a_i} \quad \text{for any countable set } \{a_i\}_i \text{ with } a_i \cap a_j = \emptyset \text{ for } i \neq j, \tag{β_4}$$

$$\chi_\emptyset \sim \mathbf{0} \quad \text{and} \quad \chi_{\mathbf{R}} \sim \mathbf{1}, \tag{γ_4}$$

then (4) gives a characterization of the Borel functions (taken to within the relation $\sim$) by these measures, i.e. the assignment $\chi \rightarrowtail \chi^f$ is one-to-one within $\sim$, and for each measure χ_a there exists an f such that $\chi_a \sim \chi_a^f$ for all a (see, for example, Varadarajan 1968, §I.4). This is the *spectral resolution* χ^f of f.

After this preparatory work, the propositions regarding measurements are easily derived. The only further concepts needed are those of the *accuracy* with which a quantity is measured and the *result* concerning a measured quantity. *The possible accuracies are in one-to-one correspondence with sets of real Borel sets.* Here it is initially undecided which sets of real Borel sets are possible representatives of an accuracy; this will be decided only when the propositions regarding measurements are established. *The possible results concerning measured quantities are in one-to-one correspondence with real Borel sets.* The extent of the representative real Borel sets is likewise determined by the propositions regarding measurements. The universal propositions regarding measurements are

$\mathbf{M}(me, f, \mathfrak{a}, a)$: measurement me of object Σ measures quantity f with accuracy $\mathfrak{a}$ and result a.

The relation **M** is mathematically described by a relation between the mathematical representatives $(\{\chi_i\}_i, \chi_{i_0}), f, \mathfrak{a}$ and a_0, defined by the conditions

$$\Sigma_{a \in \mathfrak{a}} \chi_a^f \sim \mathbf{1}; \tag{α_5}$$

for every $a \in \mathfrak{a}$ there is a subset $I_a \subseteq I$ such that

$$\chi_a^f \sim \Sigma_{i \in I_a} \chi_i; \tag{β_5}$$

$$a_0 \in \mathfrak{a}; \tag{γ_5}$$

$$\chi_{i_0} \cdot \chi_{a_0}^f \sim \chi_{i_0}. \tag{δ_5}$$

The propositions regarding measurements, thus defined, are universal in the same sense as the propositions regarding states, specified in subsection 2(d): the relation **M** is determined only by the structure of the object Σ concerned, and accordingly does not assert that a measurement of Σ has been made. But, as with the universal propositions regarding states, it provides exactly what is necessary to describe the situation which occurs if a measurement of Σ *is* made: it then states which quantities have been measured, with what accuracies, and with what results. Corresponding to (1) in subsection 2(d), we have the definition

$$Me_{me}(f, \mathfrak{a}, a) \leftrightarrow \mathbf{M}(me, f, \mathfrak{a}, a). \tag{5}$$

It is also easy to see that the measurements satisfy the characterization formula corresponding to (2) in subsection 2(d).

We must also check that the concept of measurement thus formally defined, together with the relevant propositions regarding measurements, does achieve the objective stated initially, that an individual measurement has only limited and not absolute accuracy, while measurements as a whole may have arbitrary accuracy. To see this, it is convenient to establish a relation between measurements, saying that a measurement me_1 *involves* a measurement me if

$$\mathbf{M}(me, f, \mathfrak{a}, a) \rightarrow \mathbf{M}(me_1, f, \mathfrak{a}, a) \tag{6}$$

for all f, $\mathfrak{a}$ and a. If me_1 and me are in this relation, which is clearly an ordering, then the measurement me can obviously be considered to have been made when the measurement me_1 has been made. The mathematical representation of (6), with $me = (\{\chi_i\}_{i \in I}, \chi_{i_0})$ and $me_1 = (\{\chi_j^1\}_{j \in J}, \chi_{j_0}^1)$, is as follows.

For every $i \in I$ there is a subset $J_i \subseteq J$ such that

$$\chi_i \sim \Sigma_{j \in J_i} \chi_j^1; \tag{α_6}$$

$$\chi_{j_0}^1 \cdot \chi_{i_0} \sim \chi_{j_0}^1. \tag{β_6}$$

From this it is seen that the accuracy, or rather the range, of a measurement as regards the quantities with which it deals can be made as large as is desired. Thus *there are no maximum measurements*, as is necessary if a state in exact mechanics is to be determined.

After the foregoing relatively formal discussion, something should be said regarding the content of the concept of measurement and of propositions regarding measurements. First, it must be stressed that the individual measurement me and the corresponding propositions $\mathbf{M}(me, f, \mathfrak{a}, a)$ relate to a single object Σ. The statement made by $\mathbf{M}(me, f, \mathfrak{a}, a)$ concerning Σ is, in words, that the measurement me measures the quantity f (taken relative to $\sim$) with accuracy $\mathfrak{a}$ and result a, as stated above. The result a here is in general a set of values and not the customary individual value, and accordingly it is possible to refer to the accuracy of the measurement of f; this is because the measurements are no longer absolutely accurate, and is not a point of immediate importance (this aspect is further discussed in section IV.2). But the expression that a quantity is measured with such and such a result does involve an ambiguity which is to play a certain role in quantum mechanics, and which must be disclosed even in the classical case in order to clarify the relation between the propositions regarding measurements and those regarding probabilities (see the next subsection). The ambiguity is that the measurement of a result can be regarded either as an actual acceptance of knowledge of the result or as only an objective recording of the result by means of a measuring apparatus. This distinction never gave rise to controversy in classical physics, but in quantum mechanics there was debate as to whether a measurement can ever rank as completed until it has reached the consciousness of an observer. This point will be further considered below. For classical mechanics, if understood to refer to an individual object, we shall take the first (epistemic) interpretation, since the propositions regarding probability that are shortly to be discussed will also be epistemically interpreted as propositions regarding the state of knowledge at any time. Since the measurements may change (viz. improve) this probabilistic knowledge, it is consistent to interpret the propositions regarding measurements likewise as propositions regarding newly acquired knowledge. This does not forfeit their objectivity if errors (in this case, incorrect measurements) are excluded. Then different persons may have different knowledge, but whatever knowledge a person has is correct knowledge.

Before we conclude by dealing with the theory of the measurement of statistical ensembles, it is convenient to add a further concept in measurement theory, still referring to an individual object. In the understanding of quantum mechanics, it is important to make use both of propositions which include the statement of results of measurements, and of propositions which state only which quantities are measured and with what accuracy, but not with what result. The comparison with the classical situation is particularly informative with respect to these propositions. Moreover, even in the classical case they show a clear formal analogy to certain propositions regarding measurements of statistical ensembles, and this analogy has to be considered. We shall say that two measurements *me* and me_1 are *equivalent* if

$$\mathsf{V}_a \mathbf{M}(me, f, \mathfrak{a}, a) \leftrightarrow \mathsf{V}_a \mathbf{M}(me_1, f, \mathfrak{a}, a) \tag{7}$$

for all f and $\mathfrak{a}$ (where V_a denotes "there is some a"). Thus this equivalence simply implies that the result of the measurement is disregarded. The mathematical representation of (7) is

$$\{\chi_i\}_{i\in I} \sim \{\chi_j^1\}_{j\in J}, \tag{8}$$

i.e. the element-by-element equivalence of the systems of characteristic functions which describe the measurements; the χ_{i_0} and $\chi_{j_0}^1$ which belong to *me* and me_1 respectively and which define the corresponding results no longer appear. We now treat these equivalence classes as a new type of *measurement irrespective of result*, and denote them by the variables $\dot{me}$, $\dot{me}_1$, ... They are correlated with new (universal) *propositions regarding measurements irrespective of result*:

$$\dot{\mathbf{M}}(\dot{me}, f, \mathfrak{a}) \rightharpoonup \mathsf{V}_a \mathbf{M}(me, f, \mathfrak{a}, a). \tag{9}$$

From (7) it is seen that these propositions are unambiguously defined. Their mathematical equivalence is (α_5) and (β_5); (γ_5) and (δ_5) do not occur, since they relate to the results. $\dot{\mathbf{M}}(\dot{me}, f, \mathfrak{a})$ states that *the measurement* $\dot{me}$ *measures the quantity* f *with accuracy* $\mathfrak{a}$, or in other words, according to our epistemic interpretation of the propositions $\mathbf{M}(me, f, \mathfrak{a}, a)$, that $\dot{m}e$ records certain results which, however, we choose to ignore. Thus, if the measurement $\dot{m}e$ has actually been carried out, the propositions $\dot{\mathrm{M}}(\dot{m}e, f, \mathfrak{a})$ furnish no new knowledge, but (as we shall see) there is also no loss of knowledge. This is not the case in quantum mechanics, and in order to be able to indicate this difference with sufficient clarity it is useful to define measurements irrespective of result, and the corresponding propositions, even in the classical case.

So far we have outlined a theory of measurement for an individual object in classical mechanics, but a corresponding theory for statistical ensembles is also necessary, since the propositions regarding probabilities which will appear in the next subsection are to be given both an epistemic interpretation in relation to an individual object and a statistical interpretation in relation to an ensemble of individual objects. It will be found immediately that the formalism developed above of measurements and propositions regarding measurements can be used as a theory of measurements of statistical ensembles, and needs only to be given a different interpretation. We have shown in subsections (a)–(c) that all the objects in a statistical ensemble $\{\Sigma_k\}_k$ are taken to have the same structure, to the extent that the structure has been specified in those subsections. In particular, according to subsection (c) all the Σ_k have the same range of quantities, and it is therefore meaningful to refer to a quantity f of the *ensemble* $\{\Sigma_k\}_k$. Correspondingly, we now propose to refer to the measurement of a quantity f in the *ensemble* $\{\Sigma_k\}_k$. The meaning of this is as follows. First of all, such a measurement cannot be said to have a definite result, since in general there will be a dispersion of values. The measurements that are of chief interest, however, will be those which measure the same quantities with the same accuracies in all the Σ_k. This is ensured if a *statistical measurement* of $\{\Sigma_k\}_k$ is taken to be an ensemble of individual measurements me_k of the Σ_k (in the "recording" sense) which are all mutually equivalent in the sense defined by (7), i.e. are the same measurement operation but in general give different results. This is not intended as a rigorous definition of the concept of a statistical measurement, since nobody has yet succeeded in constructing a logically defensible statistical theory of physics on the basis of concepts relating to the individual objects of an ensemble. Instead, the concept (which will be a fundamental one in our discussion) is explained only for the purpose of seeing that it may be reasonably identified in a *formal* manner with the concept of a measurement of an individual object irrespective of result. *Statistical measurements* $\dot{m}e$ *are mathematically described by systems* $\{\chi_i\}_i$ *with the properties* (α_2)–(γ_2). These measurements are correlated with *statistical propositions regarding measurements* $\dot{\mathrm{M}}(\dot{m}e, f, \mathfrak{a})$ which state that *the statistical measurement* $\dot{m}e$ *measures the quantity* f *(taken relative to* $\sim$*) with accuracy* $\mathfrak{a}$ *for all the* Σ_k. They are mathematically described by (α_5) and (β_5) in the same manner as the corresponding propositions for an individual object.

An obvious question is to ask what is the purpose of these statistical propositions regarding measurements when they occur in the specific context of an actually performed measurement $\dot{m}e$. The answer is none. The situation is exactly the same as has already been described for the corresponding individual measurements $\dot{m}e$. In the classical case, the statistical propositions regarding measurements $\dot{\mathrm{M}}(\dot{m}e, f, \mathfrak{a})$ comprise no gain but also no loss of statistical homogeneity, whereas in quantum mechanics such a loss does in general occur. In this comparison the statistical propositions regarding measurements play an important part. For an individual object, a proposition regarding a measurement, $\dot{\mathrm{M}}(\dot{m}e, f, \mathfrak{a})$, for which the measurement $\dot{m}e$ has actually been performed, always implies that a definite result has been at least recorded, even if ignored. An actual observation is present only in propositions $\mathrm{M}(me, f, \mathfrak{a}, a)$ based on a measurement me. In the statistical case also, formally analogous measurements and propositions can be arrived at if a measurement $\dot{me}$ is associated with a *selection* of the subensemble of $\{\Sigma_k\}_k$ for which this measurement has a specified result. Thus, in analogy with the measurements *me* of an individual object, which include a definite result, we can make *selective measurements* of an ensemble $\{\Sigma_k\}_k$, the selection relating to a particular one out of the various results which occur in a statistical measurement. Accordingly *the selective measurements*, like the individual measurements which include a result, *are to be described by pairs* $(\{\chi_i\}_i, \chi_{i_0})$, *in which the* χ_i *satisfy* (α_2)–(γ_2) *and* χ_{i_0} *is one of the* χ_i, namely the one which represents the result that determines the choice. The corresponding propositions regarding measurements $\mathbf{M}(me, f, \mathfrak{a}, a)$ imply that *the quantity f is measured with accuracy* $\mathfrak{a}$ *and result a in the subensemble selected by the selective measurement me.* Here "measured" means "recorded" (as distinguished above) for the individual case belonging to the subensemble. The purpose of the selection is of course to create an ensemble for which there is no dispersion of certain measured values and for which this is known to be so, but such knowledge is not part of the content of the propositions $\mathbf{M}(me, f, \mathfrak{a}, a)$.

To conclude this subsection, it must be stressed that the theory of the measurement of classical statistical ensembles, outlined above, is of course only a somewhat feeble version of what such a statistical theory of measurement might be. The reason is that this theory is only meant to make objective the epistemic theory of measurement previously given for an individual object. The process is exactly similar to the one described in the next subsection, going from probabilistic states of knowledge of an individual object to the statistical states of an ensemble of objects which make objective that knowledge. Thus we have here no epistemic theory of statistical ensembles; these are always referred to in the same manner as are individual objects in ordinary classical mechanics, i.e. completely neutrally as regards knowledge or ignorance. This does not mean that such a theory would not be valuable. The ultimate purpose of a statistical measurement is to determine the relative frequencies with which the values of a quantity occur in an ensemble. The essential point, then, is to acquire knowledge of something, and the importance of this in practical physics shows that an epistemic theory of statistical ensembles is in fact most desirable. On the other hand, such a theory (and none has yet been worked out) would not contribute much to a discussion of specifically quantum-mechanical problems. Above the level of probabilistic propositions, quantum mechanics seems no longer to differ significantly from a classical probabilistic theory. The decision is rather between the individual instance and the ensemble; the iteration would contribute nothing new.

II.3(d)(ii) Propositions regarding probabilities

The involvement of measurements and of propositions regarding them in classical mechanics, brought about in the preceding subsection, appears inadequate if we are considering only the application of this theory in thermodynamics. But it has already been stated at the beginning of subsection (d) (i) that we are interested in the opposite aspect, the comparison of classical mechanics with quantum mechanics, in particular in problems of indeterminism, and the concepts previously introduced are useful here, because in quantum mechanics the problem of the measurement process is central to the interpretation. The relationship of classical mechanics and thermodynamics must be viewed differently as regards the theory of probability: the inclusion of propositions regarding probabilities in classical mechanics was essential in defining its relation to thermodynamics. But our primary interest in the comparison with quantum mechanics also makes it necessary to include in classical mechanics, besides propositions regarding measurements, also propositions regarding probabilities. Only the two types of proposition together can form a complete analogue to the propositions regarding states in the customary formulation as given in section 2.

The concept of probability is burdened with internal problems of interpretation which cannot be fully displayed here. The history of the concept extends back to antiquity; its mathematical form dates at least from Laplace, and it has been a permanent part of physics at least since Gibbs created statistical mechanics. In the last few decades, the philosophical analysis of its basis has been pursued with great fervour on account of its steadily growing importance in science, but despite these efforts there is still no generally accepted resolution of the fundamental problems concerned. Thus it is impossible to give a full survey of this extensive area; only indications are available, which remain similar to the factors which determined the introduction of propositions regarding probabilities into classical mechanics.

The use of propositions regarding probabilities in classical statistical mechanics, created as a basis for classical phenomenological thermodynamics, has often been stated to be based on the fact that the mechanical properties of a thermodynamic system are not *known* either to a sufficient extent or with sufficient exactitude. The principal causes of this ignorance were taken to be the very large number of degrees of freedom and the microscopic nature of each individual constituent of the system. At the beginning of section 3 a number of references have been given to suggest that ordinary classical mechanics may be not adduced and modified in order to explain thermodynamic phenomena but itself subjected to critical examination. The focus of this criticism was again, however, that in any specific application the quantities used by exact mechanics to describe an object are never precisely known. The difference was that one now proceeded to assert that they *cannot* be precisely known; and this was no longer supported by pointing to any contingent circumstances such as the large number of degrees of freedom or the atomic dimensions of the basic constituent objects. It was based on the fundamental incompatibility between the way in which continuum mathematics is employed in physics and any possibility of measuring continuous quantities. In the previous subsection, this problem of the continuum for the information regarding states (other aspects will not be considered here) has been taken into account by the use of propositions regarding measurements, which replace the purely objective propositions of exact mechanics in such a way that they can no longer express the mathematically exact determination of a state. The question arises whether this procedure has not already solved the continuum problem at least to the extent that it does not itself require the use of propositions regarding probability, even if these are still demanded by other, for instance thermodynamic, factors. But the answer is that it has not. The following consequence must be considered if we accept the limited accuracy of measurement as defined in subsection d(i). If it is possible to carry out repeatedly an experiment whose aim is prediction (the standard examples are taken from ballistics) under certain initial conditions, the inaccuracy in the determination of the initial conditions will mean that the results of such a sequence of experiments show a certain amount of dispersion, in a manner which (and this is the important point) can in principle be ascertained in relation to the permissible accuracy of measurement. This becomes clear, in particular, if the measurements are not made until some time after the initial conditions have been determined, so that the original inaccuracies can become larger (cf. section IV.3). Thus a new, quantitative aspect is brought into operation, affording valuable possibilities: to find the distribution of the relative frequencies of occurrence of the various results under specified conditions, and to regard such distributions as providing a possible basis for a quantitative description of the uncertainty that exists.

It is important to understand that the two above-mentioned *basic reasons for making use of propositions regarding probabilities* (viz. inexact knowledge of initial conditions, and the resulting experimentally measurable dispersion) do not alone suffice to show *the significance of a proposition regarding probabilities*. Instead, each provides at most a guide to determining this significance, and the possession of two such guides leads to a degree of competition between them, which is a difficulty rather than a help in solving the problem of the significance. It is plausible, on the one hand, to say that a proposition regarding probabilities represents a certain state of knowledge, though an incomplete one: it is a proposition which specifies the amount of knowledge about an individual case. It might be concluded that such a proposition is made *because* something is not known precisely. This in turn implies *that* something is not known precisely. And finally a virtue is made of this necessity by stating *how* precisely it is known. In this view an important point is that the proposition regarding probabilities relates to *an individual case*, because such a case represents the characteristic situation of ignorance. But one can also take up the other notion, that of repeating an experiment under given conditions, and relate the probability of a result to the relative frequency with which this result occurs in a sequence of experiments (more precisely, the limit of this relative frequency for an infinite series of experiments). Thus, in contrast to the previous statement, the proposition regarding probabilities would be a proposition relating to *an ensemble of individual cases*, and there need be no reference to knowledge or ignorance. The second reason was described above as a consequence of the first. The dispersion occurs because the initial conditions are not exactly known. But it is possible to imagine that these conditions, so far as they are known, can be made purely objective, and that the relative frequencies, or rather their limits, are unambiguously determined by the conditions to the extent that these can be made objective.

This procedure therefore yields at least two concepts of probability, and the question then arises of which is meant when probability is mentioned in physics and, in particular, in the classical statistical mechanics here under discussion. The discussion of fundamentals during the last few decades shows that it is not easy to reach agreement as to the answer. The first of the two concepts has the disadvantage of involving a subjective component of knowledge, whose desirability is very questionable and which seems to contradict the ideal of objectivity in physics. In particular, we cannot say how a quantitative proposition regarding knowledge is to be tested on an individual case to which it is supposed to relate. But there are also difficulties in regarding probability as the limit of relative frequencies, and the most serious of these difficulties is perhaps the objectivist independence aimed at here, which makes the concept of probability invalid for expressing the entirely typical situation which occurs, in particular, as regards our knowledge of future events. The general consciousness of physicists, guided by an instinct for the dangers of prematurely adopting an extreme position, has accepted a concept of probability which stands somewhat flickeringly somewhere between the two extremes stated. The concept of ideal statistical ensembles whose states represent any existing knowledge as to the state of an individual object belonging to this ensemble, first proposed by Gibbs and later refined by Tolman (1938, Chapters I and III), certainly contributed to the stabilization of this slightly wavering attitude, but strictly speaking there are two different contents, which, by analogy with the situation already encountered in connection with propositions regarding measurements, may be termed the *epistemic* and the *statistical* interpretation of probability.

To make the discussion a little more definite, we shall continue by taking the two extreme cases as alternative bases, referring firstly to the possible *states of knowledge of an object* Σ and secondly to the possible *states of a statistical ensemble* $\{\Sigma_k\}_k$. *The states of knowledge of* Σ *or the statistical states of* $\{\Sigma_k\}_k$ *are mathematically described by Borel functions* ρ, *taken to within the equivalence relation* $\sim$, *in the phase space of* Σ *or of all the* Σ_k (see subsection (a)), *for which*

$$\rho(M) \geqslant 0, \quad \int_{\mathfrak{M}} \rho \, d\mu = 1, \tag{1}$$

μ *being the Lebesgue measure*. Where a distinction is unnecessary or inadvisable, we shall refer to both states of knowledge and statistical states as *descriptions of state*.

The states of knowledge of Σ or the statistical states of an ensemble $\{\Sigma_k\}_k$ correspond to *universal propositions regarding probabilities*, which contain the relevant degree of knowledge in the first case, or the relevant limit of relative frequencies in the second case. In either case, the individual probability is to refer to the results of possible measurements and not to properties independent of measurements, although this is not to be taken as casting doubt on the assumption that such independence exists in the classical case. We are simply taking account of the fact that the theory offers, apart from propositions regarding probabilities, only those regarding measurements. Since the probability which occurs in a proposition regarding probabilities is of course defined unambiguously by the other arguments in the proposition, we can go immediately to the corresponding probabilities themselves. Then a universal propositional function is replaced by a universal probability function **pr**, such that $\mathbf{pr}(\rho; f, a)$ *is the probability that a measurement of the quantity f will give the result a either in the state of knowledge* ρ *of an object* Σ *or in the state* ρ *of a statistical ensemble* $\{\Sigma_k\}_k$. The mathematical representation of **pr** is

$$\mathbf{pr}(\rho; f, a) = \int_{\{M: f(M) \in a\}} \rho \, d\mu. \tag{2}$$

In the mathematical probability function $\mathbf{pr}(\rho; f, a)$ or the corresponding propositions regarding probabilities $w = \mathbf{pr}(\rho; f, a)$, the states of knowledge ρ or the statistical states ρ play the same part as the measurements do in the corresponding propositions regarding measurements. Although this can be expressed by a definition

$$pr_\rho(f, a) \rightleftharpoons \mathbf{pr}(\rho; f, a) \tag{3}$$

analogous to (5) in subsection (d) (i), a proposition $w = \mathbf{pr}(\rho; f, a)$ says nothing about any actually existing states of knowledge or statistical states. These are characterized only by certain sets of propositions. A factual situation arises only when some ρ is identified as an actually existing state of knowledge or statistical state. The corresponding propositions regarding probabilities then also acquire an independent significance.

As with measurements in subsection (d) (i), so here for states of knowledge or statistical states we have to confirm that they cannot lead to the determination of a state in exact mechanics, so that again the state can be determined with perhaps arbitrary but not absolute accuracy. As will be discussed with somewhat more rigour for quantum mechanics, this is easily done by means of one of two semi-orderings:

$$\mathbf{pr}(\rho; f, a) = 1 \rightarrow \mathbf{pr}(\rho_1; f, a) = 1 \tag{4}$$

for all f and a, or

$$\mathbf{pr}(\rho; f, a) - \alpha\, \mathbf{pr}(\rho_1; f, a) \geqslant 0 \tag{5}$$

for all f and a, with $0 < \alpha \leqslant 1$. Each of (4) and (5) offers a possibility of defining, for instance in the epistemic formulation, the state of knowledge ρ_1 as being more exact than the state of knowledge ρ. We can then inquire whether there exists a *maximum* knowledge, i.e. knowledge which is essentially the most exact possible. It is easily seen that this does *not* exist with either (4) or (5): knowledge can always be increased *ad infinitum*. This is because in the form (2) and (3) we take only measures of probability in phase space which are absolutely continuous in terms of the Lebesgue measure.

II.3(e) Time dependence

In ordinary classical mechanics, the time variations of an object Σ were regarded as those of the state of Σ: the time variation of Σ was represented by a function correlating each instant t with the state M_t of Σ at that instant. Accordingly, in the epistemic formulation of classical mechanics we are concerned with the change in the existing state of knowledge concerning Σ; in the statistical formulation, with the change in the statistical state of an ensemble $\{\Sigma_k\}_k$. Here the time variation would be described by a function ρ_t. The cause of such changes is firstly, of course, just as in ordinary classical mechanics, the changes in Σ or the Σ_k *themselves*, possibly due to an intrinsic dynamics which is again to be described by a Hamiltonian H of Σ or by a function of the same type for the Σ_k (we shall assume it to be the same function for each Σ_k). In contrast to ordinary mechanics, however, we now have the additional possibility of changes caused by *measurements* of Σ or $\{\Sigma_k\}_k$. The individual measurements which include the determination of a result involve a change in the existing state of knowledge; the statistical measurements which include a selection involve a change in the statistical states (and even a change in the initial ensemble). This, together with the foregoing distinctions regarding measurements and probabilities, leads to an *epistemic* and a *statistical* formulation of classical mechanics.

For the epistemic formulation it is reasonable to suppose that a measurement me_t made at time t can change the state of knowledge ρ_t existing at that time in a way which has no relation to the change in ρ_t due to the (concealed) change in the state of Σ. For example, it may happen that previously $0 < \mathbf{pr}(\rho_t; f, a) < 1$ for a quantity f and a set of values a, but afterwards $\mathbf{M}(me_t, f, \mathfrak{a}, a)$ is true with a certain accuracy according to the result of the measurement. Then new information about Σ would have been instantaneously obtained without any change in Σ itself in the case of an ideal measurement (which will be assumed here). Thus we should expect that a measurement me_t, by the result obtained, converts a state of knowledge ρ_t valid at a time t into a well-defined state of knowledge ρ_t^+ existing "immediately after t"—and not necessarily different from ρ_t (for example, if me_t is only a check measurement), though it *may* be different from ρ_t and, if so, is a more exact state of knowledge of Σ than ρ_t was. Then ρ_t^+ is the starting-point of a further change in the state of knowledge as a result of the evolution of Σ itself. Thus we see that in the epistemic formulation the contingent conditions can be most appropriately described by a total of three functions: me_t, ρ_t, ρ_t^+. Of these, me_t is taken to be non-trivial only at discrete instants (see below), and $\rho_t^+ \neq \rho_t$ only for such an instant at which me_t is non-trivial; then ρ_t^+ has the significance stated above.

For the statistical formulation, the situation is entirely similar, as will be seen from the discussion in subsection (d) (i). A selective measurement me_t in general modifies instantaneously a statistical state ρ_t previously existing for an ensemble $\{\Sigma_k\}_k$, so as to change to a subensemble with a new state ρ_t^+ in which the result governing the selection no longer exhibits any dispersion. But, unlike the epistemic case, the object under investigation is here itself altered. In the epistemic case, new information is obtained concerning the object Σ, which itself remains unchanged; in the statistical case, the new state ρ_t^+ relates to a subensemble of the initial ensemble. The change is nevertheless still arbitrary and, like the change in the state of knowledge, does not correspond to anything in the object. Even after the selection, the previous ensemble continues to exist (at least in principle) with the previous ρ_t as the initial state; and the newly selected ensemble was in the state ρ_t^+ before it was selected, being only extracted by the measurement.

Corresponding to the functions me_t, ρ_t and ρ_t^+, which are thus seen to be essential to a particular possibility of contingent conditions in either formulation, there are contingent propositions given by (d) (i) (5) and (d) (ii) (3):

$$Me_t(f, \mathfrak{a}, a) \rightharpoondown \mathbf{M}(me_t, f, \mathfrak{a}, a), \tag{1}$$

$$w = pr_t(f, a) \rightharpoondown w = \mathbf{pr}(\rho_t; f, a), \tag{2}$$

$$w = pr_t^+(f, a) \rightharpoondown w = \mathbf{pr}(\rho_t^+; f, a), \tag{3}$$

the significance of which will not be given in words, since all the information needed to do so has already been provided. These propositions take the place of the contingent propositions regarding states (subsection 2(e), (1)) in ordinary classical mechanics, and indicate the time variation of an object or an ensemble in a new way.

After this clarification as regards content, we now come to the formal conditions on a set of three functions me_t, ρ_t and ρ_t^+, these conditions being the same for both formulations. The first condition involves one further concept. We say that the measurement me is a *trivial measurement of the quantity f* if the assertion $\mathbf{M}(me, f, \mathfrak{a}, a)$ is valid only for an $\mathfrak{a}$ consisting of a single element (which must then be equal to a). In that case, the measurement me conveys no real information concerning a. A measurement me is said to be a *trivial measurement* if it is a trivial measurement of every quantity, i.e. conveys no information at all. Mathematically, this means that in $me = (\{\chi_i\}_i, \chi_{i_0})$ the system $\{\chi_i\}_i$ consists of only one element, which must then be $\sim \mathbf{1}$ (like χ_{i_0}).

(α) In any finite range of t, me_t is trivial for all except a finite number of values of t.

(β) If $me_t = (\{(\chi_i)_t\}_i, (\chi_{i_0})_t)$, then

$$\int_{M:(\chi_{i_0})_t(M)=1} \rho_t \, d\mu > 0.$$

(γ) If me_t is trivial for all t in the open interval (t_1, t_2) with $t_2 > t_1$, we assume that

$$\rho_{t_2}(M) = \rho_{t_1}^+(U_{t_2-t_1}^{-1} M),$$

where U_t maps the phase space on itself in accordance with Hamilton's equations, subsection 2(e), (2); thus $M_\tau = U_\tau M_0$ is a solution of these equations as a function of τ for any starting point M_0. If ρ_t is differentiable, this can be replaced by the familiar continuity equation

$$\frac{\partial \rho_t}{\partial t} + \Sigma_\nu \frac{\partial \rho_t}{\partial q_\nu}\frac{\partial H}{\partial p_\nu} - \Sigma_\nu \frac{\partial \rho_t}{\partial p_\nu}\frac{\partial H}{\partial q_\nu} = 0$$

for the interval mentioned, with the initial value $\rho_{t_1}^+$.

(δ) If $me_t = (\{(\chi_i)_t\}_i, (\chi_{i_0})_t)$, then for all f and a

$$\left[\int_{\{M:\,(\chi_{i0})_t(M)\,=\,1\,\wedge\,f(M)\,\in\,a\}} \rho_t\, d\mu\right] \Big/ \left[\int_{\{M:\,(\chi_{i0})_t(M)\,=\,1\}} \rho_t\, d\mu\right] = \int_{\{M:\,f(M)\,\in\,a\}} \rho_t^+\, d\mu.$$

The condition (α) requires that non-trivial measurements lie at discrete points on the time axis. The condition (β) requires that results of measurement obtained at time t must be possible in the previous state of knowledge ρ_t or be present in the statistical state ρ_t. The condition (γ) formulates, as usual, the dynamic change; only the initial situation $\rho_{t_1}^+$, possibly modified by a preceding measurement, need be considered. Lastly, the condition (δ) states explicitly how the state of knowledge or the statistical state changes at a measurement, in the sense indicated above. The new probability is just the probability (in the usual sense) governed by the corresponding result of measurement. The consistency of all the conditions just stated is trivial for (α) and (β); for (γ) it follows from an existence theorem relating to the solutions of Hamilton's equations, and for (δ) it follows from Radon and Nikodym's theorem (see, e.g., Halmos 1950, §31). Condition (β) ensures that the denominator in (δ) is not zero.

II.4 Quantum mechanics without states for the individual object

The following formulation of quantum mechanics is related to the formulation of classical statistical mechanics chosen in the previous section, to the extent that here again there are no propositions regarding states, of the kind used in ordinary classical mechanics; they are replaced firstly by propositions regarding the outcome of *measurements* actually made at a time t, and secondly by *probabilistic* propositions regarding the outcome of measurements possible at the time t. But whereas in the classical case the reason for this change lies solely in the criticism outlined in section 3 as to the possibility of measurements in the continuum, in quantum mechanics we are concerned with the much more incisive criticism of certain ontological assumptions implied in the classical propositions regarding states, already described in a non-formal manner in Chapter I. This leads to a modification of the formalism of ordinary classical mechanics which is much more far-reaching than the formal changes resulting merely from the problem of the continuum and representing the transition from ordinary classical mechanics to the corresponding statistical mechanics. Whereas the latter is still constructed upon a classical phase space, quantum mechanics uses a complex Hilbert space, which causes a complete switch of direction.

II.4(a) Objects and statistical ensembles

In quantum mechanics, for the mathematical description of all quantum-mechanical concepts relating to an object Σ or a statistical ensemble $\{\Sigma_k\}_k$, a complex Hilbert space $\mathfrak{H}$ is used, in general having a countable infinity of dimensions, called the *Hilbert space of* Σ or $\{\Sigma_k\}_k$. The realization of the classical-mechanical phase space by a system of coordinates having a definite physical significance corresponds in quantum mechanics to a realization of the Hilbert space by a space of complex-valued functions whose variables have a definite physical significance. For example, in the quantum-mechanical analogue of the classical case mentioned in subsection 2(a), $\mathfrak{H}$ can be represented essentially by the ensemble of quadratically integrable complex-valued functions $\phi(x_1, \ldots, x_{3N})$, in which the x_ν are the position coordinates of the particles relative to a Cartesian coordinate system. The variables concerned here are therefore only one half of a classical system of canonical coordinates in phase space.

To von Neumann (1927, 1955 and elsewhere) is due the creation of the concept of the abstract Hilbert space and the "unified and, where possible and appropriate, mathematically correct representation of the new quantum mechanics" based on this concept. They are the result of a thorough analysis of the earliest versions of orthodox quantum mechanics put forward in 1925 and 1926 by Heisenberg, Born, Jordan, Dirac and Schrödinger. (A good summary of this analysis is given by von Neumann (1955, Chapter I).) Dirac gave a formulation of quantum mechanics that was not explicitly based on the concept of a Hilbert space but was subject to comparable requirements of logical rigour (for the first summary of this, see Dirac 1930). Physicists are more familiar with this than with von Neumann's formulation. The latter objected that it "did not meet the demands of mathematical rigour"; he also noted that Dirac's formulation, even if made mathematically correct, would not be equivalent to his own (von Neumann 1955, Preface). Dirac's theory does indeed operate with pure states for quantities having a continuous spectrum, which cannot occur in von Neumann's formulation. It is noteworthy that in later (unpublished) approaches von Neumann follows a line directly opposite to that of Dirac, seeking to use for the mathematical treatment of quantum mechanics structures called continuous geometries which do not permit any pure states (Birkhoff and von Neumann 1936, §15; von Neumann 1937). His first formulation would then be intermediate between these and that of Dirac. Marlow (1965) has recently applied the theory of direct integrals in an attempt to unify the Dirac and von Neumann formulations.

II.4(b) Composite objects and statistical ensembles thereof

In quantum mechanics, composite objects are mathematically described by a procedure which already displays fully the difference in mathematical structure between Hilbert space and classical phase space. It is as follows: *if* $\mathfrak{H}_1, \ldots, \mathfrak{H}_N$ *and* $\mathfrak{H}$ *are the Hilbert spaces of the objects* $\Sigma_1, \ldots, \Sigma_N$ *and* Σ, *and if* Σ *is composed of* $\Sigma_1, \ldots, \Sigma_N$ (*i.e.* $\Sigma = \Sigma_1 \otimes \ldots \otimes \Sigma_N$), *then* $\mathfrak{H}$ *is the Hilbert direct product* $\mathfrak{H}_1 \otimes \ldots \otimes \mathfrak{H}_N$. Here the Hilbert direct product of two Hilbert spaces $\mathfrak{H}_1$ and $\mathfrak{H}_2$ is defined as the topological completion of the algebraic tensor product of $\mathfrak{H}_1$ and $\mathfrak{H}_2$ relative to the (pre-Hilbert) metric uniquely defined in this product by

$$\langle \phi_1 \otimes \phi_2 | \psi_1 \otimes \psi_2 \rangle = \langle \phi_1 | \psi_1 \rangle \, \langle \phi_2 | \psi_2 \rangle.$$

The most important part of this procedure, however, is the algebraic part, since it reveals the linearity of the Hilbert space. For example, with a system of N particles in the coordinate representation, the direct products $\psi_1(\mathfrak{x}_1)\cdot\psi_2(\mathfrak{x}_2) \ldots \psi_N(\mathfrak{x}_N)$ by themselves do not form the Hilbert space of the entire system, since they do not constitute a linear manifold. The most general function $\psi(\mathfrak{x}_1, \ldots, \mathfrak{x}_N)$ for this case requires arbitrary linear combinations of these products, together with certain limit functions.

If a statistical ensemble $\{\Sigma_k\}_k$ of individual objects is considered, instead of one such object, the principle of the structural identity of the Σ_k must again be maintained. If the Σ_k are to be treated as composite objects, then this must be done for them uniformly as members of a statistical ensemble. Then, for example, $\Sigma_k = \Sigma_k^1 \otimes \ldots \otimes \Sigma_k^N$ with N the same for all k, and for a given n all the Σ_k^n have the same Hilbert space $\mathfrak{H}^n$.

This must, of course, be clearly distinguished from the particular treatment which is needed for an *individual* object Σ consisting of N structurally identical component objects Σ^n. In this case, the procedure stated initially for the formal description of Σ is, insofar as it has this composition, revoked by the priority given to additional procedures of symmetrization and antisymmetrization. It is not easy to state an axiomatically oriented form of the theory of composite objects which combines this special case with the general case, largely because an object composed of structurally similar objects is usually thought of as a *special case* of an object composed of any other objects, which occurs when the component objects are structurally alike. In both cases (general and special) this classical view corresponds to the same expression "objects composed of other objects". In quantum mechanics, however, an object composed of structurally similar objects is *not* a special case of an object composed of any other objects. For instance, the procedure for the mathematical description of the former case cannot be obtained by applying an additional condition to that for the latter case. In particular, when the component objects are structurally similar, they do not individually possess Hilbert spaces. The concepts of relationship between the component object and the composite object therefore have a quite different significance from the case of structurally different objects. In fact, symmetrization or antisymmetrization deprives the component objects of their independence to a much greater extent than will later be shown to occur in the only case considered above. Despite the interest of the case of structurally similar objects, in the sense that anomalies typical of quantum mechanics occur, it will not be discussed here, since we are concerned with the composition of objects mainly in the context of the measuring process, where the above procedure is adequate because the object and the apparatus are totally different.

The first abstract formulation of the postulate relating to composite objects is again due to von Neumann (1955, section VI.2). Concerning this postulate, Pauli (1933, section A.5) says "The way in which systems composed of a number of subsystems are described in the quantum theory is fundamentally important to that theory and most characteristic of it". London and Bauer (1939, §8) describe it as a "characteristic feature of quantum mechanics, which really embodies the very essence of the theory". Indeed, the postulate can be used to deduce particularly radical differences between quantum mechanics and classical mechanics (cf. Chapters VI and VII).

II.4(c) Quantities

Here the procedure corresponding to the two classical postulates is as follows: *the quantities corresponding to an object Σ or an ensemble $\{\Sigma_k\}_k$ are described by all the self-adjoint linear operators in the Hilbert space of Σ or $\{\Sigma_k\}_k$.*

This postulate, in the form stated, is due to von Neumann (1927, §II; 1955, sections III.1, III.3, III.5, IV.2). The decisive step, which was indeed the first step towards quantum mechanics in its narrower sense, was taken by Heisenberg (1925, §1). It was first of all extended to form matrix mechanics (Born and Jordan 1925, Introduction; Born, Heisenberg and Jordan 1926, section 1.1) and developed into an abstract operator mechanics by Dirac (1925; 1926a, b). Schrödinger's (1926) perception that his wave mechanics was mathematically equivalent to matrix mechanics formed, together with Dirac's abstract standpoint, the immediate stimulus to von Neumann's treatment in terms of abstract Hilbert space.

Von Neumann's representation of quantum-mechanical quantities by means of operators is not suitable for the treatment of these quantities in relation to propositions regarding measurements and probabilities. We shall therefore make provision now for a further mathematical representation, utilizing the spectral resolution of the operators concerned. According to the spectral theorem, the self-adjoint linear operators in $\mathfrak{H}$ are in one-to-one correspondence with all projection measures in $\mathfrak{H}$, i.e. with all mappings P which assign to a real Borel set a a projector P_a in $\mathfrak{H}$ such that

$$P_a P_b = 0 \quad \text{for} \quad a \cap b = \emptyset, \tag{α}$$

$$P_{\cup_i a_i} = \Sigma_i P_{a_i} \quad \text{for any countable set } \{a_i\}_i \text{ of Borel sets with } a_i \cap a_j = \emptyset \quad \text{for} \quad i \neq j, \tag{β}$$

$$P_\emptyset = 0 \quad \text{and} \quad P_{\mathbf{R}} = \mathbf{1}. \tag{γ}$$

The spectral resolution of the operator A is denoted by P^A. These P^A therefore correspond to the quantum-mechanical quantities.

II.4(d)(i) Propositions regarding measurements

The motivation for the use of propositions regarding measurements in quantum mechanics, to which we shall now return, is very much more abstruse than the corresponding classical procedure. There, we start from a theory which makes no reference to measurements, and, guided by the ideas that our experience in physics is largely based on measurements and that not all quantities can be measured with absolute accuracy, we derive from it a new theory in which the propositions regarding states in ordinary mechanics are replaced by propositions regarding measurements and probabilities. In quantum mechanics the situation is exactly the reverse, at least historically. The earliest view of quantum mechanics, the "Copenhagen version", was one in which the only permitted propositions about the state of an object were those regarding measurements and probabilities, and this was thought to be a necessary restriction. Chapter I has shown, in particular with reference to Bohr's views, how it was held to be due to the failure of classical propositions regarding states in the interpretation of quantum phenomena. The chief problem of interpretation in the orthodox formulation was then, however, found to be precisely the question whether these propositions must in fact necessarily occur or whether at least those regarding measurements, and possibly also those regarding probabilities, can in principle be dispensed with in this theory too, and replaced by classical propositions regarding states. The motivation for the occurrence of propositions regarding measurements would then amount ultimately (if it could be given at all) to the perception that they cannot be eliminated; and this presupposes that we know how they enter *de facto* into the orthodox formulation of quantum mechanics. The point to be noted at present is simply that one *could* then make the attempt to reintroduce classical propositions regarding states.

Another question is whether the reasons which led in subsection 3(d)(i) to a classical mechanics including propositions regarding measurements have any analogue in quantum mechanics—if not one in which classical propositions regarding states and regarding measurements are ranged together, then at least one in which there are two kinds of propositions regarding measurements: one kind which can define exact values even for quantities having a continuous spectrum, whereas the other kind cannot. Such an analogy, it may be noted here in passing, can in fact be established. Let us imagine a classical mechanics including propositions regarding measurements which in principle allow absolutely accurate measurements in the continuum. Such a formulation of mechanics is easily obtained (cf. Scheibe 1964, Chapter I). The quantum-mechanical analogue of this would be a Dirac formulation of quantum mechanics (cf. subsection (a)) including propositions regarding measurements, since here it would be possible to incorporate into the (statistical) description of the state of an object exact results of measurements of quantities having a continuous spectrum. This would be done by means of the Dirac delta functions. On the other hand, the quantum-mechanical analogue of the above-mentioned formulation of classical mechanics including propositions regarding measurements with limited accuracy is a von Neumann formulation (cf. subsection (a)) including propositions regarding measurements, of a type which will shortly be specified. The same reasons which previously led from the usual formulation of classical mechanics to the formulation including propositions regarding measurements with limited accuracy, and which of course would lead to the same thing, *mutatis mutandis*, from a formulation including measurements with absolute accuracy, would lead in quantum mechanics from the Dirac to the von Neumann formulation. The objection to exact classical mechanics stated at the beginning of this section would therefore be, when applied to quantum mechanics, an objection to the Dirac formulation, whereas the von Neumann formulation meets this objection *a priori*. This, then, would be the point at which the continuum problem would arise in quantum mechanics, and we see that the orthodox problem of measurement, which can of course be discussed in relation to the Dirac formulation also, is an entirely different matter.

If we now proceed to consider quantum-mechanical measurements and propositions regarding them, much of the initial discussion will coincide with the corresponding treatment of classical mechanics in subsection 3(d)(i), and can therefore be given very briefly. For quantum mechanics also, a distinction must be made between measurements of an individual object Σ and measurements of a statistical ensemble $\{\Sigma_k\}_k$. The former category again comprises in the first place those which include a result and which are the basis of the whole subject of measurements. These individual measurements of Σ will be formally described by pairs $(\{P_i\}_i, P_{i_0})$, where $\{P_i\}_i$ is a complete orthogonal set of projectors in the Hilbert space $\mathfrak{H}$ of Σ:

$$P_i \neq 0 \quad \text{for all} \quad i, \tag{α}$$

$$P_i P_j = 0 \quad \text{for all} \quad i \neq j, \tag{β}$$

$$\Sigma_i P_i = 1; \tag{γ}$$

P_{i_0} is one of the P_i. The corresponding *universal propositions regarding measurements*, $\mathbf{M}(me, A, \mathfrak{a}, a)$ containing a measurement me of Σ, a quantity of A of Σ, an accuracy $\mathfrak{a}$ and a result a (the two latter being described as in the classical case) state that *the measurement me of Σ measures the quantity A with accuracy $\mathfrak{a}$ and result a.* If a measurement me is described by $(\{P_i\}_{i \in I}, P_{i_0})$, then let $\mathbf{M}(me, A, \mathfrak{a}, a_0)$ be valid if and only if the following are true with P^A the spectral resolution of A:

$$\Sigma_{a \in \mathfrak{a}} P_a^A = 1, \qquad (\alpha_1)$$

for every $a \in \mathfrak{a}$ there is a subset

$$I_a \subseteq I$$

such that

$$P_a^A = \Sigma_{i \in I_a} P_i, \qquad (\beta_1)$$

$$a_0 \in \mathfrak{a}, \qquad (\gamma_1)$$

$$P_{i_0} \cdot P_{a_0}^A = P_{i_0}. \qquad (\delta_1)$$

The nature of these formal propositions regarding measurements, $\mathbf{M}(me, A, \mathfrak{a}, a)$, as universal propositions is to be understood in exactly the same way as in the classical case, and needs no further special discussion, but in relation to the ordering between measurements, whereby a measurement me_1 involves a measurement me, a difference from the classical case must be immediately pointed out. The definition is again

$$\mathbf{M}(me, A, \mathfrak{a}, a) \rightarrow \mathbf{M}(me_1, A, \mathfrak{a}, a) \qquad (1)$$

for all A, $\mathfrak{a}$ and a. The mathematical criterion is then as follows, with descriptions $(\{P_i\}_{i \in I}, P_{i_0})$ and $(\{P_j^1\}_{j \in J}, P_{j_0}^1)$:

For every $i \in I$ there is a subset $J_i \subseteq J$ such that

$$P_i = \Sigma_{\, j \in J_i} P_j^1; \qquad (\alpha_2)$$

$$P_{j_0}^1 \cdot P_{i_0} = P_{j_0}^1. \qquad (\beta_2)$$

Unlike classical statistical mechanics, however, quantum mechanics possesses *maximum measurements*, which are not "involved" by any other measurements in the sense previously defined. These are evidently just the measurements for which all the P_i have rank 1. Of course, the existence of maximum measurements does not exclude the existence of infinite sequences of measurements me_n such that me_{n+1} involves me_n but is not the same as me_n. Nor does the existence of maximum measurements imply, as it would in the classical case, that states can be determined in the classical sense. We shall be further concerned with this noteworthy point in Chapters IV, VI and VII.

As in the classical case, the significance of the quantum-mechanical propositions regarding measurements, $\mathbf{M}(me, A, \mathfrak{a}, a)$, is ambiguous: they may refer either to an actual acceptance of knowledge or merely to the recording of the result by means of a measuring apparatus. This distinction is sometimes put forward as an essential question of interpretation in orthodox quantum mechanics, but it should not be exaggerated. As has been mentioned in Chapter I, Bohr and Heisenberg emphasized that the interpretation as mere recording is compatible with the rest of their view of quantum mechanics. On the other hand, von Neumann appears to have regarded the epistemic interpretation as indispensable (cf. section VI.2 for both approaches). Our view here will be that the decision in favour of one or the other interpretation must depend on the interpretation given to propositions regarding probabilities in quantum mechanics. If, as in the present section, quantum mechanics is taken to have no states for individual objects, the propositions regarding probabilities will be interpreted either epistemically or statistically, just as in the classical case. If they are interpreted epistemically, then the propositions regarding measurements must be so too, since otherwise the reduction of states to be discussed in subsection (e) could not be meaningfully defined: this reduction of states makes an assertion regarding the change in probabilistic knowledge due to the addition of new results of measurement. But how could that be possible if no knowledge of the results is acquired? In a statistical theory, however, the propositions regarding measurements must be objectively interpreted, for similar reasons pertaining to the reduction of states. Here, objective statistical states change. The situation in a quantum mechanics with states for individual objects will be discussed in section 5.

The *measurements* (or propositions regarding them) *irrespective of result* which are especially indicative of the position in quantum mechanics can be derived by a consistent analogy with the classical case. The equivalence relation

$$\bigwedge_A \bigwedge_{\mathfrak{a}} \bigvee_a \mathbf{M}(me, A, \mathfrak{a}, a) \leftrightarrow \bigvee_a \mathbf{M}(me_1, A, \mathfrak{a}, a) \tag{2}$$

between the measurements me and me_1, whose mathematical description is

$$\{P_i\}_i = \{P_j^1\}_j \tag{3}$$

(element-to-element identity) with $me = (\{P_i\}_i, P_{i_0})$ and $me_1 = (\{P_j^1\}_j, P_{j_0}^1)$, gives the new measurements $\dot{m}e = \{P_i\}_i$, which no longer include the result. They correspond to the propositions regarding measurements, which are mathematically described by (α_1) and (β_1),

$$\dot{\mathbf{M}}(\dot{m}e, A, \mathfrak{a}) \rightleftharpoons \bigvee_a \mathbf{M}(me, A, \mathfrak{a}, a); \tag{4}$$

these state only that me measures the quantity A with accuracy $\mathfrak{a}$. Thus these propositions, for given $\dot{m}e$, yield no new information in the sense of providing a new result of measurement. Nevertheless, in marked contrast to the classical case, the mere knowledge that a measurement $\dot{m}e$ has been carried out will in general modify the previously existing probabilistic state of knowledge. This situation will be made clearer in subsection (e), and will be more precisely treated in Chapters IV and VI in particular.

Let us finally take a brief glance at statistical measurements. The ideas regarding content and form given for classical mechanics in subsection 3(d)(i) can be transferred to quantum mechanics with suitable modifications. Thus we have on the one hand *statistical measurements $\dot{m}e$ of an ensemble* $\{\Sigma_k\}_k$, corresponding formally to measurements of an individual object irrespective of the result (described by systems $\{P_i\}_i$ with the conditions (α)–(γ)). These are correlated with the statistical propositions regarding measurements, $\dot{\mathbf{M}}(\dot{m}e, A, \mathfrak{a})$, described by (α_2) and (β_2). On the other hand, we have *selective measurements me* of an ensemble $\{\Sigma_k\}_k$, corresponding formally to measurements of an individual object taking the result into account (described by pairs $(\{P_i\}_i, P_{i_0})$ with the conditions (α)–(γ) and $P_{i_0} \in \{P_i\}_i$), and propositions regarding measurements, $\mathbf{M}(me, A, \mathfrak{a}, a)$, described by (α_2)–(δ_2). Concerning the significance of these two kinds of measurements and propositions regarding them, we must first of all note what has been said in subsection 3(d)(i), and also, as previously for single measurements, be warned that quantum-mechanical measurements on the orthodox view are peculiar to themselves, as will be seen on considering the various reductions of states which relate to them.

II.4(d)(ii) Propositions regarding probabilities

In respect of the *propositions regarding probabilities*, which occur in quantum mechanics as well as the propositions regarding measurements, the first point to be stated and emphasized is that within the orthodox view certain differences of interpretation were more or less consciously brought into play from the start, but were not the subject of thorough analysis leading to a decision concerning them. Unity was achieved only in relation to the reason for the occurrence of such propositions in quantum mechanics, to the extent that the reason is not the same in classical statistical mechanics as conventionally understood. Their occurrence in quantum mechanics is not due to insufficiency of knowledge, existing in practice but avoidable in principle; it is due to a fundamental inadequacy of classical ontology as regards the concept of objects and states of objects, in the treatment of quantum phenomena. There was for long no reflection of the relationship to the inherent criticism of classical mechanics indicated in section 3, which had led to the new probabilistic formulation of classical mechanics also described in that section. Only from 1955 onwards did Born approach this topic in the papers quoted in section 3. As has already been indicated in subsection (d)(i) for the propositions regarding measurements, a von Neumann formulation of quantum mechanics takes account of this criticism *a priori* for the propositions regarding probabilities also: in the continuous spectrum of a quantity, there are no delta-function probability distributions. But this sideline of the continuum problem was likewise not the scene of any profound analysis of the concept of probability in quantum mechanics. Thus the existence of some agreement as to *the non-classical reason for the occurrence of propositions regarding probabilities* in quantum mechanics did not lead to the rejection of the two *classical interpretations of these propositions* distinguished in section 3, but only to the consideration of a third, genuinely quantum-mechanical possibility; this did not displace the other two, and all three interpretations can sometimes be found in the work of a single author. However, in any study aimed at obtaining fundamental results, they must be very carefully distinguished. In section 3, only the first two possibilities were involved.

The first interpretation is essentially parallel to the view of probability as the degree of knowledge existing with regard to an individual object, which has already been used above as a possible interpretation in classical statistical mechanics, but now the argument for this view is no longer, as in conventional classical statistical mechanics, that the positive aspect of probability is extracted from practical lack of knowledge in relation to an objective reality; it is that, of the two classical possibilities of referring to the object itself or to our knowledge of it, only the latter is available in quantum mechanics, owing to the failure of the classical propositions regarding states. According to this line of thought, the first indication of such a situation is the retreat to propositions regarding measurements, as described in subsection (d)(i), and this is now consistently extended to propositions regarding probabilities. Statements of this kind are made, for example, by Heisenberg (1930, sections II.2 and IV.1; 1951, p. 53; 1959, p. 43ff.), Wolfe (1936), von Weizsäcker (1941, p. 502ff.; 1961, section 4.b), Jeans (1943, p. 136ff.) and Süssmann (1958, p. 18); and Schrödinger, in his critical discussion of the orthodox view, took it in this way (1935a, §§6 and 7). Here it must be emphasized that the phrase used above of a different *reason* for the epistemic interpretation of the propositions regarding probabilities does not affect the *meaning* of these propositions in comparison with the classical view. We find no explicit assurances that a quantum-mechanical probability as the degree of existing knowledge is to be understood as something different from what was understood by it in the classical case. But we also find no assurances to the contrary, so that in this respect the situation remained unresolved.

Physicists, with their fondness for empiricism, were quite ready to propound the statistical interpretation in quantum mechanics alongside the epistemic interpretation just outlined, and again without any recognizable distinction of sense from the corresponding classical case. For example, Heisenberg (1930, section II.2(c)) regards the physical significance of propositions that an electron is present with a given probability at a given point in an atom as being the performance of a measurement of position in many atoms, in order to determine the relative frequency. Similar explanations are given by Weyl (1931, Chapter II, §7), Dirac (1930, §§11 and 18), Schrödinger (1935a, §2) and Pauli (1954, p. 112). These explanations are offered only in passing, but the statistical interpretation is explicitly stated, and analysed in some detail, by Born and Jordan (1930, Chapter VI, especially §57). These authors evidently follow von Neumann, who also put forward the statistical interpretation (though with some vagueness) and who first gave it a mathematical treatment suitable for quantum mechanics (1927; 1955, Chapter IV). Both Born and Jordan and von Neumann were acquainted with the pioneering work of von Mises on probability theory and used his concept of collectives, though somewhat loosely. Slater (1929) also supported the statistical interpretation, with explicit emphasis on the question of significance. Decisions in favour of this interpretation of quantum-mechanical probability on general methodological principles occur from the middle thirties onward, especially in the operational treatment by Kemble (1935; 1937, section 14b; 1938) and in the mainly objectivist treatment by Margenau (1937). But these and other authors who defended the statistical interpretation cannot be forthwith assigned to the orthodox school; they are in conflict with it on several points, some of which are vital. There is a late admission by Born, who originated (1926) the probabilistic interpretation of the Schrödinger ψ-function. In a letter to Einstein, written in 1950, he says: "The other remark concerns your interpretation of the ψ-function; it seems to me that it completely agrees with what I have been thinking all along, and what most reasonable physicists are thinking today. To say that ψ describes the 'state' of one single system is just a figure of speech, just as one might say in everyday life: 'My life expectation (at 67) is 4.3 years'. This, too, is a statement about one single system, but does not make sense empirically. For what is really meant is, of course, that you take all individuals of 67 and count the percentage of those who live for a certain length of time. This has always been my own concept of how to interpret $|\psi|^2$. Instead you propose a system of a large number of identical individuals—a statistical ensemble. It seems to me that the difference is not essential, but merely a matter of language. Or have I misunderstood you, do you mean something much more fundamental?" (Born 1971, p. 186).

For the technical details we can follow von Neumann (1927, 1955), whose mathematical method is applicable to both interpretations in one and the same manner. For the epistemic interpretation, we have an individual object Σ with a Hilbert space $\mathfrak{H}$ and a range of quantities described as in subsection (c); for the statistical interpretation, we have a statistical ensemble $\{\Sigma_k\}_k$ of a large number (in the mathematically ideal case, an infinity) of objects Σ_k, all having the same Hilbert space $\mathfrak{H}$ and the same range of quantities. The contingent possibilities corresponding to the states of classical exact mechanics are possible *states of knowledge concerning* Σ or possible *statistical states of the ensemble* $\{\Sigma_k\}_k$. *These states of knowledge or statistical states are mathematically described by the von Neumann operators in* $\mathfrak{H}$, i.e. by linear operators W defined everywhere in $\mathfrak{H}$ and such that

$$W \geqslant 0, \quad \mathrm{Tr}\,(W) = 1, \tag{1}$$

where Tr denotes the trace. The probabilistic interpretation again leads to a universal probability function **pr** for each object or ensemble, such that **pr**(W; A, a) *denotes the probability that, in a state of knowledge* W *of* Σ *or a state* W *of the ensemble* $\{\Sigma_k\}_k$, *a sufficiently accurate measurement of* A *will give a value in* a. This refers to probability as understood in either the epistemic or the statistical interpretation, and in each case the propositions regarding probabilities relate to possible measurements (as, of course, they must, on the orthodox view, there being no propositions which relate simply to the objects themselves). The mathematical description of **pr** is given by

$$\mathbf{pr}(W;\, A,\, a) = \mathrm{Tr}(WP_a^A), \tag{2}$$

where P_a^A are the spectral operators of A.

It has been shown in subsections 3(d)(i) and 4(d)(i) that the information about an object given by measurements in classical statistical mechanics can be made arbitrarily great but does not reach a maximum, whereas in quantum mechanics there is a fairly extensive range of measurements permitted by the theory which provide the maximum possible information. It has also been mentioned in subsection 3(d)(ii) that in the classical case there is no maximum information as regards probabilities. The quantum-mechanical analogue of this will now be similarly presented.

Of the two methods to be described for deriving the concept of *maximum knowledge* or the *pure statistical state*, one is due to Weyl (1931, Chapter II, §7) and the other to von Neumann (1927, §IV). The former is better suited to the epistemic interpretation, the latter to the statistical interpretation, and we shall therefore consider them in this combination. For two different states of knowledge W and W_1 concerning an object Σ, one can ask what must be the relation between them if W_1 conveys at least as much knowledge as W. The following is a plausible definition using only what is reliably known: *W_1 conveys at least as much knowledge as W* if, for all A and a,

$$\mathbf{pr}(W;\, A,\, a) = 1 \rightarrow \mathbf{pr}(W_1;\, A,\, a) = 1. \tag{3}$$

Thus there is introduced here a semi-ordering between the possible states of knowledge concerning Σ, and the mathematical equivalent of (3) is easily seen to be

$$P_{(W_1)}\, P_{(W)} = P_{(W)}\, P_{(W_1)} = P_{(W_1)} \tag{4}$$

for the carriers $P_{(W)}$ and $P_{(W_1)}$ of the operators W and W_1 (cf. Dixmier 1957, section I.4.6). In terms of this semi-ordering, W will be a state of *maximum knowledge* if the converse of (3) is true for every W_1 for which (3) itself is true, i.e. the certain knowledge existing cannot be further increased. This is easily seen to occur if and only if the corresponding operator has the form

$$W = P_\phi, \tag{5}$$

where P_ϕ is the operator projecting on the Hilbert vector ϕ. Thus the result bears an obvious similarity to that obtained for the propositions regarding measurements. In contrast to the classical case, there is here an abundance of states of maximum knowledge which, nevertheless, relate to probabilities which are neither unity nor zero.

The fairly rough semi-ordering defined by (3) can be associated with a more refined one, which will here be formulated in terms of the statistical interpretation. The essential point is that some subensembles of a statistical ensemble $\{\Sigma_k\}_k$ may have a different statistical behaviour from that of $\{\Sigma_k\}_k$ itself, and we have to ascertain the states in which these subensembles may be. By analogy with the familiar procedure in the ordinary theory of probability, in quantum mechanics we proceed by stating that *W_1 is a possible state of a subensemble of $\{\Sigma_k\}_k$ in the state W* if there is a number α such that

$$0 < \alpha \leqslant 1, \quad W - \alpha W_1 \geqslant 0 \tag{6}$$

for the corresponding von Neumann operators. This semi-ordering between W and W_1 exists if and only if there exist a number α and a von Neumann operator W_2 such that

$$0 < \alpha \leqslant 1, \quad W = \alpha W_1 + (1-\alpha) W_2. \tag{7}$$

This operator W_2 then describes the state of the subensemble in $\{\Sigma_k\}_k$ which is complementary to that described by W_1. It is easily seen that the semi-ordering between W and W_1 defined by (6) or (7) implies that defined by (3), but that the converse is not true. In this sense the new semi-ordering is more refined than the previous one, but both have the same minimum elements. A *pure state* (also referred to in the literature as a uniform or homogeneous state) *of an ensemble* $\{\Sigma_k\}_k$ is defined as a W for which every W_1 which satisfies (6) also satisfies (6) with W and W_1 interchanged. We then find that in fact W and W_1 must be equal, so that an ensemble in a pure state W is one for which every subensemble must also be in that state. At the same time it is found that the pure states are precisely described by (5), i.e. they correspond precisely to maximum knowledge.

This introduces the *purely epistemic* and *purely statistical* interpretations of the propositions regarding probabilities in quantum mechanics. The word "pure" is intended to show that states of individual objects are not yet considered. However, for lack of a better term, states of knowledge and statistical states will sometimes be referred to jointly as *descriptions of state*, since it is often unnecessary to distinguish between the two interpretations and a shorter expression is thus made available. The epistemic or statistical probabilities $\mathbf{pr}(W;\, A,\, a)$, for a given description of state W, relate implicitly to possible measurements of an individual object or statistical ensemble. In the next subsection this correlation is to be made explicit for the case where measurements are actually made. For this purpose, another derivative concept is needed, which is characteristic of the orthodox view, that of *an assemblage of states of knowledge or of statistical states*. This denotes a countable set $\{W_i,\, w_i\}_i$ of states of knowledge or statistical states W_i with corresponding probabilities w_i, the latter such that

$$w_i > 0, \quad \Sigma_i w_i = 1. \tag{8}$$

An assemblage $\{W_i,\, w_i\}_i$ with states of knowledge W_i represents the situation where just one of the states of knowledge W_i could be possessed, with probability w_i, no further measurement being necessary for this purpose, but this knowledge is refused. Similarly, an assemblage $\{W_i,\, w_i\}_i$ with statistical states W_i serves to describe a set of statistical ensembles which are in the statistical states W_i with probabilities w_i; again no further measurement would be necessary to select one of these ensembles, but yet no definite choice has been made. This concept, which at first sight appears extremely artificial, can be understood only by assigning a special position to quantum-mechanical measurement, both individual and statistical, as is done on the orthodox view. The classical approach would be simply that, if there is an assemblage $\{W_i,\, w_i\}_i$, we have *de facto* a description of state

$$W = \Sigma_i\, w_i W_i \tag{9}$$

(where W and W_i are now the representative operators), i.e. this W is equivalent to the assemblage $\{W_i,\, w_i\}_i$. In quantum mechanics, this equivalence is again valid for the propositions regarding probabilities of the outcome of measurements, based on $\{W_i,\, w_i\}_i$ or $\Sigma_i\, w_i W_i$, but *not* for the ontological status of the W_i, as will be shown in sections IV.3, VI.2 and VII.3.

II.4(e) Time dependence

In the orthodox treatment, the subject of time variations in quantum mechanics contains some remarkable features which are highly specific to that treatment. At present these features will be indicated only to the extent that is essential in order to understand the postulates given below, which govern the orthodox theory of time variations. The least difficult case is that of *purely dynamic changes* based on a Hamiltonian operator H which describes an independent dynamics in the Hilbert space of the object concerned. Here we have only to decide in favour of one "picture", i.e. decide between the Schrödinger picture and the Heisenberg picture. We shall take the Schrödinger picture, because the reduction of states in a measurement cannot be so usefully formulated in the Heisenberg picture. In the Schrödinger picture, the time variations under consideration will appear as changes of the state of knowledge of an individual object or changes of the state of a statistical ensemble, according as we use the epistemic or the statistical formulation of quantum mechanics, and the Hamiltonian operator, which in the case of a statistical ensemble is again the same for each object, will determine these variations in the usual way.

For the time variations which might be caused by a *measurement*, we have to make a clear distinction between the two formulations differentiated in subsection (d)(ii) in respect of the propositions regarding probabilities, and correlate them correctly with the views on the content of a proposition regarding a measurement, differentiated in subsection (d)(i). In the epistemic formulation, the situation is as follows. A measurement me_t made at time t can change a state of knowledge W_t existing at that time t, and this change, in accordance with the result $(P_{i_0})_t$ which is included in $me_t = (\{(P_i)_t\}_i, (P_{i_0})_t)$, will convert W_t into a well-defined new state of knowledge W_t^+, valid immediately after t. The propositions $\mathrm{M}(me_t, A, \mathfrak{a}, a)$ which correspond to me_t are then to be interpreted epistemically: those of them which are true *become actually known*, and *this* brings about the change from W_t to W_t^+. The statistical case is adjusted to this mechanism, just as in subsection 3(e). A statistical ensemble $\{\Sigma_k\}_k$ initially in the state W_t is modified by changing to the ensemble of the Σ_k for which a selective measurement me_t has given the result $(P_{i_0})_t$. Then W_t^+ is the state of the new ensemble, and the change from W_t to W_t^+ is brought about by the *selection* (compare the discussion in subsection 3(e)).

As well as this similarity between the quantum-mechanical case and the corresponding classical case, there is also a considerable difference between them. As we shall see, the change from the state of knowledge W_t to the new state of knowledge W_t^+ in quantum mechanics is in general not simply an improvement, and likewise the ensemble corresponding to W_t^+ is not simply a subensemble (in the usual sense of the word) of the ensemble corresponding to W_t. This is clear from the answer to the following problem. Let a state of knowledge or a statistical state W_t be given, and let a measurement $\dot{m}e_t = \{(P_i)_t\}_i$ be then made without taking cognizance of the result or subsequent selection. What is the situation? The answer is that there is an assemblage of states of knowledge or statistical states $W_{i,t}^+$ with certain probabilities $w_{i,t}$, among which $W_{i,t}^+$ is the state of knowledge or the statistical state which would result from a measurement if the result $(P_i)_t$ were accepted or selected. Now all would be well if it were *always* true that

$$W_t = \Sigma_i \, w_{i,t} \, W_{i,t}^+ \tag{1}$$

in accordance with the mixing procedure defined at the end of the previous subsection, (d)(ii)(9). For this would mean that $W_{i,t}^{+}$ was simply an improvement over W_t, or that it was the new statistical state of a subensemble of the previous ensemble in the state W_t, and hence that the measurement *with* acceptance or selection is nothing more than acceptance or selection of what already existed before the measurement. But the relation (1) is in general *not* valid in quantum mechanics. This is the clearest evidence that the quantum-mechanical concept of measurement is totally different from its classical counterpart. In particular, it shows that the change from W_t to $W_{i,t}^{+}$ does not come about *only* through the acceptance or selection of a result of measurement, which has been mentioned in the previous paragraph as one of its causes; cf. sections IV.3, VI.2 and VII.3.

The conditions imposed by the dynamics and by measurements on the interrelated time variations of an individual object or a statistical ensemble $\{\Sigma_k\}_k$ can again be most simply stated as conditions on a total of three time-dependent functions: me_t, W_t and W_t^{+}. Here me_t is an epistemically interpreted individual measurement and result, or a statistical selective measurement. The case of a measurement $\dot{m}e_t$ without acceptance or selection of the result will be obtained as a consequence of the following postulates.

(α) In any finite range of t, me_t is trivial (in the same sense as in the classical case; see subsection 3(e)) for all except a finite number of values of t.

(β) If $me_t = (\{(P_i)_t\}_i, (P_{i_0})_t)$, then $\mathrm{Tr}\,(W_t(P_{i_0})_t) > 0$.

(γ) If me_t is trivial for all t in the open interval (t_1, t_2) with $t_2 > t_1$, we assume that

$$W_{t_2} = U_{t_2-t_1} \cdot W_{t_1}^{+} \cdot U_{t_2-t_1}^{-1},$$

where U_τ is defined in terms of the Hamiltonian operator H by

$$U_\tau = e^{-(i/\hbar)H\tau}.$$

The corresponding differential equation is here

$$i\hbar \dot{W}_t = HW_t - W_t H,$$

with the initial value $W_{t_1}^{+}$.

(δ) If $me_t = (\{(P_i)_t\}_i, (P_{i_0})_t)$, then

$$W_t^{+} = (P_{i_0})_t\, W_t(P_{i_0})_t / \mathrm{Tr}\,(W_t(P_{i_0})_t).$$

The consistency of (α)–(δ) is trivial except for (γ), and there it is ensured by a familiar procedure for constructing U_τ from H. The postulate (δ) in this general form was first enunciated by Lüders (1951). For the case of a maximum measurement it originates with von Neumann; see section V.4 for further details. The change specified in (δ) is called *reduction of states having regard to the result*. There is, of course, strictly speaking no reduction of an individual state here; this will occur only in the next section. The misnomer has in the past given rise to considerable misunderstanding, but can be tolerated here, since the whole discussion of orthodox quantum mechanics has been so presented that the differences of content are made fully explicit.

Let us finally consider the case where only a measurement $\dot{m}e_t$ without acceptance (or, in the statistical case, without selection) is made. The situation is clear on the basis of the full description already given in subsection (d)(i) (and in 3(d)(i)) of the relation between such a measurement and a measurement me_t. In $\dot{m}e_t = \{(P_i)_t\}_i$, the $(P_i)_t$ represent the *a priori* possible (and mutually exclusive) results. These possibilities may be restricted by an existing description of state W_t: $\mathrm{Tr}(W_t(P_i)_t)$ represents the probability of occurrence of $(P_i)_t$, which must be greater than zero; cf. (β). If the measurement then takes place, in the epistemic interpretation just one of the $(P_i)_t$ having $\mathrm{Tr}(W_t(P_i)_t) > 0$ has occurred, but we do not know which; only the probability $\mathrm{Tr}(W_t(P_i)_t)$ is known. With the same probability it is known that the state of knowledge would be

$$W_{i,t}^{+} = (P_i)_t\, W_t(P_i)_t/\mathrm{Tr}(W_t(P_i)_t) \tag{2}$$

if the result had been accepted and had been $(P_i)_t$; cf. (δ). Thus there exists the assemblage

$$\{W_{i,t}^{+}, \mathrm{Tr}(W_t(P_i)_t)\}_{i:\mathrm{Tr}(W_t(P_i)_t)>0} \tag{3}$$

with the $W_{i,t}^{+}$ defined in (2). In the statistical case, after the (statistical) measurement $\dot{m}e_t$, all the $(P_i)_t$ having $\mathrm{Tr}(W_t(P_i)_t) > 0$ have occurred, with those probabilities (here equivalent to relative frequencies), but no selection has been made. The ensembles defined by the results are in the states (2). Thus we again have the assemblage (3), this time with a statistical interpretation. For further predictions, in accordance with ordinary probability theory only the new description of state

$$W_t^{\times} = \Sigma_i (P_i)_t\, W_t(P_i)_t \tag{4}$$

is relevant, since for all quantities A and sets of values a we must have

$$\mathbf{pr}(W_t^{\times}; A, a) = \Sigma_{i:\mathrm{Tr}(W_t(P_i)_t)>0}\, \mathrm{Tr}(W_t(P_i)_t)\, \mathbf{pr}(W_{i,t}^{+}; A, a), \tag{5}$$

which is equivalent to (4); see (d)(ii)(2). The change from W_t to the assemblage of descriptions of state (3) or the single description of state (4) is called *strong* and *weak reduction of states* respectively (without reference to the result). The significance of this distinction between strong and weak reduction of states, which is necessary because $W_t^{\times}$ may be different from W_t, will be further discussed in sections IV.3, VI.2 and VII.3. These also, like the reduction of states (δ), give no reduction of the state of an individual object, which is not under discussion here. What they do give has been conceptually established by the foregoing analysis and will be further considered below.

II.5 Quantum mechanics with states for the individual object

As has already been mentioned in the introduction to this chapter, the customary presentations of orthodox quantum mechanics are permeated by a lack of clarity regarding the quantum-mechanical concept of a state. The state of an individual object is spoken of in quantum mechanics as in classical physics, but at the same time it is not clear what such states actually are and how they relate to the other, specifically quantum-mechanical concepts used, in particular those which pertain to the probability concept. This situation will not be made the grounds for a detailed analysis and criticism, for it can easily be put right in a provisional, albeit still incomplete, manner. In the preceding section, for example, two versions of quantum mechanics have been given which do not contain the concept of the state of an individual object, but can nevertheless be properly considered as orthodox versions. This already leads on to the demonstration that the concept is not necessary in an orthodox interpretation. In the present section we shall attempt a formulation of orthodox quantum mechanics in which it is brought in as a fundamental concept. This will be called the *ontic formulation of quantum mechanics*; the question whether it differs essentially from the formulation without individual states remains to be investigated. We shall certainly find it ultimately necessary to extend in an epistemic or statistical manner this ontic formulation, which refers primarily to individual states but *not* to states of knowledge or statistical states.

This will allow a comparison with the formulations given in the preceding section. For brevity, the division into subsections used in sections 2 to 4 will not be made explicit, but the order of procedure will be the same.

We begin, therefore, with the ontic formulation itself. The comments made in subsections 4(a)–(c) remain valid as regards the concepts of object, composite object and quantity; and the same is true of the concept of measurement and the corresponding proposition regarding measurement (with and without acceptance of the result), as defined for an individual object in subsection 4(d)(i), except that here it is appropriate to add that an individual measurement with its result must be interpreted objectively, i.e. as recording the result obtained by means of the apparatus. The first point of difference relates to the descriptions of state and the corresponding propositions regarding probabilities. The concept of the *state of an individual object* Σ makes its appearance here. *Such states are described mathematically by the vectors of the Hilbert space* $\mathfrak{H}$ *of* Σ, taken to within a complex factor. There is again a universal probability function **pr**, such that $\mathbf{pr}(\phi; A, a)$ is *the probability that, in a state* ϕ *of* Σ, *a suitable measurement of the quantity A shows its value to be* (*objectively*) *in a*. This function **pr** has the mathematical form

$$\mathbf{pr}(\phi; A, a) = \|P_a^A \phi\|^2, \tag{1}$$

which is a special case of (2) in subsection 4(d)(ii) with $W = P_\phi$. But this relation is at present purely formal, and we have to ascertain which kind of probability is now under discussion.

The earliest form was the idea of an ontology of "blurred reality", as it was termed by Schrödinger (1935a, §§4 and 5). In the state ϕ of Σ, not all quantities of Σ are equally real as regards their value. We do not have the classical alternative that the value is or is not such-and-such; the probabilities specify the *degree of realization* of the values in the state ϕ. However, Schrödinger's terminology of "blurredness" of the quantum-mechanical quantities, though the first to be used, was later joined by others. London and Bauer (1939, §3) refer to " 'potential' probabilities which become valid only when a measurement is actually made". Thus we have here to understand the reality of a quantum-mechanical object as a sum of quantitatively considered possibilities of exhibiting its properties *if* suitable measurements are made, without any indication of what these properties will be *until* a measurement is made. This was later very clearly put by Bohm (1951, §6.9): "Thus, before the electron has interacted with a measuring apparatus, the wave function defines two important kinds of probability; namely, the probability of a given position and the probability of a given momentum. But the wave function by itself does not tell us which of these two mutually incompatible probability functions is the appropriate one. This question can be answered only when we specify whether the electron interacts with a position-measuring device or with a momentum-measuring device. We conclude that, although the wave function certainly contains the most complete possible description of the electron that can be obtained by referring to variables belonging to the electron alone, this description is incapable of defining the general form . . . in which the electron will manifest itself. We are, therefore, again led to interpret momentum and position . . . as incompletely defined potentialities latent in the electron and brought out more fully only by interaction with a suitable measuring apparatus." Margenau has since 1949 held a view which may safely be said to be similar (first published in Margenau 1949, §3; 1950, §§8.2 and 17.3); and Heisenberg, at least in his later publications (1959, pp. 42 and 53; 1961), has written in the same sense and regarded the concept of probability in the quantum theory as a certain extension of the Aristotelian concept of possibility: "The concept that events are not determined in a peremptory manner, but that the possibility or 'tendency' for an event to take place has a kind of reality—a certain intermediate layer of reality, halfway between the massive reality of matter and the intellectual reality of the idea or the image—this concept plays a decisive role in Aristotle's philosophy. In modern quantum theory this concept takes on a new form; it is formulated quantitatively as probability and subjected to mathematically expressible laws of nature" (Heisenberg 1961, p. 9f.). Heisenberg too regarded the relationship between the state of an object and the measurements that can be made of the object in the same way as appears in the above quotations from London and Bauer and from Bohm; this had essentially been in use since the early days of the Copenhagen interpretation. Bohr's warning (cf. Chapter I) that the state of a quantum-mechanical object is not defined in the absence of a measuring system to define it, can be dealt with by means of a qualification of the term "is defined" which makes use of the distinction between the actual and the possible. As an ensemble of possibilities, the state is defined by the object alone; as an actual thing, it is defined only by a measurement which imposes the transition from the possible to the actual (Heisenberg 1959, p. 53ff.). Thus this view ensures for the quantum-mechanical state, as an ensemble of probabilities which are intrinsic tendencies of behaviour, the maximum degree of independence that is compatible with Bohr's requirement stated above.

Finally, let us consider time variations of the object Σ. These are of course changes in the state of Σ, so that there will certainly be a function ϕ_t for the description, and this function will vary, firstly, because of an intrinsic dynamics H of Σ. The noteworthy part, however, will again be the change due to a measurement me_t. For the ontic formulation of orthodox quantum mechanics, this becomes evident even from the mere question whether and how a *measurement* me_t can change the *state* ϕ_t of an individual object. For, so long as classical ideas are combined with these concepts of measurement and state, and it is assumed (as in the orthodox view) that measurement does not cause any essentially avoidable perturbations, the question is indeed an absurd one. The details will be left till later; at present we may note that, in the orthodox sense, ϕ_t and me_t define a new state ϕ_t^+ (which in particular depends on the result implied in me_t) as the state which exists immediately after the measurement. It is also clear that the propositions regarding measurements must be objectively interpreted, there being no reference to states of knowledge in the concept of state.

The formal conditions on the three functions me_t, ϕ_t and ϕ_t^+ are now easily obtained:

(α) (as in subsection 4(e), (α)).

(β) If $me_t = (\{(P_i)_t\}_i, (P_{i_0})_t)$, then $\|(P_{i_0})_t\, \phi_t\|^2 > 0$.

(γ) With the same condition as in subsection 4(e), (γ),

$$\phi_{t_2} = U_{t_2 - t_1} \cdot \phi_{t_1}^+$$

or the corresponding Schrödinger equation

$$i\hbar\dot{\phi}_t = H\phi_t$$

is valid, with the initial state $\phi_{t_1}^+$.

(δ) With the same condition as in subsection 4(e), (δ),

$$\phi_t^+ = \frac{(P_{i_0})_t\, \phi_t}{\|(P_{i_0})_t\, \phi_t\|}.$$

Here (δ) is the *true reduction of states*, as distinct from the corresponding process in subsection 4(e), (δ), in which states of knowledge or statistical states are reduced. The present (δ) is the natural generalization of the "reduction of a wave packet" in a measurement of position. Although the content of this (δ) is totally different from that in subsection 4(e), there is a relationship between them, which will shortly be specified.

This completes the ontic formulation of orthodox quantum mechanics. Although it is obviously self-supporting, an epistemic or statistical extension of it is desirable. This is shown even by the question as to the situation which exists after a measurement $\dot{m}e = \{P_i\}_i$ without acceptance of the result, if the object Σ was previously in the state ϕ. One might wish to say that there exists an assemblage of states $P_i\phi/\|P_i\phi\|$ which occur with probabilities $\|P_i\phi\|^2$. This is not yet permissible, however, since the probabilities $\|P_i\phi\|^2$ are no longer the ontic probabilities, the only ones so far defined; these determine the meaning of the propositions regarding states, $w = \mathbf{pr}(\phi;\, A,\, a)$, and nothing else. We must not be deceived by the existence of the *numerical* equalities $\|P_i\phi\|^2 = \mathbf{pr}(\phi;\, A,\, a_i)$, according to (1), for quantities A and sets of values a_i such that $P^A_{a_i} = P_i$ for all i. There is a fundamental difference of *content* here in the orthodox view. *Before* the measurement, it is a question of probabilities, to be interpreted ontically, for the behaviour of the object upon measurement. *After* the measurement, it is a question of probabilities, to be interpreted epistemically or statistically, relating to results of measurement that are already objectivized.

The extension will be carried out in such a way as to incorporate (1) the whole of the purely epistemic or purely statistical quantum mechanics given in section 4, (2) the ontic formulation of quantum mechanics just described, (3) a connection (still to be established) between these two theories. According as the epistemic or the statistical theory is chosen for (1), the result is called the *epistemic* or the *statistical extension* of the ontic version. It will be clear from the preceding discussion that the decision has to be taken in respect of the propositions regarding probabilities. The connection sought consists, for the concepts defined in subsections 4(a)–(d)(i), simply in a trivial identification of the concepts relating to an individual object, while the statistical concepts so far introduced (for the statistical extension) are initially left untouched. There is then an unavoidable question of the nature of the *description of state* in the desired extension, seeing that the theory now contains states of individual objects.

Here there are evidently two competing possibilities, as has been very clearly shown by Süssmann (1958). One is the use of (countable) assemblages $\{\phi_i,\, w_i\}_i$ of states ϕ_i of the object Σ, or of the objects Σ_k in a statistical ensemble, in which the probabilities w_i, which are to be interpreted epistemically or statistically, satisfy the conditions

$$w_i > 0, \quad \Sigma_i\, w_i = 1 \tag{2}$$

and would represent the degree of knowledge of the occurrence of the state ϕ_i of Σ, or the approximate relative frequency of the occurrence of ϕ_i in the ensemble $\{\Sigma_k\}_k$. The use of such assemblages would make explicit reference to the individual states which characterize the ontic formulation, but there would be no immediate way of deriving, from an assemblage of states, propositions regarding the results of measurements of quantities. The probabilities w_i and the probabilities $\|P^A_a\, \phi_i\|^2$ have different meanings and cannot be forthwith combined in the required propositions. Secondly, besides the assemblages of individual states, provided by the ontic formulation, the purely epistemic or purely statistical formulation provides for the descriptions of an individual state the states of knowledge W concerning Σ, or statistical states W of $\{\Sigma_k\}_k$. Here the reverse situation occurs. Propositions regarding the results of measurements exist for these states. But they do not contain the individual states explicitly, and what they assert regarding these is initially unknown. The required connection between the two theories must evidently be precisely such as to eliminate this asymmetry.

The first step is to provide the assemblages of individual states with the missing propositions regarding the results of measurements of quantities. To do this, we must relate the ontically interpreted probabilities to the epistemically or statistically interpreted probabilities, for example as follows.

(OE) If it is known that the object Σ is in the state ϕ, then the ontically interpreted probability $\|P_a^A\phi\|^2$ for any quantity A and set of values a is *numerically* equal to the epistemically interpreted probability of obtaining the result a in a suitable measurement of A.

(OS) If all the Σ_k in a statistical ensemble $\{\Sigma_k\}_k$ are in the same state ϕ, then the ontic probability $\|P_a^A\phi\|^2$ for any A and a is *numerically* equal to the statistical probability ($\approx$ relative frequency) of obtaining the result a in a suitable measurement of A.

Thus the relationship is established for the special case of an assemblage comprising only a single state ϕ. It follows at once from (OE) or (OS) that, in this special case, there then exists the state of knowledge P_ϕ or the statistical state P_ϕ. These are therefore *unambiguously* assigned to the *known* state ϕ of Σ or to the state ϕ which is the *same* for all Σ_k in the ensemble. The general case, however, follows also: if there exists any assemblage $\{\phi_i, w_i\}_i$ of individual states, application of (OE) or (OS) and the argument just given to each of the ϕ_i converts this assemblage into an assemblage of states of (maximum) knowledge or (pure) statistical states (cf. the end of subsection 4(d)(ii)), for which the required probability, in the sense of ordinary probability theory, is then given by

$$\Sigma_i\, w_i\, \|P_a^A\, \phi_i\|^2. \tag{3}$$

Since

$$\mathrm{Tr}((\Sigma_i\, w_i\, P_{\phi_i})P_a^A) = \Sigma_i\, w_i \|P_a^A\, \phi_i\|^2, \tag{4}$$

the state of knowledge or the statistical state

$$W = \Sigma_i\, w_i\, P_{\phi_i} \tag{5}$$

is *unambiguously* assigned to the assemblage $\{\phi_i, w_i\}_i$: this state of knowledge or this statistical state exists if the assemblage $\{\phi_i, w_i\}_i$ of individual states exists. A probability arising in the form (3) is evidently envisaged by Heisenberg's statement (1959, pp. 44f., 53f.) that subjective and objective elements are intermingled in the quantum-mechanical probabilities. A probability (3) is in fact a mixture of the probabilities $\|P_a^A\, \phi_i\|^2$ with the probabilities w_i. The subjective-objective distinction intended by Heisenberg is not that between epistemically and statistically interpreted probabilities, but *that between either one of these and the ontically interpreted probabilities.* The latter are represented in (3) by the $\|P_a^A\, \phi_i\|^2$, though by means of (OE) or (OS), not directly. The former are given by the mixing probabilities w_i. Thus there is a distinction to be made between avoidably and unavoidably occurring probabilities (see also Born and Jordan 1930, §57; Pauli 1954).

The postulates (OE) and (OS) therefore provide a complete answer to the question of epistemic or statistical propositions regarding the results of measurements for a given assemblage of states. We now come to the other question, that of the information conveyed by an epistemic or statistical description of state W as to the existence of an individual state. Let us first assume that W is a maximum or pure state: $W = P_\phi$. If such a description of state exists together with an assemblage $\{\phi_i, w_i\}_i$ of individual states, then according to (OE) or (OS) and the argument in the previous paragraph we must have $P_\phi = \Sigma_i\, w_i\, P_{\phi i}$, whence it follows immediately that $\{\phi_i, w_i\}_i$ contains only one state, namely ϕ, with probability unity. This leads to the following postulates.

(OE′) In a maximum state of knowledge P_ϕ of Σ, Σ is in the state ϕ and is known to be in that state.

(OS′) If $\{\Sigma_k\}_k$ is in the pure statistical state P_ϕ, then all the Σ_k are in the state ϕ.

From this it then follows that, if there exists the description of state P_ϕ, the epistemically or statistically interpreted probabilities $\|P_a^A\phi\|^2$ are numerically equal to the ontic probabilities which belong to the state ϕ. Thus (OE′) and (OS′) simply invert (OE) and (OS) respectively.

Next, let the description of state W be no longer maximum or pure. Then W does not unambiguously define an assemblage of individual states. There are certainly always assemblages $\{\phi_i, w_i\}_i$ which belong to W in the sense of (OE) or (OS), i.e. give the same probabilities, since every von Neumann operator W has a representation in the form (5). But, firstly, this representation is not unambiguous if W is not maximum or pure; in fact, it then always has infinite ambiguity. Secondly, the individual Σ or Σ_k need not be in any state, and therefore none of the assemblages which (5) affords for W need exist. This is one of the most noteworthy features of quantum mechanics, "the one that enforces its entire departure from classical lines of thought" (Schrödinger 1935b, §1), and will be further discussed in Chapters VI and VII. For the present we shall simply note that (OE) and (OE′), or (OS) and (OS′), exhaust the possibilities of coupling the purely epistemic or purely statistical quantum mechanics to this ontic version.

Finally, we have to inquire as to the extension of the theory of time variations based on (α)–(δ). It is immediately clear how an assemblage of individual states will vary dynamically. From (γ),

$$\left.\begin{aligned} \phi_{\mu,t_2} &= U_{t_2-t_1}\,\phi^+_{\mu,t_1}, \\ w_{\mu,t_2} &= w^+_{\mu,t_1} \end{aligned}\right\} \tag{6}$$

is the assemblage at time $t_2 > t_1$ if the assemblage $\{\phi^+_{\mu,t_1}, w^+_{\mu,t_1}\}_\mu$ existed at time t_1 and no measurements have taken place in the interval (t_1, t_2). This transition is also easily seen to be compatible with the transition from $W^+_{t_1}$ to W_{t_2}, subsection 4(e), (γ):

$$\Sigma_\mu\, w_{\mu,t_2}\, P_{\phi_{\mu,t_2}} = U_{t_2-t_1}\,(\Sigma_\mu\, w^+_{\mu,t_1}\, P^+_{\phi_{\mu,t_1}})\, U^{-1}_{t_2-t_1}. \tag{7}$$

Thus (γ) in subsection 4(e) follows for the descriptions of state defined by the two assemblages in the sense of (5).

The treatment of the reduction of states is somewhat more complicated. To simplify the formulae, the t subscript will be omitted. Let $me = (\{P_i\}_i, P_j)$ be the measurement and $\{\phi_\lambda, w_\lambda\}_\lambda$ the assemblage initially present. Then, from (β), $\|P_j\,\phi_\lambda\|^2 > 0$ for at least one λ. From (δ), the states of the new assemblage all have the form $P_j\,\phi_\lambda/\|P_j\,\phi_\lambda\|$. Let the different states among these be $\phi^+_{j\mu}$. We ask for the probability with which $\phi^+_{j\mu}$ occurs in the new assemblage. In the previous assemblage, ϕ_λ is present with probability w_λ. *If* ϕ_λ exists, it becomes $P_j\,\phi_\lambda/\|P_j\,\phi_\lambda\|$ with probability $\|P_j\,\phi_\lambda\|^2$. The *unconditional* probability of this transition is therefore $w_\lambda \cdot \|P_j \phi_\lambda\|^2$. The unconditional probability of the occurrence of $\phi^+_{j\mu}$ is therefore

$$w^\times_{j\mu} = \sum_{\lambda:\, P_j\phi_\lambda \sim \phi^+_{j\mu}} w_\lambda \|P_j \phi_\lambda\|^2. \tag{8}$$

P_j occurs with the probability $\Sigma_\lambda w_\lambda \|P_j \phi_\lambda\|^2 > 0$ (see above). Thus the probability, *conditional* upon this result, of the occurrence of $\phi^+_{j\mu}$ in the new assemblage, is

$$w^+_{j\mu} = \frac{\Sigma_{\lambda:\, P_j\phi_\lambda \sim \phi^+_{j\mu}} w_\lambda \|P_j \phi_\lambda\|^2}{\Sigma_\lambda w_\lambda \|P_j \phi_\lambda\|^2}. \tag{9}$$

$\{\phi^+_{j\mu}, w^+_{j\mu}\}_\mu$ is the new assemblage, taking account of the result P_j. With the measurement $\dot{m}e = \{P_j\}_j$, which does not take account of the result, the reduction to P_j is simply absent, i.e. we have the new assemblage $\{\phi^+_{j\mu}, w^\times_{j\mu}\}_{j\mu}$, with $w^\times_{j\mu}$ given by (8). Here again there is compatibility with the reductions of state according to subsection 4(d)(ii): $\{\phi_\lambda, w_\lambda\}_\lambda$ corresponds, through (OE) or (OS), to an assemblage of states of knowledge or statistical states. The way in which this is changed by measurements has been given in subsection 4(e) only for the special case where $\{\phi_\lambda, w_\lambda\}_\lambda$ is trivial. Thus the first generalization of the trivial case is that just obtained for the particular case of maximum knowledge or pure statistical states. For a trivial $\{\phi_\lambda, w_\lambda\}_\lambda = \{\phi\}$, $\phi^\times_{j\mu}$ becomes simply $P_j\phi/\|P_j\phi\|$, (8) becomes $\|P_j\phi\|^2$, and (9) becomes unity. Thus the reduction taking account of the result now gives $P_j\phi/\|P_j\phi\|$, in accordance with 4(e), (δ), and the strong reduction without taking account of the result gives $\{P_j\phi/\|P_j\phi\|, \|P_j\phi\|^2\}_{j:\|P_j\phi\|^2>0}$, in agreement with 4(e), (3). The corresponding results for the purely epistemic or purely statistical descriptions of state, related to the assemblages of states by (5), are

$$\Sigma_\mu w^+_{j\mu} P_{\phi^+_{j\mu}} = \frac{P_j\, (\Sigma_\lambda w_\lambda P_{\phi_\lambda})\, P_j}{\mathrm{Tr}((\Sigma_\lambda w_\lambda P_{\phi_\lambda})\, P_j)} \tag{10}$$

for the measurement taking account of the result, and

$$\Sigma_{j\mu} w^\times_{j\mu} P_{\phi^+_{j\mu}} = \Sigma_j P_j\, (\Sigma_\lambda w_\lambda P_{\phi_\lambda})\, P_j \tag{11}$$

for the measurement not taking account of the result, as may be confirmed by simple calculation. Here also, therefore, the reductions defined in subsection 4(e) are a consequence of (δ) and (OE) or (OS).

CHAPTER III

Orthodox Quantum Mechanics in Hilbert Space: Formulation in Terms of Properties

---

THE formulations of the three theories of mechanics given in Chapter II include among their basic physical concepts that of a quantity of an object. In Chapter III we shall show how this concept can be replaced by that of a *property* of an object. This change brings about a logical simplification of the formulation of a physical theory which is useful in certain fundamental studies. The possibility of such a replacement in quantum mechanics was known to von Neumann (1955, section III.5). An elegant and rigorous treatment has recently been given by Mackey (1963, §2–2), who uses propositions regarding probabilities to support the relationship of content between quantities and properties. In the following it will be shown that propositions regarding measurements can also be used for this purpose. In sections 1–3, the new formulation of the three theories of mechanics will be stated, and in section 4 it will be shown to be equivalent to the corresponding formulations in Chapter II, sections 2–5. The method of formulating a physical theory described in section II.1 will be retained.

III.1 Classical mechanics

A phase space is assigned to the object Σ, as in section II.2, and composite objects are described by the Cartesian product of the phase spaces of the component objects, as in subsection II.2(b). The concept of a quantity of Σ is, however, replaced by that of the (contingent) *properties of* Σ. *These are described by all Borel sets in the phase space of* Σ, in one-to-one correlation. The states of Σ are described, as in subsection II.2(d), by points in the phase space of Σ. The relation between properties and states is now given by a universal propositional function $S_*(M, E)$, which has the meaning that Σ in the state M has the property E. $S_*(M, E)$ is mathematically represented by $M \in E$, i.e. Σ *in the state* M *has the property* E if and only if the phase point corresponding to M is in the Borel set corresponding to E in phase space. Lastly, time variations of state are subject to the same conditions as in subsection II.2(e): together with the phase space we have the Hamiltonian function, and the variations M_t are subject to Hamilton's equations.

III.2 Classical statistical mechanics

In classical statistical mechanics also, the first change as compared with the formulation in section II.3 is to replace the concept of a quantity of an object Σ by that of a *property of* Σ. The classification of Borel functions in the phase space $\mathfrak{M}$ of Σ corresponds to a classification of Borel sets in $\mathfrak{M}$. Two Borel sets E and F in $\mathfrak{M}$ are said to be *equivalent* (denoted by $E \sim F$)

D

if their symmetric difference $(E \cap F^\perp) \cup (E^\perp \cap F)$ has Lebesgue measure zero ($\cap$, $\cup$, and $^\perp$ denoting the intersection, union and complement in $\mathfrak{M}$). *The properties of* Σ *are now described by these classes of Borel sets.* Here, to simplify the notation, we shall again represent a class by one of its members, and must accordingly bear in mind that all statements thus formulated in relation to $\sim$ are independent of the particular member chosen.

The concept of a measurement of Σ, subsection II.3(d)(i), is retained, but measurements are now described by pairs $me = (\{E_i\}_i, E_{i_0})$, in which the Borel sets E_i taken to within $\sim$ satisfy the conditions

(α) $E_i \nsim \emptyset$,

(β) $E_i \cap E_j \sim \emptyset$ for $i \neq j$,

(γ) $\cup_i E_i \sim \mathfrak{M}$.

In this case there are two kinds of propositions regarding measurements. $\mathbf{M}^1_*(me, E)$ has the meaning that *the property E is a possible result of the measurement me.* $\mathbf{M}^2_*(me, E)$ has the meaning that *the property E is an actual result of the measurement me.* The relation between $(\{E_i\}_{i\in I}, E_{i_0})$ and E which describes $\mathbf{M}^1_*$ is to be valid if and only if

(α_1) there is a subset $I_E \subseteq I$ belonging to E, such that $E \sim \cup_{i\in I_E} E_i$;

the relation which describes $\mathbf{M}^2_*$ is to be valid if and only if (α_1) holds and also

(β_1) $E_{i_0} \cap E \sim E_{i_0}$.

In accordance with the procedure in subsection II.3(d)(i), two measurements *me* and me_1 would be said to be *equivalent* if they had the same properties as possible results, in terms of the propositional function $\mathbf{M}^1_*$. In this way the measurements $\dot{m}e$ arise, in which the result is ignored; their description is then of the form $\dot{m}e = \{E_i\}_i$. They correspond to the propositions $\dot{\mathbf{M}}^1_*(\dot{m}e, E)$ that *a property E is a possible result of the measurement* $\dot{m}e$, and these propositions are described only by (α_1). So far, the measurements relate to an individual object. The change to statistical measurements with or without selection has exactly the same content as in subsection II.3(d)(i), and formally they correspond to the measurements $\dot{m}e$ and *me* just defined, together with their mathematical descriptions.

The description of states of knowledge of Σ or the statistical states of an ensemble $\{\Sigma_k\}_k$ is taken to be the same as in subsection II.3(d)(ii). The corresponding universal probability function $\mathbf{pr}_*(\rho; E)$ gives the probability (to be appropriately interpreted) that, *assuming the state of knowledge or statistical state* ρ, *the property E will be found when a suitable measurement is made.* The mathematical representation of $\mathbf{pr}_*$ is

$$\mathbf{pr}_*(\rho; E) = \int_{\{M:\ M \in E\}} \rho \, d\mu. \tag{1}$$

As regards the time variations, the functions me_t, ρ_t and ρ_t^+ are now subject to the conditions

(α_2) as (α) in subsection II.3(e),

(β_2) if $\mathbf{M}^2_*(me_t, E)$, then

$$\int_{\{M:\ M \in E\}} \rho_t \, d\mu > 0,$$

(γ_2) as (γ) in subsection II.3(e),

(δ_2) if $me_t = (\{(E_i)_t\}_i, (E_{i_0})_t)$, then for all E

$$\int_{\{M \in E_{i0} \cap E\}_M} \rho_t \, d\mu \Big/ \int_{\{M \in E_{i0}\}_M} \rho_t \, d\mu = \int_{\{M \in E\}_M} \rho_t^+ \, d\mu.$$

III.3 Quantum mechanics

Here again, the differences from the formulation in section II.4 begin with the introduction of the concept of a *property* of an object Σ in place of the concept of a quantity. *The properties of Σ are described by all the (closed) subspaces of the Hilbert space $\mathfrak{H}$ of Σ.* As convenient, we may also describe the properties not by subspaces of but by projectors in $\mathfrak{H}$ which project on these subspaces.

The concept of a measurement of Σ, subsection II.4(d)(i), will be retained, but *we now describe the measurements by pairs* $me = (\{E_i\}_i, E_{i_0})$, the subspaces $E_i \subseteq \mathfrak{H}$ satisfying the conditions

(α) $E_i \neq \{0\}$ (the null space in $\mathfrak{H}$),

(β) $E_i \perp E_j$ for $i \neq j$ ($\perp$ = perpendicular to),

(γ) $\overline{\Sigma}_i E_i = \mathfrak{H}$ ($\overline{\Sigma}_i$ = closed algebraic sum).

As in section 2, there are two kinds of propositions regarding measurements, $\mathbf{M}^1_*(me, E)$ and $\mathbf{M}^2_*(me, E)$, which respectively have the meanings that *the property E is a possible and an actual result of the measurement me.* The describing relations between $(\{E_i\}_{i\in I}, E_{i_0})$ and E satisfy the following conditions:

(α_1) there is a subset $I_E \subseteq I$ belonging to E, such that

$$E = \overline{\Sigma}_{i\in I_E} E_i,$$

or this together with

(β_1) $E_{i_0} \subseteq E$ ($\subseteq$ = is contained in).

Here again, as in section 2, we define measurements $\dot{m}e$ in which the result is ignored, and the corresponding propositions. We can then proceed as before to the two types of statistical measurements.

For the possible states of knowledge W concerning an object Σ, or the states W of a statistical ensemble, described as in subsection II.4(d)(ii), or the states ϕ of an object Σ (section II.5) (according as an epistemic, statistical or ontic formulation is used), we have again a universal probability function $\mathbf{pr}_*(W; E)$ or, in the third case, $\mathbf{pr}_*(\phi; E)$, whose values are the variously interpreted *probabilities that the property E is found when a suitable measurement is made.* $\mathbf{pr}_*(W; E)$ and $\mathbf{pr}_*(\phi; E)$ are mathematically described by

$$\mathbf{pr}_*(W; E) = \mathrm{Tr}(WP_E), \tag{1}$$

where P_E is the operator which projects on the subspace pertaining to E, and

$$\mathbf{pr}_*(\phi; E) = \|P_E\phi\|^2. \tag{2}$$

For the time variations we again follow subsection II.4(e) (or section II.5). For the epistemic and statistical interpretations this gives

(α_2) analogous to (α) in subsection II.4(e),

(β_2) if $\mathbf{M}^2_*(me_t, E)$, then $\mathrm{Tr}(W_t P_E) > 0$,

(γ_2) analogous to (γ) in subsection II.4(e),

(δ_2) if $me_t = (\{(E_i)_t\}_i, (E_{i_0})_t)$, then

$$W_t^+ = \frac{P_{E_{i0}} W_t P_{E_{i0}}}{\mathrm{Tr}(W_t P_{E_{i0}})}.$$

Correspondingly, for the ontic interpretation the conditions are

(α_3) analogous to (α) in section II.5,

(β_3) if $M_*^2(me_t, E)$, then $\|P_E\,\phi_t\|^2 > 0$,

(γ_3) analogous to (γ) in section II.5,

(δ_3) if $me_t = (\{(E_i)_t\}_i, (E_{io})_t)$, then

$$\phi_t^+ = \frac{P_{E_{io}}\phi_t}{\|P_{E_{io}}\phi_t\|}.$$

III.4 Proofs of equivalence

In classical mechanics, the equivalence of the property formulation (section 1) and the quantity formulation (section II.2) can be shown as follows. Starting from the latter, we consider all sets of states E having the form

$$E = \{S(M, f, \alpha)\}_M, \tag{1}$$

where f is a quantity and α a value. Let these sets of states be the *properties* of the object Σ concerned. We correlate with the property (1) the Borel set E' of all points M' in the phase space $\mathfrak{M}$ of Σ for which the Borel function f' which describes f satisfies the equation $f'(M') = \alpha$:

$$E' = \{f'(M') = \alpha\}_{M' \in \mathfrak{M}}. \tag{2}$$

This correlation is independent of the f and α used in (1) to define E. The correlation is a one-to-one relation between all the properties of Σ and all the Borel sets in $\mathfrak{M}$. We choose it as the description of properties, required in accordance with section 1. The S_* in section 1 is such that

$$S_*(M, E) \leftrightarrow M \in E. \tag{3}$$

$M \in E$ if and only if $M' \in E'$. Thus S_* is described by

$$S'_*(M', E') \leftrightarrow M' \in E', \tag{4}$$

as required in section 1. This completes the essence of the proof. In the proof, a difference of notation has been maintained between the physical entities and the mathematical entities which represent them, i.e. these are not identified in the usual manner. The same will be done in all similar subsequent proofs, since the identification convention would lead to misunderstandings.

Now let the property formulation be the starting-point. First, we assign to each property E the set of states

$$E^* = \{S_*(M, E)\}_M. \tag{5}$$

We then consider real functions f in the space of states for which

(α) for every set of values (= real Borel set) a, the set of states $f^{-1}(a)$ is one of the sets (5), and therefore represents a property.

Let these functions be the *quantities* of Σ. We assign to the quantity f the function

$$f'(M') = f(M) \tag{6}$$

in $\mathfrak{M}$, where M' is the phase point representing the state M. It is easy to see that in this way all Borel functions in $\mathfrak{M}$ are put in one-to-one relation with all quantities of Σ, as required.

From (α), for a quantity f and every value α the set of states $f^{-1}(\{\alpha\})$ is a set (5). The correlation (5) is one-to-one, and therefore $f^{-1}(\{\alpha\})$ corresponds to a unique property $E_{(f,\alpha)}$ which represents it:

$$f^{-1}(\{\alpha\}) = \{\mathbf{S}_*(M, E_{(f,\alpha)})\}_M. \tag{7}$$

We thus define

$$\mathbf{S}(M, f, \alpha) \rightharpoonup \mathbf{S}_*(M, E_{(f,\alpha)}). \tag{8}$$

From (7) and (8), $f(M) = \alpha$ if and only if $\mathbf{S}(M, f, \alpha)$; therefore, from (6), $f'(M') = \alpha$ if and only if $\mathbf{S}(M, f, \alpha)$. Thus S is described by

$$\mathbf{S}'(M', f', \alpha) \rightharpoonup f'(M') = \alpha, \tag{9}$$

as it should be.

The proofs of equivalence for classical statistical mechanics and quantum mechanics are so similar in general procedure that they need be given only for the latter. We again start firstly from the quantity formulation. In order to derive the property concept, some definitions are needed. A pair (A, a) consisting of a quantity A and a set of values a will be called a *pre-property*. A pre-property (A, a) uniquely defines a property, namely the property that is the actual result of a measurement *me* if and only if *me* measures the quantity A with the result a and appropriate accuracy; but different pre-properties may yield the same property. We must therefore construct on this basis an equivalence relation between pre-properties. Let the pre-property (A, a) be called a *possible* or an *actual result* of the measurement *me* if there exists respectively an accuracy $\mathfrak{a}$ such that $a \in \mathfrak{a}$, and a set of values a_1 such that $\mathbf{M}(me, A, \mathfrak{a}, a_1)$ or only an accuracy $\mathfrak{a}$ such that $\mathbf{M}(me, A, \mathfrak{a}, a)$. We also say that a pre-property (A, a) *includes* a pre-property (B, b) if both are possible results of a suitable measurement *me*, and if (B, b) is an actual result for every measurement *me* such that (A, a) is an actual result and (B, b) a possible result. (In classical statistical mechanics, the first part of this definition is redundant, since the measurement in question always exists, but in quantum mechanics this is not so.) Lastly, (A, a) and (B, b) are said to be *equivalent* if each includes the other. We now define the *properties of* Σ as the equivalence classes of pre-properties of Σ:

$$E = \{(A, a)\}^{\approx}. \tag{10}$$

Two pre-properties (A, a) and (B, b) are equivalent if and only if

$$P_a^A = P_b^B \tag{11}$$

in respect of the spectral resolutions belonging to A and B. Thus, if a property (10) is specified, it corresponds to a unique subspace

$$E' = \{P_a^A \phi = \phi\}_{\phi \in \mathfrak{H}} \tag{12}$$

of the Hilbert space $\mathfrak{H}$ of Σ. This assignment is in fact one-to-one, and relates to every subspace of $\mathfrak{H}$. It can therefore be used as the required description of properties.

For *measurements* of Σ there is simply a different representation in the property formulation, but it is obtainable in an obvious manner from that of the quantity formulation, being given immediately by the canonical correlation between all projection operators and all subspaces of $\mathfrak{H}$. The propositions regarding measurements of Σ are defined as follows: the property E is a possible or an actual result of the measurement *me*, i.e. $\mathbf{M}_*^1(me, E)$ or $\mathbf{M}_*^2(me, E)$ is valid, if and only if a representative pre-property (A, a) is a possible or an actual result of *me*. Then (10) gives

$$\mathbf{M}_*^1(me, E) \leftrightarrow \bigvee_{\mathfrak{a}} \bigvee_{a_1} . \mathbf{M}(me, A, \mathfrak{a}, a_1) \wedge a_1 \in \mathfrak{a}, \tag{13}$$

$$\mathbf{M}_*^2(me, E) \leftrightarrow \bigvee_{\mathfrak{a}} . \mathbf{M}(me, A, \mathfrak{a}, a). \tag{14}$$

This definition is independent of the choice of the (A, a) which represents E. The conditions $(\alpha_1)-(\delta_1)$ in section II.4(d)(i) easily lead to (α_1), or (α_1) and (β_1), section 3, for the describing relations.

For states of knowledge, statistical states or ontic states, the property formulation remains unchanged. The corresponding probability functions are obtained, using (10), from

$$\mathbf{pr}_*(W; E) = \mathbf{pr}(W; A, a) \tag{15}$$

or, for the ontic formulation, from

$$\mathbf{pr}_*(\phi; E) = \mathbf{pr}(\phi; A, a); \tag{16}$$

on account of (12), this immediately gives the correct description (1) or (2) (section 3). This completes the essentials of the proof.

Now let the property formulation be the starting-point. To derive the concept of a quantity, some definitions are again needed. We say that the property E *includes* the property F if both are possible results of a suitable measurement *me*, and if F is an actual result for every measurement such that E is an actual result and F a possible result. (In the classical case, the first part of this definition is again redundant.) This relationship holds if and only if

$$E' \subseteq F' \tag{17}$$

is valid for the subspaces in Hilbert space which describe E and F. If now a set $\{E_\mu\}_\mu$ of properties is given, we may seek a property E which is included by every E_μ and which also includes every property which is included by every E_μ. Such an E is called an *upper bound* of the E_μ and denoted by $\cup_\mu E_\mu$. According to the mathematical description of the properties of subspaces of $\mathfrak{H}$, and (17), there is always just one upper bound of the E_μ, and it is described by the closed sum

$$(\cup_\mu E_\mu)' = \overline{\Sigma}_\mu E'_\mu \tag{18}$$

of the subspaces which describe the E_μ. Together with this concept, we also need the relation whereby the property E *excludes* the property F. This is said to occur if and only if there exists a measurement for which E and F are possible results, and if F is *not* an actual result for every measurement such that E is an actual result and F a possible result. (In the classical case, the first part of this definition is again redundant.) The mathematical description here is that the describing subspaces E' and F' are orthogonal:

$$E' \perp F'. \tag{19}$$

The exclusion relation will also be denoted by $E \perp F$. We finally require the definition of the two *trivial properties* $E_\cap$ and $E_\cup$, which are respectively actual results of no measurement and of every measurement. They correspond in Hilbert space to

$$E'_\cap = \{0\}, \quad E'_\cup = \mathfrak{H}, \tag{20}$$

i.e. in particular, they always exist as *one* property. The concept of a quantity can now be specified as follows. A quantity is a mapping A which assigns to every real Borel set a a property $A(a)$ such that

(α_1) $A(a) \perp A(b)$ if $a \cap b = \emptyset$,

(β_1) $A(\cup_\mu a_\mu) = \cup_\mu A(a_\mu)$ if the a_μ form a sequence of pairwise disjoint sets,

(γ_1) $A(\emptyset) = E_\cap$, $A(\mathbf{R}) = E_\cup$.

Strictly, (β_1) should state only that $A(\cup_\mu a_\mu)$ is an upper bound of the $A(a_\mu)$, and (γ_1) only that $A(\emptyset)$ and $A(\mathbf{R})$ are the result of no measurement and of every measurement respectively, but we have included in the definition itself the existence and uniqueness which follow from the mathematical representation. The transition from the properties to the describing subspaces, and from these to the corresponding projection operators, shows by means of the spectral theorem that the quantities thus defined are in one-to-one relationship with all the self-adjoint operators, as they should be.

The description of measurements, required in the quantity formulation, is again obtained from the description given by the property formulation simply by means of the canonical transition from subspaces to projection operators in $\mathfrak{H}$. To derive the propositions regarding measurements, we first define an *alternative* as a set $\{E_i\}_i$ of properties such that

(α_2) $E_i \neq 0$,

(β_2) $E_i \perp E_j$ for $i \neq j$,

(γ_2) $\cup_i E_i = E_{\mathsf{U}}$.

The mathematical description is given by (18)–(20). In (γ_2), as in (β_1) and (γ_1), the statement should initially be only that every upper bound of the E_i is a result of every measurement; (γ_2) is obtained from the mathematical representation of E_i.

Finally, we have to derive the probability functions and their mathematical representation from the concepts of state of knowledge, statistical state or ontic state, which are the same in the two formulations. The definition is simply

$$\mathbf{pr}(W; A, a) \rightharpoondown \mathbf{pr}_*(W; A(a)), \tag{21}$$

or

$$\mathbf{pr}(\phi; A, a) \rightharpoondown \mathbf{pr}_*(\phi; A(a)). \tag{22}$$

The required mathematical representation is obtained from (1) or (2) in section 3, since (α_1)–(γ_1) show that the property $A(a)$ is described by the subspace which corresponds to P_a^A.

The manner of the above proof of the equivalence of different formulations of a physical theory indicates the sense in which two such formulations are here said to be equivalent. First of all, it must be recalled that these formulations are both of the type specified in section II.1: both involve systems of physical concepts $\mathfrak{P}_1$ or $\mathfrak{P}_2$, on the one hand, and systems of mathematical concepts $\mathfrak{M}_1$ and $\mathfrak{M}_2$ on the other, and the two are related in that the former is represented in the latter (symbolized by $\mathfrak{P}_1 \rightarrowtail \mathfrak{M}_1$ or $\mathfrak{P}_2 \rightarrowtail \mathfrak{M}_2$). The proof of equivalence was obtained as follows. $\mathfrak{P}_2$ and $\mathfrak{P}_1$ were respectively obtained from the first and second formulation. This was permissible only by using $\mathfrak{P}_1$ and $\mathfrak{P}_2$ respectively (together with universal procedures of logic and mathematics, of course), not by using $\mathfrak{M}_1$ and $\mathfrak{P}_1 \rightarrowtail \mathfrak{M}_1$, or $\mathfrak{M}_2$ and $\mathfrak{P}_2 \rightarrowtail \mathfrak{M}_2$, and certainly not by using $\mathfrak{M}_2$ and $\mathfrak{P}_2 \rightarrowtail \mathfrak{M}_2$, or $\mathfrak{M}_1$ and $\mathfrak{P}_1 \rightarrowtail \mathfrak{M}_1$. It is only in proving that $\mathfrak{P}_2$ and $\mathfrak{P}_1$ have the respective representations $\mathfrak{P}_2 \rightarrowtail \mathfrak{M}_2$ and $\mathfrak{P}_1 \rightarrowtail \mathfrak{M}_1$ that we can and must of course use also the representations $\mathfrak{P}_1 \rightarrowtail \mathfrak{M}_1$ and $\mathfrak{P}_2 \rightarrowtail \mathfrak{M}_2$ respectively. This procedure is the basis for the force of the assertion of equivalence: it implies that $\mathfrak{P}_2$ or $\mathfrak{P}_1$ can be derived by purely logical means from $\mathfrak{P}_1$ or $\mathfrak{P}_2$ respectively, and that the representation $\mathfrak{P}_2 \rightarrowtail \mathfrak{M}_2$ or $\mathfrak{P}_1 \rightarrowtail \mathfrak{M}_1$ then follows by using the other representation, $\mathfrak{P}_1 \rightarrowtail \mathfrak{M}_1$ or $\mathfrak{P}_2 \rightarrowtail \mathfrak{M}_2$ respectively.

CHAPTER IV

Determinism and Indeterminism

In this chapter we deal with conclusions relating to the mechanical theories formulated in Chapters II and III, and chiefly to the question of whether they are deterministic or indeterministic. The problem thus brought to the forefront carries an unusual burden of philosophical tradition, and is in any case intrinsically complicated. We must therefore note first of all that only certain aspects of it can be discussed here. The first such aspect (section 1) is based on the distinction between contingent propositions and assertions of laws in a physical theory. The distinction will not be generally established here (see Scheibe 1964), but it may be taken as accepted that theories which involve in an essential manner the changes of an object with time contain both contingent propositions which express the possibility of a time behaviour of an object, and assertions of laws which restrict the ensemble of all such possibilities. If a dynamical theory has such a structure, it is found that the description of any particular possible time behaviour is redundant, in the sense that even a very few of the contingent propositions in this description determine, and indeed logically determine, by their truth values together with the relevant laws, the truth value of all the other contingent propositions in the description. If a physical theory is regarded as deterministic when such a situation can be shown to exist in it, we may expect that quantum mechanics, like classical mechanics, is thoroughly deterministic. If, on the other hand, restrictions are imposed, for example that the determining contingent propositions all refer to the same instant, whereas the propositions to be determined refer to some other instant, then the situation in quantum mechanics is such that it can be called deterministic only with certain qualifications. But this situation is nevertheless exactly analogous to that in classical statistical mechanics, so that this aspect does not lead to any profound difference between quantum mechanics and classical mechanics.

The aspect just mentioned is intrinsic to the theory, in that a decision whether quantum mechanics is deterministic or indeterministic does not result from a direct comparison of the descriptions of state in quantum mechanics with the corresponding descriptions in classical mechanics to show, for instance, that one of them involves probabilities and measurements and the other does not. If the statement that quantum mechanics is indeterministic is based on such a comparison, this is evidently done from a different aspect, which will be discussed in sections 2 and 3. The chief task there will be to develop certain differences between quantum mechanics and both exact and statistical classical mechanics. The latter also has been formulated in section II.3 with contingent propositions regarding measurements and probabilities, and it will have to be investigated how far the probabilistic nature of quantum mechanics proves an indeterminism distinct from the classical case. When only the situation at a single instant is considered, such a distinction can be based on the occurrence of maximum probabilistic descriptions of state, and consequently of a finite number of quantities not simultaneously measurable, and also of further related exclusion principles, of which the

Heisenberg relations are the most familiar. These relationships give rise to problems which will be discussed in section 2.

In that section we consider only the features of contingent propositions in quantum mechanics valid at a particular instant, as compared with their classical counterparts; in section 3 the changes with time will also be included, from essentially the same standpoint as in section 2. For the time changes of probabilistic descriptions of state which result from dynamics, there appears in quantum mechanics the peculiarity of the interference of superposed maximum descriptions of state. The reduction of states caused by acts of measurement also leads to a characteristic uncertainty; in direct contrast to the classical case, the dynamic evolution of the descriptions of state is interrupted by the mere performance of a measurement, regardless of the acceptance or selection of a result.

The literature on indeterminism in physics is almost impossible to survey. In accordance with the highly selective nature of the present studies of orthodox quantum mechanics, reference will be made only to the following sources for further reading. The work of Stegmüller (1969) has some points of contact with the discussion in section 1. The problems of absence of simultaneous measurability and those of the Heisenberg relations (section 2) have recently been treated strictly separately, in accordance with the anti-orthodox trend. Recent publications, which will provide further references, are those of She and Heffner (1966), Prugovečki (1967), Park and Margenau (1968), and Dombrowski (1969). Section 3 is concerned principally with the possible orthodox views on the reduction of states. This is part of the topic of the measurement process in quantum mechanics, which is the subject of a very extensive literature criticizing the orthodox approach. Some of this literature will be cited in Chapter VI. For section IV.3, especial mention should be made of the Margenau school's critique, including recent papers by Margenau (1958, 1963a, b) and Park (1968). Lastly, the analysis by Scott (1968) has certain affinities with the examples given in section 4.

IV.1 Indeterminism and laws

On the customary view, ordinary classical mechanics is a typical deterministic theory; on the orthodox interpretation, quantum mechanics is a typical indeterministic theory. Somewhere between these two extremes lies classical statistical mechanics, with some deterministic and some indeterministic features. The problem is the precise meaning of thus generally labelling physical theories as deterministic or indeterministic. In the present section, this problem will be discussed only in respect of certain laws relating contingent propositions in all three mechanical theories.

Let us first consider ordinary classical mechanics. Here we can prove a theorem which appears at first sight to formulate, at least implicitly, the deterministic nature of classical mechanics; that is, which contains a reason for saying that classical mechanics is deterministic even if no general definition of a deterministic theory has been given. This theorem is:

R_{cl}. Let there be a classical-mechanical object Σ with phase space $\mathfrak{M}$ and Hamiltonian function H; let *on* denote some propositional function with a time t, a quantity f and a value α as arguments, which describes a possible time behaviour of Σ as specified in section II.2, i.e. which represents a change of state subject to Hamiltonian equations belonging to the given H as in subsection II.2.e, equations (1) and (2). Then the following are valid.

(α) For every time t_0, every set of canonically conjugate quantities $p_1, \ldots, q_n$, and every set of possible values $\alpha_1, \ldots, \beta_n$ of these quantities, there is just one propositional function *on* for which the propositions $on_{t_0}(p_1, \alpha_1), \ldots, on_{t_0}(q_n, \beta_n)$ are true.

(β) For every propositional function *on* there exist $t_0, p_1, \ldots, q_n$ and $\alpha_1, \ldots, \beta_n$ as specified in (α), such that the propositions mentioned in (α) are true.

The significance of this theorem R_{cl} is evidently that it reduces the ensemble of propositions $on_t(f, \alpha)$ describing any time behaviour of Σ to a very small subset. For the finite set of propositions mentioned in (α) in relation to the data $t_0, p_1, \ldots, q_n, \alpha_1, \ldots, \beta_n$, we *may* require that all these propositions are true, in the sense that there is then always a propositional function *on* such that these particular assertions are true. But if this requirement *is* imposed, there exists only one such *on*; thus the *finite* number of propositions of the form $on_t(f, \alpha)$ stated by (α) to be true determine for *every* time t, quantity f and value α (and determine logically in combination with certain relevant laws) whether $on_t(f, \alpha)$ with these arguments is true or not. The further assertion (β), which in the present case is trivially true, ensures that the reducing parameters $t_0, p_1, \ldots, q_n, \alpha_1, \ldots, \beta_n$ together with the reducing condition given in (α) have a sufficient range of operation, i.e. that the reduction embraces every possible propositional function *on* and therefore every possible behaviour of Σ.

From this viewpoint, R_{cl} is therefore a solution to the following reduction problem of classical mechanics.

RP_{cl}. Under the assumptions stated in R_{cl}, find a set $\mathfrak{R}$ of reducing data and a reducing assignment $\mathfrak{r} \rightarrowtail \Delta\mathfrak{r}$ which assigns to every $\mathfrak{r} \in \mathfrak{R}$ a minimal subset $\Delta\mathfrak{r}$ of the region of definition of a propositional function *on*, in such a way that

(α) for any given $\mathfrak{r} \in \mathfrak{R}$ there is just one *on* that is true for every set of three arguments in $\Delta\mathfrak{r}$;

(β) for any given propositional function *on* there exists some $\mathfrak{r} \in \mathfrak{R}$ such that *on* is true for every set of three arguments in $\Delta\mathfrak{r}$.

It is clear that this general problem has more than one solution, and accordingly that the above theorem R_{cl} is also only one of several possible answers. In R_{cl}, $\mathfrak{R}$ is evidently the set of all triplets of time instant, canonical system of quantities, and possible values for these, and the assignment $\mathfrak{r} \rightarrowtail \Delta\mathfrak{r}$ consists in assigning to $t_0, p_1, \ldots, q_n, \alpha_1, \ldots, \beta_n$ ($\equiv \mathfrak{r}$) the set $\{(t_0, p_1, \alpha_1), \ldots, (t_0, q_n, \beta_n)\}$ as a subset $\Delta\mathfrak{r}$ of the range of definition of a propositional function *on*. It is also clear, however, that R_{cl} is a particularly far-reaching solution of RP_{cl}: every reducing set of propositions (defined by a $\Delta\mathfrak{r}$) is finite and involves only one time.

This last remark leads to the question why classical mechanics is said to be deterministic *because* its reduction problem RP_{cl} is soluble by R_{cl}. Taking the word "deterministic" in its narrowest sense, this nomenclature has certainly been applied because R_{cl} gives the entire time variation of a classical-mechanical system Σ in terms of propositions valid for Σ at a single time. If the word "deterministic" is taken in its widest sense, approximately meaning "governed by laws", it suffices to point out that R_{cl} is a non-trivial solution of RP_{cl}; for every solution of the reduction problem must be based on the fact that the solution makes use of certain laws of mechanics, in the sense that from them the truth value of certain contingent propositions about Σ is logically determined by the truth values of certain other such propositions. Without seeking to specify here a position with respect to the resulting range of possible applications of the word "deterministic", we should mention at this point that the assessment of the indeterministic nature of classical statistical mechanics, and still more of quantum mechanics, will depend greatly on which such position is taken.

In addition to this intrinsic ambiguity, it should also be pointed out here that there are other ways of approaching the problem of determinism and indeterminism, besides the viewpoint adopted in this section. When proceeding to classical statistical mechanics and to quantum mechanics, we shall, in respect of our present viewpoint, merely have to ask how a reduction problem analogous to RP_{cl} can be formulated in these theories, and what are its solutions. The main point of comparison in this analogy will be that in these two theories also there exist contingent propositions, namely those regarding measurements and regarding probabilities, and that there are again laws which are logically related to the contingent propositions. We have then to ascertain the interrelation of these in the reduction problem, finding that, just as R_{cl} reveals certain relationships between the time-dependent propositions regarding states in ordinary classical mechanics, analogous solutions of the reduction problem in classical statistical mechanics and quantum mechanics indicate certain relationships between *their* contingent propositions. But nothing is discovered, at least directly, about the relationship between these propositions and the classical propositions regarding states.

Here we encounter a new aspect, the relationship between contingent propositions in two different theories, in contrast to the relationship between contingent propositions in the same theory, which have alone been considered until now (as in the reduction problem). The distinction between these two approaches is often not made with sufficient clarity. In the present context, this is shown, for example, by the fact that many interpreters of quantum mechanics (mainly in the orthodox camp) consider the appearance of propositions regarding probabilities in this theory as justifying the view that quantum mechanics is indeterministic; whereas another group of interpreters (often among the adherents of dialectical materialism) consider the deterministic behaviour of the time dependence (in a sense to be further specified below) of these propositions regarding probabilities as affording preference to the view that quantum mechanics is deterministic. There is in reality no contradiction here, of course, but only an unjustifiable equivocation which results from ignoring just the distinction mentioned above. If quantum mechanics is held to be indeterministic because propositions regarding probabilities play an essential role (or any role) in it, this is evidently in the sense of evaluating the relationship between the quantum-mechanical propositions regarding probabilities and the classical propositions regarding states. The conclusion could not be disproved by anything concerning the mutual relationship of the quantum-mechanical propositions (whether or not this may be similar to the corresponding relationship of the classical propositions). Likewise, if quantum mechanics is held to be deterministic because the time dependence of the propositions regarding probabilities is deterministic, this is evidently in the sense of evaluating the mutual relationship between these propositions only. The conclusion is unaffected by emphasizing any difference between the propositions regarding probabilities and the classical propositions regarding states.

The two viewpoints in question are evidently such that, if an appropriate basis of comparison is assumed, then e.g. a distinction between the mutual relationship of the contingent propositions in quantum mechanics and that of the classical propositions regarding states leads indirectly to a conclusion as to the relationship between the two types of proposition. For example, if it were found that the time dependence of the quantum-mechanical propositions regarding probabilities (in the sense specified above, of a reduction of these propositions to a single time) is deterministic (which in fact is true only under certain restrictions), then anyone who described quantum mechanics as indeterministic, because of the propositions regarding probabilities which occur in it, would have to propound some other basis for understanding this indeterminism, e.g. that of a direct comparison between the quantum-mechanical propositions regarding probabilities and the classical propositions regarding states, as already mentioned. On the other hand, if it were found that reduction to a single time is not possible for the time dependence of the quantum-mechanical propositions, this in itself would be sufficient grounds for considering that quantum mechanics involves indeterminism. The situation as a whole will be more complicated than can be shown by discussing a single aspect; the only purpose here has been to describe the ideas underlying the reduction problem (with which alone this section deals) sufficiently clearly to provide a basis for indicating how it might be compared with other possibilities in analysing the subject of determinism and indeterminism.

Let us now consider the reduction problem for classical statistical mechanics and quantum mechanics. The problem and its solution are found to be similar for both theories, so that it will be sufficient to give the discussion explicitly for quantum mechanics only, calling attention to any differences from classical statistical mechanics. As in all the treatments of orthodox quantum mechanics distinguished in Chapter II, the contingent relationships of a single object, or of a statistical ensemble (which will have to be modified in the course of time on account of any measurements that are made), are fully represented only by several contingent functions. The mere fact that, unlike classical mechanics, there are here several functions and not one function, suggests that the solutions of the reduction problem will be less straightforward than the solution R_{cl} of RP_{cl}, for example. For our purposes, however, it is both clearer and adequate to specify only parts of the complete solutions. There is no fundamental problem in putting these parts together, and this need not be discussed in detail here. The idea of breaking down the problem makes necessary a more liberal formulation of the problem, so as to include also partial functions derived from the contingent functions to be reduced. We shall use the following formulation.

RP_{qu}. Reduce an ensemble of groups $(\mathfrak{F}_1, \ldots, \mathfrak{F}_n)$ of n contingent functions $\mathfrak{F}_\nu$, which are subject to certain conditions, in particular those given in sections II.4 and II.5 as general requirements in quantum mechanics with a given Hilbert space $\mathfrak{H}$ and (where appropriate) a given Hamiltonian operator H, but possibly also other conditions which are themselves contingent. The problem is to find a set $\mathfrak{R}$ of reducing data and a reducing assignment $\mathfrak{r} \rightarrowtail (\mathfrak{F}_1|\Delta\mathfrak{r}^1, \ldots, \mathfrak{F}_n|\Delta\mathfrak{r}^n)$, where $\Delta\mathfrak{r}^\nu$ is a (minimal) subset of the region of definition of $\mathfrak{F}_\nu$, and $\mathfrak{F}_\nu|\Delta\mathfrak{r}^\nu$ is a function in $\Delta\mathfrak{r}^\nu$, in such a way that

(α) for any $\mathfrak{r} \in \mathfrak{R}$ there is just one group $(\mathfrak{F}_1, \ldots, \mathfrak{F}_n)$ for which each function $\mathfrak{F}_\nu$ coincides with $\mathfrak{F}_\nu|\Delta\mathfrak{r}^\nu$ in $\Delta\mathfrak{r}^\nu$;

(β) for each group $(\mathfrak{F}_1, \ldots, \mathfrak{F}_n)$ there exists some $\mathfrak{r} \in \mathfrak{R}$ such that each function $\mathfrak{F}_\nu$ coincides with $\mathfrak{F}_\nu|\Delta\mathfrak{r}^\nu$ in $\Delta\mathfrak{r}^\nu$.

In contrast to RP_{cl}, we here speak simply of functions, not of propositional functions only. This difference is unimportant, however, and merely allows us to consider probability functions directly. All the functions which occur can be converted in an obvious manner into propositional functions. Moreover the reduction here is not only to subranges $\Delta\mathfrak{r}^\nu$ of the ranges of definition of the $\mathfrak{F}_\nu$, but to functions that are to be specified there. For propositional functions this is a generalization of the case treated in RP_{cl}, to the extent that there the reduction condition was that the propositional functions to be reduced should be true throughout the reducing subranges of their respective ranges of definition, whereas here we allow them to be true for some arguments and false for other arguments.

As a first example of a solution of RP_{qu}, we shall give a case which concerns only propositions regarding measurements, to be made at a single time, and for which the time behaviour therefore need not be made explicit.

R^1_{qu}. Let the set of propositional functions to be reduced be the ensemble of all propositional functions M regarding measurements and dependent on a quantity A, an accuracy $\mathfrak{a}$ and a set of values a, for which there is a maximum measurement *me* with $M(A, \mathfrak{a}, a) \leftrightarrow \mathbf{M}(me; A, \mathfrak{a}, a)$ for all A, $\mathfrak{a}$ and a, with $\mathbf{M}$ the universal propositional function regarding measurements for a single object, used in subsection II.4(d)(i).

(α_1) For every quantity A_0 having a pure non-degenerate point spectrum and every eigenvalue α_0 of A_0 there is just one propositional function M such that the proposition $M(A_0, \{\alpha\}_\alpha, \{\alpha_0\})$ is true with $\{\alpha\}_\alpha$ the spectrum of A_0.

(β_1) For each of the propositional functions M there is a quantity A_0 having a pure non-degenerate point spectrum and an eigenvalue α_0 such that $M(A_0, \{\alpha\}_\alpha, \{\alpha_0\})$ is true.

This is a reduction of each set of propositions $M(A, \mathfrak{a}, a)$ with given M to a single true proposition. The part of (α_1) which refers to uniqueness states that, if a maximum measurement has been made of a quantity A_0 having a pure non-degenerate point spectrum, with the result α_0, then for every quantity A, every accuracy $\mathfrak{a}$ and every result a it is established whether this measurement has or has not measured the quantity A with accuracy $\mathfrak{a}$ and result a. From (β_1) it is evident that the reducing set of all pairs of the type (A_0, α_0), thus selected, is not at the same time suited to a corresponding reduction of all propositions regarding measurements, including non-maximum measurements. From $M(A_0, \{\alpha\}_\alpha, \{\alpha_0\})$, because of the equivalence to $\mathbf{M}(me, A_0, \{\alpha\}_\alpha, \{\alpha_0\})$, it follows that the measurement *me* is maximum. Hence, if $\mathbf{M}$ pertains to a non-maximum measurement, the existence statement in (β_1) is false. This illuminates the significance of the added statement, which often is trivially true. No generalization of R^1_{qu} to all propositions regarding measurements is known.

The theorem R^1_{qu} corresponds exactly to the following theorem about propositions regarding probabilities:

R^2_{qu}. Let the set of functions to be reduced be the set of all probability functions p, depending on a quantity A and a set of values a, for which there is a state of maximum knowledge or a pure statistical state or simply a state ϕ (according to the basic formulation used) such that $p(A, a) = \mathbf{pr}(\phi; A, a)$ for all A and a, $\mathbf{pr}$ being the universal probability function defined in subsection II.4(d)(ii) or section II.5.

(α_2) For every quantity A_0 having a pure non-degenerate point spectrum and every eigenvalue α_0 of A_0 there is just one probability function p such that $p(A_0, \{\alpha_0\}) = 1$.

(β_2) For every probability function p there is a quantity A_0 having a pure non-degenerate point spectrum and an eigenvalue α_0 of A_0 such that $p(A_0, \{\alpha_0\}) = 1$.

The discussion of R^1_{qu} above applies to R^2_{qu} with appropriate modifications, and therefore need not be repeated. Neither of the two reductions is valid for classical statistical mechanics, since the latter has no maximum measurements and no states of maximum knowledge or pure statistical states (cf. subsections II.3(d)(i) and (ii)).

R^1_{qu} and R^2_{qu} refer only to the situation which can exist at a single time. They are therefore useful in solving the problem of the determinism of quantum mechanics only if a fairly wide class of law-like relations are said to be deterministic, and not just those which establish unambiguous connections between states at different times (cf. the distinction made previously). In a certain sense, this broad application of the word "deterministic" is entirely justified. For example, in R^2_{qu} certain probabilities "determine" other probabilities relating to the same instant. Similarly, we shall see in R^3_{qu} that (with some limitations) certain probabilities relating to a time t_0 "determine" other probabilities relating to a later time t_1. If now we ask what, in the two cases, is the cause of something's being determined, then there is a possible answer common to both cases, as follows: the data concerned can be supplemented by further data in such a way that a purely logical determination results. These further data will of course be part of a whole theory. For example, if we know from the theory of the quantum-mechanical harmonic oscillator that the energy of such an oscillator has its lowest eigenvalue at time t_0, it is logically impossible for the position probability in the range a at that time to be other than $\|P_a^X \phi_0\|^2$, where ϕ_0 describes the ground state. On the same supposition, it is likewise impossible for the energy of the *closed* system at a later time t_1 to be other than the lowest energy eigenvalue. This shared property of simultaneous and non-simultaneous determination is of course a very general one. The question whether cases of determination with time variation have some specific additional feature, in comparison with those where time variation does not occur, need not be discussed here; such cases have to be considered in quantum mechanics also, so as to be able to compare them with the corresponding classical case contained in R_{cl}.

A first partial solution to the reduction problem for time variations is:

R_{qu}^3. We have to reduce groups of three functions *Me, pr, pr*$^+$ which describe, as in subsection II.4(e), the time behaviour of an object or statistical ensemble together with the measurements performed and results obtained in the time interval $[t_1, t_2]$. It will also be assumed that *Me* is trivial throughout $[t_1, t_2]$.

(α_3) For every $t_0 \in [t_1, t_2]$ and every description of state W_0 there is just one group (Me, pr, pr^+) of the type just defined, such that $pr_{t_0}(A, a) = \mathbf{pr}(W_0; A, a)$ for all A and a.

(β_3) For every group (Me, pr, pr^+) of the type just defined, there is a $t_0 \in [t_1, t_2]$ and a description of state W_0 such that the reduction condition specified in (α_3) is satisfied.

This theorem, which is easily proved from the conditions stated in section II.4, contains the determinism, as regards time, of the quantum-mechanical propositions regarding probabilities, which occurs for a specified Hamiltonian operator. The first point to note is that here the specified W_0 connects *all* the probabilities existing at time t_0 with the given data, and the reduction therefore takes place *only* in respect of time.

What is particularly important, however, is that according to R_{qu}^3 nothing is determined for the time t simply by the data relating only to the specified time t_0. Such a determination might be supposed to occur because in (α_3) the specification (of the description of state W_0) appears only for t_0. Actually, however, the ensemble of functions to be reduced is limited *a priori*, as regards *Me*, to the case where these functions are trivial throughout the time interval $[t_1, t_2]$. If this restriction is omitted, we cannot say that a W_0 assumed for t_0 implies anything about a time other than t_0. This is, broadly speaking, because the probabilities at one time are determined by the probabilities existing at another time (apart from dynamics) only through measurements made and results obtained in the intervening period, and (1) making a measurement at a time t is not dependent on making a measurement at any other time, (2) even if the making of measurements at various times were included in the data, their results would still be indeterminate. In orthodox quantum mechanics, therefore, there is strictly speaking no determinism in the narrower sense, nor is there in the modifications of classical mechanics described in section II.3. The most that can be said is that there is determinism in respect of propositions regarding probabilities, provided that no measurement intervenes, and precisely this is said in R_{qu}^3.

Although this fundamental situation remains unchanged, we can of course examine the mechanism of mutual dependence of contingent propositions further than is done in R^3_{qu}, by making more general the special assumption about the propositions regarding measurements. As an example, the following is the next step in this direction, allowing at least one non-trivial measurement:

R^4_{qu}. We have again to reduce groups of three functions *Me*, *pr*, pr^+, which are subject to the same conditions as in R^3_{qu}, but the additional assumption is weakened to the extent that *Me* is allowed to be non-trivial at not more than one specified t_0' in the specified interval $[t_1, t_2]$.

(α_4) For every $t_0 \in [t_1, t_2]$ with $t_0 < t_0'$ and every description of state W_0 and measurement me_0' whose result P'_{i_0} is such that

$$\mathrm{Tr}(U_{t_0'-t_0} W_0 U^{-1}_{t_0'-t_0} \cdot P'_{i_0}) > 0, \tag{1}$$

there is just one group (*Me*, *pr*, pr^+) of the type defined above, such that

$$\left.\begin{aligned} pr_{t_0}(A, a) &= \mathbf{pr}(W_0; A, a) \\ Me_{t_0'}(A, \mathfrak{a}, a) &\leftrightarrow \mathbf{M}(me_0'; A, \mathfrak{a}, a) \end{aligned}\right\} \tag{2}$$

for all arguments A, $\mathfrak{a}$ and a.

(β_4) For every group (*Me*, *pr*, pr^+) of the type defined above there is a $t_0 \in [t_1, t_2]$ with $t_0 < t_0'$, and a W_0 and me_0' satisfying (1), such that the reduction condition (2) is satisfied.

An important point is that R^4_{qu} makes reference to a direction of time. If $t_0 < t_0'$, then (γ) and (δ) in subsection II.4(e) in fact make *pr* and pr^+ unambiguous at times after t_0'. But if $t_0 > t_0'$, these functions would not be unambiguous at times before t_0', since the specified description of state W_0 would refer to a time later than (or simultaneous with) the time of me_0', and then the description of state before the measurement could not be reconstructed. If $t_0 < t_0'$ we have the expected (though not ambiguous) dependence of the description of state after t_0' on the result given by the measurement. If the result were not included, we could not deduce anything regarding the times after t_0' from the W_0 specified for t_0.

The discussion given in this section may be very briefly summarized as follows. In the widest sense of the word "deterministic", which refers only to the establishment of certain laws whereby certain contingent propositions as data determine (logically) other contingent propositions, i.e. the latter are reduced to the former, classical statistical mechanics and quantum mechanics are just as deterministic as ordinary classical mechanics. In the narrower sense of the word "deterministic", where the reduction just mentioned allows the whole evolution to be brought back to a single time, these theories are not deterministic. No specific difference between quantum mechanics and classical statistical mechanics has appeared. In the next two sections we shall discover such differences on the basis of other means of assessing indeterminism.

IV.2 Indeterminism, non-simultaneous measurability, and Heisenberg relations

At the time when quantum mechanics was established, Heisenberg drew attention to the fact that the indeterminism of quantum mechanics lies in its *instantaneous* descriptions of state: "We have not assumed that the quantum theory, unlike the classical theory, is an essentially statistical theory in the sense that only statistical conclusions can be drawn from exactly specified data. . . . But in the incisive formulation of the law of causality: 'If we know the present exactly, we can calculate the future' it is the premise and not the conclusion which is false. It is in principle impossible to discover every determining component of the present" (Heisenberg 1927, p. 197). Schrödinger too emphasized this point: "Note that no reference has been made to time variations. It would not help to allow the model to change in a totally 'unclassical' manner, for example by 'jumping'. This will not work even for one instant" (Schrödinger 1935a, §4). Evidently Heisenberg's words allude chiefly to two quite different kinds of problem. In the classical causality law (here given in the epistemic formulation) we have the reduction problem discussed in section 1, and in particular in the narrower sense of the reduction of time variations to a single time. But the assertion that in this law "it is the premise and not the conclusion which is false" refers not to the problem of relative determinations but to the problem of the sense in which, and possibly the reason why, the premise of the law is false in quantum mechanics. In section 1 we have seen how the reduction problem can be stated and solved even in quantum mechanics, and how quantum mechanics is found to be indeterministic as regards this problem: the reduction of time variations to a single time is impossible. The point of the quotations above, however, is to refer to the other problem, that of the peculiarity of instantaneous descriptions of state in quantum mechanics, and an indeterminism which may be related to this. We have to discuss not the question of how propositions valid at one time can be reduced to propositions valid at another time or how the latter determine the former but the fact that the descriptions of state regarded as possible at a particular time are themselves incomplete in the sense of a direct comparison with the corresponding situation in exact classical mechanics.

In fact the uncertainty of propositions regarding probabilities can be quite clearly distinguished from the certainty of classical propositions regarding states. A deduction from the postulates of sections II.4 and II.5, which has already been mentioned in passing but now appears very explicitly, is that any state of knowledge, any statistical state and any individual state involves genuine propositions regarding probabilities, i.e. propositions for which the probabilities are neither zero nor unity. Similarly, no single measurement can provide a complete state characterization, i.e. one which assigns definite values to all quantities. For this reason also one *can* say that quantum mechanics is indeterministic, provided that due notice is taken of the warning in section 1 that the term is then used in a different sense from the one described in that section and repeated above. The occurrence of propositions regarding probabilities and incomplete propositions regarding measurements may be more profound than the irreducibility of time variations to a single time, in that one is inclined to say that the latter is due to the former. But this relative assessment does not for the moment take us much further, since in the new sense classical statistical mechanics is also indeterministic: here too we deduce from the postulates in section II.3 that any state of knowledge and any statistical state involves genuine probabilities, and that no measurement is complete. Thus both theories are indeterministic in the sense of section 1 and in the sense just defined. There can still exist the question as to the relative justification of the two kinds of indeterminism, but this will tell us nothing about any difference between quantum mechanics and its classical counterpart: both theories are, as regards the criteria so far advanced, equally far removed from exact classical mechanics.

Let us look, then, for specific differences. One is that in the epistemic formulation of quantum mechanics there are states of maximum knowledge, in its statistical formulation there are corresponding pure statistical states, and in both formulations there are maximum measurements, whereas none of these exists in the corresponding formulations of classical statistical mechanics; cf. subsections II.3(d)(i), (ii), II.4(d)(i), (ii). These descriptions of state are not remarkable in themselves, but in the present case they are incomplete in the sense defined above, and a state of maximum knowledge which is incomplete (because probabilistic) *is* remarkable: no improvement of it is allowed by the theory, although there is a probabilistic incompleteness. Similarly for the pure statistical states, which are not dispersion-free. From classical probability theory it is known that randomly selected subensembles are always in the same state as the original ensemble. But subensembles selected on the basis of results of measurements may be in a different state which behaves in some respect more uniformly than the original state. In quantum mechanics, however, for a pure statistical state there is *no* subensemble differing in its statistical behaviour from the original ensemble. Lastly, there are maximum measurements which are incomplete, and this also is classically quite unintelligible.

The occurrence of maximum, but classically incomplete, measurements and descriptions of state in the epistemic and statistical treatments of quantum mechanics foreshadows an ontological peculiarity which does not appear explicitly in these formulations but which plays an important part in the ontic treatment of quantum mechanics. The latter relates primarily to states of individual objects, not to states of knowledge or statistical ensembles. The corresponding propositions regarding states, however, are probabilistic (in contrast to classical mechanics) and are ontically interpreted in the sense that they express an indeterminacy as to the *actual* conditions existing in the state of an *individual* object (cf. section II.5). This indeterminacy is an essential part of reality as conceived in the ontic interpretation, and the corresponding propositions regarding probabilities express one such reality completely. The indeterminism in question, characterized by the features described above, thus acquires a positive aspect in the ontic interpretation.

We come now to two further features which distinguish quantum mechanics from classical mechanics. The occurrence of maximum but incomplete measurements (of an individual object) must have the consequence that two such measurements are related in a classically impossible way: they cannot be made simultaneously, at least if (as in the orthodox view) the results of measurements are made the basis of new descriptions of state. A corresponding anomaly is to be expected in consequence of the occurrence of states of maximum knowledge or pure statistical states, which are likewise incomplete because probabilistic. In the Copenhagen interpretation of quantum mechanics, these features are embraced by the concept of non-simultaneous measurability of quantities or by generalizations of the Heisenberg relations. There has, however, been some lack of precision in the orthodox definition of these concepts and of their physical interpretation, especially as it is not always made clear whether this definition is to be based on a consistent Hilbert-space formulation of quantum mechanics or on a dualistic theory using both classical mechanics and field theory. This situation has given rise to confusion and criticism. Actually, a more exact formal analysis of the (von Neumann) Hilbert-space formulation shows that the features in question can be more clearly differentiated than might appear at first sight, and that they can be treated by a variety of theoretically distinct concepts, even if some of the distinctions are really academic. In the following this will be at least indicated; the explicit analysis will again be given only for quantum mechanics, and the classical results mentioned where they differ.

The differentiation of measurements from descriptions of state, introduced and made explicit in Chapter II, is the most important point to be borne in mind here. Let us first consider *measurements*. The meaning of saying that a measurement me_1 involves a measurement me has been given in subsection II.4(d)(i), (1). Both the measurements themselves and this relation between them included taking account of the results of measurements. We shall now ignore the results, and say that the measurement $\dot{m}e_1$ (ignoring the result; subsection II.4(d)(i), (2)) involves the measurement $\dot{m}e$ apart from the result if

$$\dot{\mathbf{M}}(\dot{m}e, A, \mathfrak{a}) \to \dot{\mathbf{M}}(\dot{m}e_1, A, \mathfrak{a}) \tag{1}$$

for all quantities A and accuracies $\mathfrak{a}$; cf. subsection II.4(d)(i), (4). The mathematical criterion for this is just (α_2) in subsection II.4(d)(i). If now $\mathfrak{M}$ is a set of measurements $\dot{m}e$ (ignoring the result), then we say that *the measurements in* $\mathfrak{M}$ *can be made simultaneously* if there is a measurement $\dot{m}e_1$ which involves all $\dot{m}e \in \mathfrak{M}$ apart from the result. This definition signifies that all $\dot{m}e \in \mathfrak{M}$ are made when $\dot{m}e_1$ is made. It can easily be shown that in classical statistical mechanics any finite number of measurements can be made simultaneously, but in quantum mechanics, for every non-trivial measurement, there is another measurement which cannot be made simultaneously with it. In quantum mechanics the mathematical description of simultaneous measurability for a *finite* $\mathfrak{M}$ is that

(α_1) $P_i P_j^1 = P_j^1 P_i$ for any two $\{P_i\}_i, \{P_j^1\}_j \in \mathfrak{M}$.

In classical statistical mechanics too, however, there are *infinite* sets of measurements which cannot be made simultaneously, a trivial example being the set of all measurements. This "weak" non-simultaneous feasibility of measurements is imposed already by the exclusion of measurements having absolute accuracy in the continuum (subsection II.3(d)(i)). The occurrence in quantum mechanics of finite sets of measurements which cannot be made simultaneously is evidently much more far-reaching and is unrelated to the continuum problem,

This demonstration of the distinction between quantum mechanics and classical statistical mechanics reveals the real core of the concept of non-simultaneous measurability. The topic is, however, usually discussed in terms of the simultaneous measurability of quantities, not in terms of the simultaneous making of measurements. This involves technical difficulties related to the multiplicity of the spectra of quantities, so that a variety of concepts of simultaneous measurability of quantities correspond to the one concept, just defined, of measurements which can be made simultaneously. In the formulation of quantum mechanics given in sections II.4 and II.5, the question whether two specified quantities can be simultaneously measured does not even have a theoretically unambiguous meaning, quite apart from ambiguities which may result from the interpretation. This is due to the explicit introduction of the concept of measurement in subsection II.4(d)(i), as distinct from the concept of a quantity. Von Neumann, who does not use an explicit concept of measurement, originally defined two quantities as being simultaneously measurable if there is a third quantity on which both are functionally dependent, and whose measurement consequently provides a simultaneous measurement of both the first two quantities (von Neumann 1927, §II). It must now be remembered, however, that according to the discussion in subsection II.4(d)(i) a quantity does not unambiguously define a measurement that measures it. Any measurement measures a quantity only with a certain accuracy; at best we can select from all measurements of a quantity having a point spectrum the measurement which measures that quantity most accurately. For continuous quantities this is not possible. Von Neumann (1955, section III.3) later gave a treatment of the problem which takes account of these facts, but which is burdened by other difficulties that we need not consider here. Using the basis provided by subsection II.4(d)(i), we can proceed as follows.

Let A be a quantity and $\dot{m}e$ a measurement. We assign unambiguously to these data a number $\Delta_{\dot{m}e}A$ in the range from zero to $+\infty$, representing the unsharpness with which the quantity A is measured by the measurement $\dot{m}e$. To do so, we first consider any accuracy $\mathfrak{a}_0$ such that $\dot{\mathbf{M}}(\dot{m}e, A, \mathfrak{a}_0)$ is valid. Such $\mathfrak{a}_0$ always exist, since this assertion can be made trivial, e.g. by putting $\mathfrak{a}_0 = \mathbf{R}$. Each $a \in \mathfrak{a}_0$ has an unambiguous length, viz. the lower bound of the length of all intervals which contain a. The upper bound of all such lengths can be unambiguously assigned to $\mathfrak{a}_0$. Let it be $\lambda(\mathfrak{a}_0)$. We now vary $\mathfrak{a}_0$, i.e. keep A and $\dot{m}e$ fixed and consider the non-empty set of all $\mathfrak{a}_0$ for which $\dot{\mathbf{M}}(\dot{m}e, A, \mathfrak{a}_0)$ is valid. Let the lower bound of all $\lambda(\mathfrak{a}_0)$ be $\Delta_{\dot{m}e}A$. This can be defined for any pair A and $\dot{m}e$. The definition evidently implies no special relation between A and $\dot{m}e$ needed to obtain $\Delta_{\dot{m}e}A$; on the contrary, it has the merit of providing relations between A and $\dot{m}e$.

In this sense we now define A as being *measured with absolute accuracy* by $\dot{m}e$ if $\Delta_{\dot{m}e}A = 0$, and say that A is *measured with limited accuracy* by $\dot{m}e$ if A is trivial, i.e. $\sim \lambda\mathbf{1}$, or if $\Delta_{\dot{m}e}A$ is less than the length of the spectrum of A. If the spectrum of A is infinitely long, this condition is satisfied if and only if $\Delta_{\dot{m}e}A$ is finite. If $\Delta_{\dot{m}e}A$ is itself infinite, it provides no further differentiation, and we have to say that A is *measured non-trivially* by $\dot{m}e$ if A is trivial, i.e. $\sim \lambda\mathbf{1}$, or if $\dot{\mathbf{M}}(\dot{m}e, A, \mathfrak{a})$ is valid for at least one $\mathfrak{a}$ consisting of at least two non-empty sets of values. In this case a true decision is clearly made even if $\Delta_{\dot{m}e}A = +\infty$. If A is measured with absolute accuracy by $\dot{m}e$, it is also measured with limited accuracy, and the latter similarly implies that it is measured non-trivially.

Let $\mathfrak{G}$ be a set of quantities. We say that the quantities in $\mathfrak{G}$ are *simultaneously measurable with absolute accuracy* if there exists a measurement $\dot{m}e$ such that all $A \in \mathfrak{G}$ are measured with absolute accuracy by $\dot{m}e$. The quantities in $\mathfrak{G}$ are *simultaneously measurable with arbitrary accuracy* if, for every positive function C in $\mathfrak{G}$ (i.e. such that $C_A > 0$ for all $A \in \mathfrak{G}$), there exists a measurement $\dot{m}e$ such that $\Delta_{\dot{m}e}A > C_A$ for all $A \in \mathfrak{G}$. The quantities in $\mathfrak{G}$ are *simultaneously measurable with limited accuracy* if there exists a measurement $\dot{m}e$ such that all $A \in \mathfrak{G}$ are measured with limited accuracy by $\dot{m}e$. Lastly, the quantities in $\mathfrak{G}$ are *simultaneously non-trivially measurable* if there exists a measurement $\dot{m}e$ such that all $A \in \mathfrak{G}$ are non-trivially measured by $\dot{m}e$. Each of these four concepts could be weakened by requiring the existence of the measurement $\dot{m}e$ only for a finite number of $A \in \mathfrak{G}$ (though these could otherwise be specified arbitrarily). This is called a *finite weakening* of the concepts.

We can begin to illustrate these concepts by listing a number of conclusions which follow from the definition, starting with the concept of quantities simultaneously measurable with absolute accuracy. A single quantity can be measured with absolute accuracy if and only if its operator has a pure point spectrum. A finite number of quantities can be simultaneously measured with absolute accuracy if and only if their operators have pure point spectra and commute pairwise (in the sense that the projectors of their spectral resolutions commute pairwise). Hence it follows that in quantum mechanics there are sets of as few as two quantities which are individually measurable with absolute accuracy, but not simultaneously. Standard examples are two components of spin or angular momentum. (It is trivially true that two quantities are not simultaneously measurable with absolute accuracy if one or both individually are not measurable with absolute accuracy.) In classical statistical mechanics, a finite number of quantities measurable with absolute accuracy are always simultaneously measurable with absolute accuracy. But there are infinite sets of quantities measurable with absolute accuracy which are not simultaneously measurable with absolute accuracy, for instance the set of all quantities measurable with absolute accuracy. If the finite weakening is included, then in quantum mechanics the quantities in any set $\mathfrak{G}$ are simultaneously measurable with absolute accuracy if and only if all the quantities in $\mathfrak{G}$ have pure point spectra and have operators which commute pairwise. In classical mechanics, every $\mathfrak{G}$ consisting of quantities having a pure point spectrum would then be a set of quantities simultaneously measurable with absolute accuracy.

A set of quantities simultaneously measurable with absolute accuracy is also a set of quantities simultaneously measurable with arbitrary accuracy, but the converse is not true. Any individual quantity can be measured with arbitrary accuracy, but not necessarily with absolute accuracy, for example not if it has a continuous spectrum. A finite number of quantities are simultaneously measurable with arbitrary accuracy if and only if their operators commute pairwise. Hence it follows that quantum mechanics likewise contains sets of two quantities which are not simultaneously measurable with arbitrary accuracy. The standard examples here are the conjugate position and momentum coordinates of a particle. If this brings to mind the Heisenberg relations, it must be noted that these are not at present under discussion. What is asserted is that the quantities in question are not simultaneously measurable with arbitrary accuracy, *as defined above*. In classical statistical mechanics, any finite number of quantities are simultaneously measurable with arbitrary accuracy, but there are again infinite sets for which this is not true, e.g. the set of all quantities. If the finite weakening is again included, then in quantum mechanics the commutability of the operators is generally necessary and sufficient for this kind of simultaneous measurability, while in the classical case there is then no restriction: any set of quantities can be simultaneously measured with arbitrary accuracy if the finite weakening applies. Whether the finite weakening is included or not, the concept of quantities which are simultaneously measurable with arbitrary accuracy implies no *relational* restriction in comparison with that of quantities which are simultaneously measurable with absolute accuracy: if the quantities in a set $\mathfrak{G}$ are individually measurable with absolute accuracy and together simultaneously measurable with arbitrary accuracy, then they are simultaneously measurable with absolute accuracy.

Every set of quantities simultaneously measurable with arbitrary accuracy is a set of quantities simultaneously measurable with limited accuracy, but the converse is not true. Examples of relevant cases will be given in section VII.2 in connection with the EPR paradox. It does not seem possible to construct an example from the usual quantities in quantum mechanics. Similar results are valid when we go finally to the quantities which are simultaneously non-trivially measurable. The analogous classical relations can again be analysed and compared with those of quantum mechanics, and we now see how these concepts of simultaneous measurability lead to a host of individual results which illuminate the finer distinctions resulting from the variety of spectra that occur. The most far-reaching result affecting quantum mechanics would be that the position and momentum coordinates of a particle furnish examples of quantities which are not even simultaneously non-trivially measurable, i.e. for which *no* simultaneous information can be obtained by measurement. This result is, of course, tied to the orthodox concept of measurement (quite apart from all matters of detail), as defined in subsection II.4(d)(i) and for which in particular measurements which reduce states (subsections II.4(e) and II.5) can be used to obtain new information relating to the future. This remark must not, however, be misconceived as implying that the non-simultaneous measurability of position and momentum coordinates is a consequence of the relationship between measurements and reductions of state. All that is asserted is that both are consequences of the concept of measurement here introduced on the basis of the orthodox view.

We now place alongside the hierarchy of concepts of simultaneous measurability of quantities described above another hierarchy of concepts referring not to measurements but to *descriptions of state*; these likewise can express an instantaneous indeterminism. When Heisenberg says, in the passage quoted at the beginning of this section, that "it is in principle impossible to discover every determining component of the present", there is still an ambiguity as to whether this inability relates to measurements or to the achievable descriptions of state. In orthodox quantum mechanics, it is true, the reduction of states is such as to ensure that measurements and descriptions of state agree in what can be or is achieved by them. But as pure concepts measurements and descriptions of state are basically different, and this must be taken into account. For all concepts which depend on descriptions of state, the various treatments of quantum mechanics distinguished in Chapter II should strictly be dealt with separately. In order to avoid having to use too laborious a terminology, the following concepts will be described in terms appropriate only to the epistemic treatment of quantum mechanics (without individual states), but the other two treatments (statistical and ontic) given in Chapter II will be kept in mind.

Let W be a state of knowledge and A a quantity of an object. As previously we assigned to a measurement $\dot{me}$ and a quantity A a number $\Delta_{\dot{me}}A$ which is a measure of the unsharpness with which A is measured by $\dot{me}$, we shall now define a number $\Delta^*_W A$ which is a measure of the *unsharpness* with which A is *known* when the state of knowledge is W. For this purpose we might use the range of dispersion as in ordinary probability theory, especially having regard to the statistical case. There is, however, another measure which is more closely related to the unsharpness $\Delta_{\dot{me}}A$ and more suitable for an initial comparison, namely the lower bound of the lengths of all sets of values a for which $\mathbf{pr}(W; A, a) = 1$. Let this number be $\Delta^*_W A$. As above, we can then give successive definitions: A is *known with absolute accuracy in* W if $\Delta^*_W A = 0$, and *known with limited accuracy* if $\Delta^*_W A$ is less than the length of the spectrum of A for a non-trivial A. For a spectrum of infinite length, the weakest case cannot be dealt with by means of $\Delta^*_W A$. We can then say that A is *known non-trivially in* W if there exists a set of values a with $\mathbf{pr}(W; A, a) = 1$ and $\mathbf{pr}(W_1; A, \mathbf{R}-a) = 1$ for a suitably chosen state of knowledge W_1. In this case one thus always has more knowledge than corresponds to the *a priori* knowledge of A.

Now let $\mathfrak{G}$ be a set of quantities. The $A \in \mathfrak{G}$ can be *simultaneously known with absolute accuracy* if there is a state of maximum knowledge ϕ such that all $A \in \mathfrak{G}$ are known with absolute accuracy in ϕ. They can be *simultaneously known with arbitrary accuracy* if, for every positive function C in $\mathfrak{G}$, there is a ϕ such that $\Delta^*_\phi A < C_A$ for all $A \in \mathfrak{G}$. They can be *simultaneously known with limited accuracy* or *non-trivially* if there is a ϕ such that all $A \in \mathfrak{G}$ are known in ϕ with limited accuracy or non-trivially, respectively. These concepts are evidently inclusive in the order given. It is easy to find examples to show that they are not necessarily inclusive in the opposite order. For the first pair, we need only take the set of functions of a quantity having a continuous spectrum. For the second pair, we again refer to section VII.2. For the third pair, we use the same construction as in that section, but with spectra of infinite length and with continuous functions having infinitely degenerate eigenvalues. Finally, in that construction there can be two orthonormal bases of which no two subsets span a common subspace. These easily yield two quantities which cannot be simultaneously non-trivially known. The position and momentum coordinates of a particle also provide quantities which cannot be simultaneously non-trivially known. For example, if for a certain ϕ there is a momentum distribution which is 1 in a true part of momentum space and 0 elsewhere, the same ϕ gives a position distribution which does not have the probability 1 for any true part of position space.

There is a cross-link between the concepts of simultaneous measurability and those of the possibility of simultaneous knowledge, in that the members of the former series include the corresponding members of the latter series. The proof is simple, and will be illustrated by means of the second concept in each series. Let the $A \in \mathfrak{G}$ be simultaneously measurable with arbitrary accuracy, and let C be a positive function in $\mathfrak{G}$. Then there exists a measurement $me = \{P_i\}_i$ with $\Delta_{\dot{me}} A < C_A$ for all $A \in \mathfrak{G}$. By the definition of $\Delta_{\dot{me}} A$ (see above), there then exists for each A an accuracy $\mathfrak{a}$ with $\dot{\mathbf{M}}(\dot{me}, A, \mathfrak{a})$ and $\lambda(\mathfrak{a}) < C_A$, since $\Delta_{\dot{me}} A$ is the lower bound of the $\lambda(\mathfrak{a})$. For all $a \in \mathfrak{a}$ we then have $l(a) < C_A$, since $\lambda(\mathfrak{a})$ is the upper bound of the lengths $l(a)$. We now take some j and a corresponding ϕ such that $P_j \phi = \phi$. The $\mathfrak{a}$ selected for A defines a resolution of $\{P_i\}_i$ in the form

$$P_a^A = \Sigma_{i \in I_a} P_i.$$

P_j occurs in one of these sums, and, for the corresponding a, $\mathbf{pr}(\phi; A, a) = 1$. Since for this a in particular $l(a) < C_A$ and $\Delta_\phi^* A \leqslant l(a)$, it follows that $\Delta_\phi^* A < C_A$, as was to be proved.

The considerable freedom in the choice of ϕ here shows that the relationships just proved by way of example cannot be inverted. The most familiar example is given by the three angular-momentum components of the electron in the hydrogen atom. As Condon (1929) noted, these cannot be simultaneously measured with absolute accuracy, but they are simultaneously known when the energy is known and the total angular momentum is known to be zero (both with absolute accuracy). The requirement of simultaneous measurability therefore goes further than that of the corresponding possibility of simultaneous knowledge, and accordingly the impossibility of simultaneous knowledge is more unusual than the impossibility of simultaneous measurement. In particular, the impossibility of position and momentum coordinates being simultaneously non-trivially known is the most far-reaching anomaly of quantum mechanics that can be imagined in this connection.

The unsharpness $\Delta_W^* A$ so far considered is not generally used. Its definition refers directly to the possibility that the probability $\mathbf{pr}(W; A, a) = 1$ for a certain set of values a whose length is less than that of the spectrum of the quantity A. In such a case, this number can therefore be used as a measure (in the epistemic interpretation) of the certain knowledge of a quantity. A probability distribution may be non-trivial for the whole spectrum and yet have a considerable concentration around one point. The unsharpness $\Delta_W^* A$ is not capable of representing such properties; the ordinary dispersion is then more suitable. This is, like $\Delta_W^* A$, a functional of the description of state and of the quantity concerned, and its interpretation is therefore unambiguously established by that of the description of state. In the epistemic interpretation, if there is a state of knowledge W concerning an object Σ, and if A is a quantity of Σ, then

$$\mathrm{Exp}(W, A) = \int_{-\infty}^{\infty} \alpha \, d(\mathbf{pr}(W; A, (-\infty, \alpha])) \tag{2}$$

is the familiar *expectation value* of A when the state of knowledge is W, $\mathbf{pr}(W; A, a)$ being the probability function corresponding to W. From (2) in subsection II.4(d)(ii) we then have

$$\mathrm{Exp}(W, A) = \mathrm{Tr}(WA). \tag{3}$$

(The mathematical difficulties in defining the right-hand side of (3) for arbitrary self-adjoint operators A will be ignored here. For bounded A, (3) can always be justified (Dixmier 1957, section I.6.1). For unbounded A, $\mathrm{Tr}(WA)$ need not exist.) Using the expectation value, we then define the (ordinary) *uncertainty* $\Delta_W A$ of A when the state of knowledge is W:

$$(\Delta_W A)^2 = \text{Exp}(W, (A - \text{Exp}(A))^2), \tag{4}$$

and this with (3) gives immediately

$$(\Delta_W A)^2 = \text{Tr}(WA^2) - (\text{Tr}(WA))^2. \tag{5}$$

For the position coordinates q and momentum coordinates p of a particle, Heisenberg (1930, section II.1) has interpreted $\Delta_W q$ and $\Delta_W p$ as the unsharpness of the knowledge of q or p in W. With the same formalism, $\Delta_W A$ can, of course, also be defined for the statistical formulation of quantum mechanics, and then signifies simply the *dispersion* of A about the *mean value* (2) for the relevant ensemble in the (statistical) state W. Finally, if ϕ is a state of an object Σ in the ontic interpretation, one can call (2) for $W = P_\phi$ the *fictitious value* of A in the state ϕ (fictitious because according to this interpretation A does not in general have a well-defined value, and the assignment must therefore be regarded as fictitious). Similarly, for $W = P_\phi$ (4) may be called the (ontic) *indeterminacy* of A in the state ϕ. It is a measure of the extent to which the value of A actually occurs. A fairly consistent interpretation of $\Delta_\phi A$ according to the ontic interpretation of quantum-mechanical probabilities is given by Süssmann (1963, especially §§1.b and 14.f).

The uncertainty $\Delta_W A$ can, like $\Delta_W^* A$ above, be made the basis of definitions which provide (for example, in the epistemic interpretation) new concepts of the possibility of simultaneous knowledge of quantities. For the cases of simultaneous knowledge of quantities A in a set $\mathfrak{G}$ with absolute, arbitrary or limited accuracy, we can use the previous treatment verbatim, simply replacing $\Delta_W^* A$ by $\Delta_W A$. The corresponding concepts will be distinguished from those obtained with $\Delta_W^* A$ by adding the phrase "in the wider sense", since the three concepts previously defined imply the corresponding concepts. This follows immediately from the inequality

$$\Delta_W A \leqslant \Delta_W^* A, \tag{6}$$

which may be proved as follows. For given W and A we first assume additionally that $\text{Exp}(W, A) = 0$. Then, let a be a set of values having $\mathbf{pr}(W; A, a) = 1$. Since $\mathbf{pr}(W; A, \mathbf{R}) = 1$, such an a always exists. Let $[\alpha, \beta]$ be the smallest (perhaps infinite) interval which covers a. Then $\mathbf{pr}(W; A, [\alpha, \beta]) = 1$, and so

$$\begin{aligned} \text{Exp}(W, A^2) &= \textstyle\int_{-\infty}^{\infty} \lambda^2 d\,(\mathbf{pr}(W; A, (-\infty, \lambda])) \\ &= \textstyle\int_{\alpha}^{\beta} \lambda^2 d\,(\mathbf{pr}(W; A\,(-\infty, \lambda])) \\ &= \lambda_1^2 \end{aligned}$$

with $\alpha \leqslant \lambda_1 \leqslant \beta$. Because of the special assumption that $\text{Exp}(W, A) = 0$, we have $\alpha \leqslant 0 \leqslant \beta$, and hence $\lambda_1^2 \leqslant (\beta - \alpha)^2$. For the same reason (cf. (4)) $(\Delta_W A)^2 = \text{Exp}(W, A^2)$. Finally, $\beta - \alpha$ is simply the length $l(a)$ of the set of values a. Thus we have $(\Delta_W A)^2 \leqslant l^2(a)$, and by taking the lower bound we get (6) for the special case $\text{Exp}(W, A) = 0$. In the general case, we change from A to $A - \text{Exp}(W, A)\mathbf{1} = A'$. Since $\text{Exp}(W, A') = 0$, (6) is valid for A' and W, and since this change obviously does not alter $\Delta_W A$ or $\Delta_W^* A$, (6) is valid in general.

From (6) we have immediately the result already stated: if the quantities in a set $\mathfrak{G}$ can be simultaneously known with absolute or arbitrary or limited accuracy, then the same is true in the wider sense. Together with earlier results, this shows *a fortiori* that, if the quantities in a set $\mathfrak{G}$ can be simultaneously measured with absolute or arbitrary or limited accuracy, they can also be simultaneously known with corresponding accuracy in the wider sense. In the descriptions of state, the various cases of possible simultaneous knowledge in the narrower and wider senses behave differently as regards the reversal of the implication stated above. For absolute accuracy it can be reversed, for limited accuracy it cannot. The latter fact is illustrated by the important case of the position and momentum coordinates of a particle. Two conjugate position and momentum coordinates cannot be simultaneously known with arbitrary accuracy (and therefore certainly not with absolute accuracy) in the wider sense. This is what is usually regarded as a direct deduction from the Heisenberg relations

$$\Delta_W p \cdot \Delta_W q \geqslant \tfrac{1}{2}\hbar. \tag{7}$$

But p and q *can* be simultaneously known with limited accuracy in the wider sense: it is sufficient for both $\Delta_W p$ and $\Delta_W q$ to be finite (for a spectrum of infinite length), and this can be achieved in a well-known manner. We have seen already that p and q cannot be simultaneously known with limited accuracy in the narrower sense. The latter point was probably first noted by von Neumann (1929, Introduction, §2). Since p and q moreover cannot even be simultaneously non-trivially known, we now see that the possibility of simultaneous knowledge with limited accuracy in the wider sense is very liberally formulated. On the orthodox view also this possibility of simultaneous knowledge of position and momentum does exist, and it may perhaps exist for any finite number of quantum-mechanical quantities.

The foregoing discussion has introduced a fair number of very general concepts which in the literature often appear only as a murky cloud within which the Heisenberg relations (7) may be perceptible. If we now take these relations as the basis and look back, the situation is as follows. The significance of the relations (7) depends first of all on that of $\Delta_W p$ and $\Delta_W q$. The answer to the latter question is easily envisaged from what everyone agrees on concerning the derivation of (7) in the formulation of orthodox quantum mechanics based on the Hilbert-space formalism: that this derivation must in principle follow a path first stated by Weyl (1931, Chapter II, §7), following a suggestion by Pauli. Omitting what is unimportant to our discussion, we can regard as belonging to this derivation the definition of $\Delta_W p$ and $\Delta_W q$ as given above (4) (with A replaced by the position or momentum operator, and taking the square root). Thus $\Delta_W p$ and $\Delta_W q$ are defined as the standard deviations (in the sense of mathematical probability theory) corresponding to a particular existing momentum or position distribution $\mathrm{Tr}(WP_a^p)$ or $\mathrm{Tr}(WP_a^q)$, and therefore depend on the existing description of state W as well as on p and q. Their interpretation thus leads back to the interpretation of the quantum-mechanical probabilities, and we then have a special case of the three possibilities already generally discussed above: $\Delta_W p$ or $\Delta_W q$ is either the inaccuracy in the state of knowledge of the momentum or position when the probabilistic state of knowledge is W, or the statistical dispersion of the momentum or position in the state W of a statistical ensemble, or the (objective) indeterminacy of these quantities in an ontically interpreted state $W = P_\phi$ of an individual object. From the Heisenberg relations (7) themselves, it then follows immediately that, according to the interpretation used, p and q cannot have values simultaneously known with arbitrary accuracy or dispersions which are simultaneously arbitrarily small or values simultaneously defined with arbitrary accuracy, all *in the wider sense*. Hence there follows the corresponding denials of possibility *in the narrower sense*, where the definition (see above) involves not the standard deviations $\Delta_W p$, $\Delta_W q$ but instead the unsharpness values $\Delta_W^* p$, $\Delta_W^* q$ defined previously. From this new circumstance, it then follows that p and q also cannot be simultaneously *measured* with arbitrary accuracy—a different statement, which involves not the concept of description of state but that of measurement. Corresponding to this line of argument, we have from each of the three statements another which refers not to arbitrary accuracy but to absolute accuracy. These six statements about positions and momenta follow, other things being equal, from the Heisenberg relations (7). But their content differs as between corresponding pairs, and also of course from that of (7). In addition, there are four further statements valid in relation to position and momentum coordinates in the present context: they cannot be simultaneously known non-trivially or with limited accuracy (in the narrower sense), and they are not simultaneously measurable non-trivially or with limited accuracy. These statements, however, are not consequences of the Heisenberg relations. Their proof depends on the relationship of position and momentum as Fourier transforms.

IV.3 Indeterminism and changes with time

The special features of quantum mechanics described in section 2 are joined by new ones as soon as we include time variations. These new features can be roughly divided into those which result simply from the dynamics of an object without the intervention of measurement, and those which depend entirely on the making of measurements. They will be discussed in that order.

From the time behaviour of propositions regarding probabilities, which (as we have seen in section 1) is deterministic on the assumption that there are no measurements, it is possible to derive a new viewpoint from which this behaviour appears indeterministic, both in classical statistical mechanics and in quantum mechanics. In the latter, the phenomenon concerned is known as the spreading of wave packets. This is generally an ergodic time behaviour of the probability functions. The phenomenon has been known in essence since the work of Duhem (1906) in relation to the critique of exact classical mechanics which is the basis of the classical statistical mechanics formulated in section II.3. The probabilistic aspect was certainly clearly stated by Koopman and von Neumann (1932, p. 256): "We will show that under this hypothesis all the initially observed properties of the system are obliterated by the lapse of time: the method of elementary mechanics of computing the final from the initial state must be replaced by the methods of the theory of probability. Contrary to the case of classical statistical mechanics, this situation is not dependent upon the system's having an enormous number of degrees of freedom: this number may perfectly well reduce to two". Born and Brillouin in particular regarded this situation as important in respect of quantum mechanics, holding that classical mechanics is itself indeterministic (Born 1955a, c, 1958, 1959, 1961; Born and Hooton 1955; Brillouin 1964, Part II; see also the discussion between von Laue (1955) and Born (1955b)). In quantum mechanics itself, any ergodic behaviour of the probability functions and the related indeterministic aspect is of minor importance, since here the indeterminism is much more unpleasantly noticeable in another form, viz. in measurements. However, a gradual spreading of the ψ-function in space, *when its phase is taken into account*, is very significant in quantum mechanics also, since it causes the familiar interference effects. This phenomenon cannot be expressed in the terms just used, in that these do not belong to the language of quantum mechanics specified in sections II.4 and II.5. We shall return to this point in section 4, and here merely indicate the relevant position in classical mechanics.

This position can be stated most briefly as follows. The probabilistic description of state for a mechanical object Σ (in the epistemic interpretation) is subject—in the absence of measurements—to a time determinism on the basis of the dynamics of Σ; that is, a probability distribution pr_{t_1} valid at time t_1 is transformed by the dynamics into a definite other distribution pr_{t_2} at time t_2. In this transformation, all the information contained in pr_{t_1} is preserved. But, in the following sense, there may be a decrease of the information restricted to a selected quantity f of Σ. It may happen that pr_{t_1} (f, a) is a delta function or nearly so; then we know for certain or almost for certain that Σ possesses a property $\{(f, a_0)\}^{\approx}$ and does not possess any property excluded by this. But pr_{t_2} may still give positive probabilities for all a. For instance, if f is a position coordinate for a system of free particles, it not only can but must happen that, after knowing f initially with fair accuracy, we know nothing about it when a certain time has elapsed. The standard example is that of a single particle moving on a finite segment of the x axis and elastically reflected at the ends of the segment. For such cases of increasing uncertainty of position, the reason is, of course, obviously the initial uncertainty of momentum. The only difficulty is that this uncertainty cannot be disposed of within the possibilities allowed here. If the momenta can be known only with arbitrary, not absolute, accuracy on the basis of the initial distribution (as is true here), the consequences as regards position are unalterable. The investigations mentioned above conclude that this spreading is the rule and not the exception, as regards both the dynamics and the particular quantity selected. An epistemically regarded determinism with quantitatively precise predictions of the contingent properties of an individual object is thus seen to be in general an illusion.

Let us now go on to the time variations of the probability functions due to measurements. It has been shown in section 1 that it is the dependence of these variations on measurements which makes it impossible to reduce the contingent propositions to a single time as in ordinary classical mechanics. Although classical statistical mechanics and quantum mechanics thus proved to be indeterministic in this sense, it is not too serious a matter, if the dependence in question can be made sufficiently understandable. For the classical case this is not difficult. In the epistemic interpretation, we say that a later state of knowledge depends not only on an earlier state of knowledge and on the relevant dynamics, but also on the new information acquired in the interval; and this new information is not determined by the earlier state of knowledge and the inherent dynamics of the object concerned, in accordance with the principles of any epistemology governed by what is humanly possible. Correspondingly, for the statistical interpretation, we do not wonder why and in what sense the new statistical state depends on the old state *and* on the result of measurement. We go from the initially given ensemble to another which is a selection from it defined by that result. It must be realized, however, that the situation is finally settled on the basis that the variations under discussion are *conservative* ones: states of certain knowledge are not lost, and values of quantities free from dispersion remain so, as follows immediately from (δ) in subsection II.3(e).

The situation is different in quantum mechanics; it will first be explained in the purely epistemic formulation. We have seen in subsection II.4(d)(ii) and section IV.2 that in this case there are states of knowledge such that they cannot be improved (in a sense that has been defined previously). They are described by the statistical operators, which have the form $W = P_\phi$, but these do not combine with the dynamics to decide future states of knowledge regardless of measurements which take place during the intervening time. This is because they too refer to truly probabilistic states of knowledge. For many quantities they do not comprise a reliable prediction as to values. The missing knowledge can be acquired by measurement, but not all at once, of course; we have to decide which quantities to measure whose values are still unknown. Then, as already stated in general, the mechanism of the reduction of states ((δ) in subsection II.4(e)) together with the result of the measurement leads unambiguously to a new state of knowledge $W^+ = P_{\phi+}$. The remarkable combination of maximumness and incompleteness in the original state of knowledge $W = P_\phi$ means, however, that the new state of knowledge is not an ordinary improvement of the old one. In fact the state of knowledge $W = P_\phi$ gave a reliable prediction of the value of any quantity for which ϕ is an eigenvector; in $W^+ = P_{\phi+}$, this information is lost, which must necessarily have ontological consequences. The mechanism of the reduction of states, which still operates flawlessly here—the unambiguous determination of a new state of knowledge W^+ by an earlier state W and a new result of measurement—and its epistemic interpretation which is here under discussion—the abrupt alteration of an existing state of knowledge by new information—must not be allowed to conceal the fact that the dependence of the state of knowledge W^+ on a result of measurement, when W is a state of maximum knowledge, is fundamentally different from the classical position, and heralds a new type of ontological indeterminism. We shall see in the subsequent analysis that it would be misleading to make a comparison (by way of actual apologetics) between the quantum-mechanical reduction of states in the epistemic interpretation and the corresponding process in classical statistical mechanics, in that in both "our knowledge of the system does change discontinuously" (Heisenberg 1930, section II.2(c); 1959, p. 38; Born 1955c, section I.3). The difference, emphasized above, between the classical and the quantum-mechanical reduction of states appears even in the epistemic interpretation of probability, as is clearly expressed by Pauli (1933, section A.1): "The influence of the momentum (or position) measuring apparatus on the system is such that . . . the earlier knowledge of the position (or momentum) cannot afterwards be used to predict the results of subsequent measurements of position (or momenta). . . . It will appear that this 'complementarity' has no analogue in the classical theory of gases, although the latter also makes use of statistical laws. That theory does not contain the statement, which depends on the finiteness of the quantum of action, that measurements on a system must sometimes necessarily destroy the knowledge of the system gained by earlier measurements, i.e. make this knowledge unusable. . . . As already mentioned, this does away with the possibility of making physical phenomena unambiguously objective, and therefore with the possibility of a causal description of these phenomena in space and time." The point has also been expressly mentioned by Schrödinger (1935a, §9): "There is no ψ-function which represents exactly the same propositions as another, and further propositions too. Hence, if the ψ-function of a system changes, either spontaneously or as a result of measurements, there must always be propositions missing from the new function which were present in the previous one. The catalogue cannot simply receive new entries; there must also be deletions. Now knowledge can be *gained*, but not *forfeited*. The deletions must therefore mean that propositions previously true are now false. A true proposition can become false only by a change in the *object* to which it refers."

The situation is entirely similar in the purely statistical interpretation of quantum mechanics. The states of maximum knowledge of an individual object Σ correspond to the pure states of an ensemble $\{\Sigma_k\}_k$ (cf. subsection II.4(d)(ii)). If $\{\Sigma_k\}_k$ is in the pure state $W = P_\phi$, it contains no subensemble in a state different from P_ϕ, but there may be many quantities showing dispersion. If now a measurement $(\{P_i\}_i, P_{i_0})$ is made on the ensemble, with $P_i\phi \nsim \phi$ for all i, then the ensemble defined by the reduction of states for i_0 is in the state $P_{i_0}\phi/\|P_{i_0}\phi\| \neq \phi$. Thus this ensemble is not, in the usual sense, a subensemble of $\{\Sigma_k\}_k$ in the state ϕ. Every quantity for which ϕ is a non-degenerate eigenvector shows dispersion, whereas such a quantity does not show dispersion in the state ϕ of $\{\Sigma_k\}_k$, nor therefore in any subensemble.

Here again, as already in the epistemic interpretation, we are apparently forced to the conclusion that the measurement must cause something to happen to the individual objects Σ_k, although this has only a statistical consequence in the present context.

In the ontic view of quantum mechanics, dynamics and reduction of states ensure, according to section II.5, that a later state of Σ is unambiguously determined by the initial state, the dynamics of Σ, and the results of measurements made during the intervening time. Here, however, the dependence of the later state on the results of measurements appears immediately as an anomaly of content. We have to say that the state of Σ is altered by a measurement, and in such a way that the new state is determined by the result itself, and is therefore largely indeterminate before the measurement. In the epistemic and statistical interpretations it was at least possible to explain initially why the probabilities are altered by measurement, although even then there were difficulties on closer examination. In the ontic interpretation, neither states of knowledge nor ensembles are concerned, but the objectively determined state of an individual object. How can this be said to alter when the result of a measurement is obtained?

The answer can be approached by reconsidering the situation, already indicated at the end of subsection II.4(e) and section II.5, which occurs in the reduction of states by a measurement whose result is ignored. In classical statistical mechanics it is regarded as entirely obvious that a state of knowledge or statistical state ϕ is unchanged by the mere act of making a measurement on Σ or $\{\Sigma_k\}_k$, without taking account of the result. This is taken to be obvious because, in the ideal case of a measurement completely free from perturbation, which we shall always assume here, the measurement does not cause any change in the object, and therefore our state of knowledge of the object, or the statistical state, can change only when the result of the measurement is taken into account. This assumption always made tacitly can be deduced by means of the changes which then occur. We shall do so on the basis of the formulation given in section III.2. If the state of knowledge or the statistical state corresponds to the probability function pr, and if the measurement $(\{E_i\}_i, E_{i_0})$ is made, then the subsequent probability function for any possible result E_i is

$$pr^+_{E_i}(F) = pr(E_i \cap F)/pr(E_i). \qquad (1)$$

Since this case in turn occurs with probability $pr(E_i)$ and some one of the E_i is certainly obtained as the result of the measurement, the existing state of knowledge if we do not know which E_i occurs or the statistical state without subsequent selection is

$$pr^+(F) = \Sigma_i pr(E_i) pr^+_{E_i}(F); \qquad (2)$$

and it can be shown immediately that

$$pr^+(F) = pr(F), \qquad (3)$$

i.e. no change has occurred. In particular, if two measurements are considered, it makes no difference which is performed, as regards the event associated with each of these measurements and the consequent effect on the object: from (3), the two weak reductions of states (2) associated with the measurements are the same. The only difference concerns the resulting knowledge of the object. In these circumstances there is in particular no need for a distinction between the strong and weak reductions of states.

The situation is quite different in quantum mechanics. As has already been mentioned at the end of subsection II.4(e), the weak reduction of states corresponding to the case (2) does not restore the former description of state: the mere performance of a measurement alters the description of state. For the ontic formulation also, in section II.5, equation (11), we in general have

$$\Sigma_{j\mu} w^{+}_{j\mu} P_{\phi+j\mu} \neq \Sigma_\lambda w_\lambda P_{\phi\lambda}.$$

In every case this is most clearly seen if, before the measurement, there existed a state of maximum knowledge or a pure statistical state ϕ or (in the ontic formulation) an individual state ϕ which was known or the same for all the objects in an ensemble. If then a maximum measurement $\dot{m}e = \{\phi_i\}_i$ is made, the weak reduction of states gives

$$W^\times = \Sigma_i |\langle\phi|\phi_i\rangle|^2 P_{\phi_i}, \tag{4}$$

and it is evident that in general $W^\times \neq P_\phi$. This is to be regarded as the most important difference from the classical case. It has the consequence that a distinction must be made between the strong and weak reductions of states: only the former fully reflects the situation defined by the process of measurement. A further consequence is that any two different measurements (and not only any two which cannot be made simultaneously) may have different effects on the object itself, not only on the existing state of knowledge of it. The strong reduction of states must therefore be distinguished in a very significant way from the mere acceptance or selection of a result of measurement. As Heisenberg pointed out as early as 1930, the process of measurement must be "divided into two sharply distinguished parts, the first of which is to subject the system to a physically real external interaction which alters the course of events. . . . In consequence, the system is brought into an assemblage of states, in general infinite in number. . . . The second part of the measurement selects as the actual state one particular state out of the infinity of states in the assemblage. This second part does not itself influence the course of the process, but merely alters our knowledge of the actual state of affairs" (Heisenberg 1930 (German edition), section IV.2). These words clearly relate to the epistemic extension of the ontic interpretation of quantum mechanics, since they refer both to individual states and to states of knowledge. The following points are to be successively noted in respect of the various formulations distinguished previously.

The measurement is first of all "an interaction which alters the course of events". If, in the purely epistemic or purely statistical interpretation, there is initially a state of knowledge or statistical state W, and if the measurement $\dot{m}e = \{P_i\}_i$ is made (individually or statistically; cf. subsection II.4(d)(i)), we have afterwards, first of all, the assemblage

$$\left\{\frac{P_i W P_i}{\mathrm{Tr}(WP_i)}, \mathrm{Tr}(WP_i)\right\}_{i:\mathrm{Tr}(WP_i)>0} \tag{5}$$

of states of knowledge or statistical states $P_i W P_i/\mathrm{Tr}(WP_i)$ with the probabilities $\mathrm{Tr}(WP_i)$, to be interpreted epistemically or statistically, as the case may be; this is the strong reduction of states (cf. the end of subsection II.4(e)). The change from W to the assemblage (5) expresses the real part of the process of measurement, the "interaction which can be neither neglected nor surveyed" (Bohr), which arises from the classical treatment of the measuring apparatus "with the quantum of action essentially excluded" (cf. Chapter I). This strong reduction of states cannot be replaced by the corresponding weak reduction, i.e. the change from W to

$$W^\times = \Sigma_i P_i W P_i, \tag{6}$$

since, as we shall see in sections VII.2 and VII.3, this replacement would generate too many quantities which immediately after the measurement had a definite value in $W^\times$, even if this were unknown or showed dispersion (likewise in $W^{\times}$). The difference between (5) and (6) is precisely that in (5), by the separate appearance of the $P_iWP_i/\mathrm{Tr}(WP_i)$, certain aspects of reality are expressed which in (6) are intermingled with other possibilities. The new aspects of reality brought about by the measurement are, however, not yet expressed even in (5), insofar as the individual object might be affected by the change. This can, of course, occur only in the ontic formulation, where, in the simplest case (cf. (8) in section II.5), the state ϕ of the individual object is converted by the real part of the measurement into the assemblage

$$\left\{\frac{P_i\phi}{\|P_i\phi\|}, \|P_i\phi\|^2\right\}_{i:\|P_i\phi\|^2 > 0} \tag{7}$$

of new individual states $P_i\phi$, which occur with the probabilities $\|P_i\phi\|^2$, to be interpreted epistemically or statistically as the case may be. More precisely, if the state ϕ of Σ is known before the measurement, then after completion of the real part of the measurement we know from (7) that Σ is in one of the states $P_i\phi/\|P_i\phi\|$, with probability $\|P_i\phi\|^2$. If we have a corresponding ensemble $\{\Sigma_k\}_k$, whose Σ_k are all in the state ϕ before the measurement, the (statistical) measurement transfers this ensemble into a new situation in which the states of the individual Σ_k are distributed in accordance with (7). The *individual* change of state does not consist in the fact that quantities which previously had definite values are caused by the measurement to take other values which are again objectively definite, but not known in advance because the measurement cannot be sufficiently surveyed. What the change *does* consist in is specified by the ontic interpretation of the concept of state. The object changes its potential tendencies to react in various ways to measurements—the degree of objective indeterminacy as regards the results of possible measurements. For the quantities actually measured, this is a transition from the possible to the actual, as in Heisenberg's discussion. In the ontic formulation too, we have finally to note that the weak reduction of states corresponding to (7) in general allows no conclusion as to the state in which the object is: when the components of (7) are combined in the form

$$W^{\times} = \Sigma_i\|P_i\phi\|^2 P_{P_i\phi}, \tag{8}$$

the aspects of reality defined by the measurement are again blurred.

The real part of the process of measurement is also evidenced by the fact that it is not a matter of indifference which of two different measurements is made, even if they can be made simultaneously. To see this, it is sufficient to use the weak reduction of states, for which

(α) if the weak reduction of states relating to a measurement $\dot{m}e$ converts every description of state into the same description as does the weak reduction of states relating to a measurement $\dot{m}e'$, then the two measurements are the same.

We have to show that the equation

$$\Sigma_i P_i W P_i = \Sigma_k Q_k W Q_k$$

for all W implies that $\{P_i\}_i$ and $\{Q_k\}_k$ are equal element by element. The proof can be expressed in such a way that the conclusion is reached in terms of cases $W = P_\phi$. Let some P_{i_0} be chosen. If $P_{i_0}\phi = \phi$, then with $W = P_\phi$ the above equation gives

$$P_\phi = \Sigma_k\|Q_k\phi\|^2 P_{Q_k\phi},$$

whence $Q_{k_0}\phi = \phi$ for just one k_0, $Q_k\phi = 0$ for all $k \neq k_0$. It is easily seen that within the eigenvectors ϕ of P_{i_0} this k_0 is independent of ϕ. Thus $P_{i_0}Q_{k_0} = Q_{k_0}P_{i_0} = P_{i_0}$ with a suitable k_0 for given i_0, and hence $\{Q_k\}_k$ is a coarsening of $\{P_i\}$. The converse is also true, from the same argument with P_i and Q_k interchanged. This proves (α). From (α) it follows that any two *different* measurements may lead to *complementary* phenomena as defined by Bohr. There is always some ϕ which leads to different weakly reduced descriptions of state. If one phenomenon is defined by ϕ, $\dot{m}e = \{P_i\}_i$ and $W^+ = \Sigma_i \|P_i\phi\|^2 P_{P_i\phi}$, then a change in the arrangement of the experiment so as to make the measurement $\dot{m}e' = \{Q_k\}_k$ would give the phenomenon ϕ, $\dot{m}e'$ and $W^{+\prime} = \Sigma_k \|Q_k\phi\|^2 P_{Q_k\phi}$; this would be different from the other, because $W^{+\prime} \neq W^+$, even if $\dot{m}e$ and $\dot{m}e'$ can be made simultaneously. In particular, even in this case the two measurements could be said to be mutually exclusive in Bohr's terminology. The difference between measurements that can be made simultaneously, those that cannot be made simultaneously, and certain extreme cases of the latter would then consist, roughly speaking, in the multiplicity of initial states that lead to the same results in the weak reduction of states: there would be respectively many, fewer, and no such initial states. Thus, for measurements that might be made simultaneously, we might have a complete orthogonal set of maximum descriptions of state of the kind in question (a common orthonormal basis through $\{P_i\}_i$ and $\{Q_k\}_k$). On the other hand, it might well be that no initial state of this kind exists, if the P_i describe position ranges and the Q_k momentum ranges of a particle. There seem to have been no precise analyses of this problem, nor has the question of incorporating the Bohr concept of the phenomenon into a Hilbert-space formulation of quantum mechanics yet been discussed; here we too shall not do more than indicate its existence.

The second step in measurement, as distinguished by Heisenberg, "does not itself influence the course of the process, but merely alters our knowledge of the actual state of affairs". Clearly it is this second step which is familiar from classical physics, where it is (in our epistemic or statistical view of classical mechanics) the only one involving any change, viz. the reduction of the state of knowledge or transition to a subensemble, in accordance with a particular result of measurement. In quantum mechanics the second step, which is sharply separated from the first, is also of a straightforward classical kind. It is simply a selection among the possibilities shown separately in (5) or (7) which has actually occurred: "The observation itself changes the probability function discontinuously; it selects of all possible events the actual one that has taken place. Since through the observation our knowledge of the system has changed discontinuously, its mathematical representation also has undergone the discontinuous change and we speak of a 'quantum jump'. If the old adage 'Natura non facit saltus' is used as a basis for criticism of quantum theory, we can reply that certainly our knowledge can change suddenly and that this fact justifies the use of the term 'quantum jump'" (Heisenberg 1959, p. 54). When the situation is stated in this manner, it must be realized that the sharp distinction between the two steps has been lost again, and misunderstandings may therefore occur. The actual "quantum jump" takes place as the first or real part of the measurement process. Its discontinuity is shown by the difference between the new description given by the strong reduction of states and the description which existed before the measurement. The second step is a "classical jump". Otherwise, "quantum jumps" would have occurred whenever anyone learnt something new, and quantum mechanics would not have been something new (see also Heisenberg 1955, p. 27).

IV.4 Illustrations

In section I.3, the initial presentation of Bohr's interpretation of quantum mechanics was illustrated by means of two diffraction experiments. We have now presented the orthodox view, making considerable use of the Hilbert-space formalism of quantum mechanics, and an illustration in terms of the same experiments is therefore appropriate here. Only the outline of a treatment on this new basis will be given. The reader is invited to use the examples to go through all the aspects mentioned in Chapters II and IV. This applies in particular to the three interpretations of quantum-mechanical propositions regarding states and to all the concepts derived from these interpretations.

First, let us consider again the one-hole experiment. For simplicity, the analysis will be restricted to two spatial dimensions (cf. Fig. 1, section I.3). Let the screens S and S_1 be theoretically infinite in both directions of the y axis, i.e. in practice much longer than the width of the slit in S. We shall also assume that the process can be roughly divided into four successive stages in time:

(1) behaviour of the particle before interacting with S,
(2) short-period interaction with S (passage through S),
(3) behaviour of the particle between interaction with S and interaction with S_1,
(4) interaction with S_1 (impinging on S_1).

These four stages will be discussed successively, in order to note their quantum-mechanical description.

(1) The description of state here is often taken to be a plane wave with an exact momentum in the positive x direction. This is, strictly speaking, impossible in a von Neumann formulation of quantum mechanics. Let us therefore consider a description of state ϕ which only approximates to this case and is valid for the time immediately preceding the passage of the particle through S. Without selecting any particular ϕ, we may state the essential conditions which it must satisfy. The expectation values must satisfy

$$\left.\begin{aligned} \operatorname{Exp}(\phi, p_x) &= \overline{p_x} > 0, \\ \operatorname{Exp}(\phi, p_y) &= 0, \end{aligned}\right\} \tag{1}$$

and the ordinary dispersions must satisfy

$$\left.\begin{aligned} \Delta_\phi p_x &\ll \overline{p_x}, \\ \Delta_\phi p_y &\approx 0. \end{aligned}\right\} \tag{2}$$

A further restriction could be imposed by assuming that

$$\Delta^*_\phi p_x, \Delta^*_\phi p_y < +\infty, \tag{3}$$

where $\Delta^*_\phi p_x$ and $\Delta^*_\phi p_y$ are the unsharpness values defined in section 2, whose relationship to the ordinary dispersions (see equation (6) in section 2) must be noted. With the assumptions (3), the momentum is restricted to a finite region of momentum space. As regards the position of the particle, this would imply that no part (and certainly not an infinite part) of position space can be excluded, and in particular that the particle may be in the region $x > 0$, i.e. to the right of S. In order to approach the (classical) idea of a passage of the particle, initially to the left of S, through the slit in S, we can also require that

$$\mathbf{pr}(\phi; X, \{x > 0\}_x) \approx 0, \tag{4}$$

without contradicting (1)–(3). If (3) is omitted we can even have equality in (4). In every case the consequence of (1)–(4) is that the position in the region $x < 0$ is largely indeterminate. The content of the conditions (1)–(4) for the three interpretations of ϕ (epistemic, statistical and ontic; cf. Chapter II) should be examined by the reader.

(2) The interaction of the particle with the screen S can take place in three basically different ways. It can be regarded as a two-body problem in a purely quantum-mechanical form; this would require in particular a quantum-mechanical treatment of S. We can also try to describe the action of S on the particle by a Hamiltonian operator, thus obtaining a Schrödinger problem for the particle alone. Finally, the process may be regarded as a measurement followed by a reduction of states. It is important to emphasize, as in subsection I.2(e), that none of these possibilities is excluded by the orthodox view or, in particular, by Bohr's view. The only possibility that is excluded is that the last of the three treatments can be completely replaced by one of the other two (cf. sections 2 and 3). We shall take the direct treatment as a measurement process, in order to be able to illustrate such a procedure in terms of the experiment in question, using the orthodox formulation. The situation then involves first of all a measurement $\dot{me}$ without reference to any result. If S occupies the space $a = [-x_0, +x_0]$ in the x direction and the slit in S occupies the space $b = [-y_0, +y_0]$ in the y direction, then $\dot{me}$ is described by

$$\{P_a^X P_b^Y, P_a^X P_{\mathbf{R}-b}^Y, P_{\mathbf{R}-a}^X\}. \tag{5}$$

This comprises the following possibilities:

$P_a^X P_b^Y$ passage through the slit in S

$P_a^X P_{\mathbf{R}-b}^Y$ absorption by S

$P_{\mathbf{R}-a}^X$ no reaction with S

The measurement $\dot{me}$ leads to a strong reduction of ϕ in which all these possibilities are preserved, although classically separated. They will not be written here, since the experimental arrangement distinguishes *de facto* the component represented by $P_a^X P_b^Y$: for an interaction, e.g., by striking the second screen S_1, the only particles possible in practice are those for which $P_a^X P_b^Y$ occurs. The others are either absorbed by S ($P_a^X P_{\mathbf{R}-b}^Y$) or are still to the left of S ($P_{\mathbf{R}-a}^X$), since the fraction of $P_{\mathbf{R}-a}^X\phi$ to the right of S is very small on account of (4). Thus we have *de facto* the selective measurement me described by

$$(\{P_a^X P_b^Y, P_a^X P_{\mathbf{R}-b}^Y, P_{\mathbf{R}-a}^X\}, P_a^X P_b^Y). \tag{6}$$

For a particle which travels onwards, the new description of state must therefore be

$$\phi^+ = P_a^X P_b^Y \phi / \| P_a^X P_b^Y \phi \|, \tag{7}$$

and this leads to

$$\left.\begin{aligned} \phi^+(x, y) &= \phi(x, y) \text{ for } x \in a,\ y \in b, \\ &= 0 \text{ otherwise.} \end{aligned}\right\} \tag{8}$$

The properties of position and momentum have interchanged roles in this description of state, as compared with ϕ. It is true that the unsharpness values of position, defined in section 2, are

$$\Delta_{\dot{me}} X = \Delta_{\dot{me}} Y = +\infty. \tag{9}$$

For, if the result $P_a^X P_b^Y$ is ignored, there remain the two other possibilities $P_a^X P_{\mathbf{R}-b}^Y$ and $P_{\mathbf{R}-a}^X$ and therefore an infinite-length set of values both for Y (from $\mathbf{R}-b$) and for X (from $\mathbf{R}-a$), within which no decision is possible. It is also possible, however, to take account of the result when defining the unsharpness in measurement. For the general case $me = (\{P_i\}_i, P_{i_0})$, this $\Delta_{me} A$ is to be regarded as the lower bound of the lengths of all sets of values a such that

(α_1) $\mathsf{V}_{\mathfrak{a}}\mathbf{M}(me; A, \mathfrak{a}, a)$,

(β_1) $P_a^A P_{i_0} = P_{i_0}$.

E

Then evidently

$$\Delta_{me} A \leqslant \Delta_{\dot{me}} A, \tag{10}$$

and in our particular case

$$\Delta_{me} X = 2x_0, \Delta_{me} Y = 2y_0, \tag{11}$$

$2x_0$ being the thickness of S and $2y_0$ the slit width (see above). This provides estimates of the two unsharpness values relating to the description of state (cf. section 2). In general we have:

(α_2) if $me = (\{P_i\}_i, P_{i_0})$ and $P_{i_0}\phi = \phi$,

then $\Delta_\phi^* A \leqslant \Delta_{me} A$.

From (6) in section 2, we therefore have in particular here

$$\Delta_{\phi+} X \leqslant \Delta_{\phi+}^* X \leqslant 2x_0, \Delta_{\phi+} Y \leqslant \Delta_{\phi+}^* Y \leqslant 2y_0. \tag{12}$$

Hence it follows that, depending on the interpretation, the momentum is known with only low accuracy, or shows a high dispersion, or has become extremely indeterminate. Since $\Delta_{\phi+}^* X$ and $\Delta_{\phi+}^* Y$ are finite, all momentum values are, strictly speaking, possible, in both the x and the y direction. This causes the spatial dispersion of the particles which occurs in stage (3). To calculate this explicitly, it is best to approximate ϕ^+ as a function of x and y by a function which is sufficiently many times differentiable, unlike the function ϕ^+ itself, which drops discontinuously to zero outside the rectangle defined by a and b. A simple form is (Tomonaga 1966, §61)

$$\phi^+(x, y) \approx \frac{1}{\delta\sqrt{\pi}} \exp\left(-\frac{x^2+y^2}{2\delta^2}\right) \exp\left(\frac{i}{\hbar} \bar{p}_x x\right), \tag{13}$$

with

$$(\Delta_{\phi+} X)^2 \approx (\Delta_{\phi+} Y)^2 \approx \tfrac{1}{2}\delta^2. \tag{14}$$

Here it would be assumed in particular (concerning ϕ) that the reduction of states does not alter the expectation values of the momentum:

$$\left.\begin{aligned} \text{Exp}\,(\phi^+, p_x) &= \text{Exp}\,(\phi, p_x) = \overline{p_x} > 0, \\ \text{Exp}\,(\phi^+, p_y) &= \text{Exp}\,(\phi, p_y) = 0. \end{aligned}\right\} \tag{15}$$

(3) In the third stage the particle is to be regarded as free, and behaving in accordance with the force-free Schrödinger equation

$$\frac{\hbar}{i}\frac{\partial\psi}{\partial t} = \frac{\hbar^2}{2m}\left(\frac{\partial^2}{\partial x^2}+\frac{\partial^2}{\partial y^2}\right)\psi, \tag{16}$$

with the initial condition $\psi = \phi^+$. The spreading of the wave packets here causes the spatial dispersion just mentioned. On the assumption (15), we have essentially a forward scattering in the positive x direction, symmetrical about the x axis. With x_1 (> 0) as the distance between S_1 and S, the centre of the wave packet (16) reaches the screen S_1 at the time

$$\overline{t_1} = m x_1 / \overline{p_x}. \tag{17}$$

If we assume that x_1 is sufficiently large:

$$\frac{\hbar}{\overline{p_x}} \cdot \frac{x_1}{\delta} \gg \delta, \tag{18}$$

then at the time $\overline{t_1}$ ψ produces on the screen S_1 approximately the probability density

$$|\phi_1(x_1, y)|^2 \approx \frac{1}{\eta\pi} \exp\left(-\frac{y^2}{2\eta^2}\right), \tag{19}$$

where

$$\eta = \frac{1}{\sqrt{2}} \frac{\hbar}{\overline{p_x}} \frac{x_1}{\delta} \tag{20}$$

and again

$$(\Delta_{\phi_1} Y)^2 \approx \tfrac{1}{2}\eta^2. \tag{21}$$

From (18) and (13) we then have

$$\Delta_{\phi_1} Y \gg \Delta_{\phi+} Y \tag{22}$$

(cf. Tomonaga 1966, §61). This expresses the spatial dispersion in the y direction, initially as a prediction based on the quantum-mechanical description of state.

(4) We can now test this prediction by making a final measurement: the incidence of the particles on the screen S_1 can be regarded as such a measurement. If no definite result is considered, this measurement $\dot{m}e_1$ would be described by

$$\{P^X_{a_1} P^Y_{b_{1,i}}, 1-P^X_{a_1}\}_i, \tag{23}$$

where $a_1 = [x_1 - x_0, x_1 + x_0]$ denotes the space occupied by S_1 in the x direction and $\{b_{1,i}\}_i$ is a division of the y axis which corresponds to the accuracy of measurement.

If the above presentation of the one-hole experiment, which is largely based on the quantum-mechanical formalism, is compared with the more qualitative heuristic presentations indicated in section I.3, the following points in particular emerge. The heuristic presentations do not limit themselves to either a stationary or a time-dependent treatment of the problem, although there appears to be a certain preference for the former. An explicit and rigorous stationary treatment is given by Beck and Nussenzveig (1958, which see for further references). Our decision has been for at least an approximate time-dependent presentation, since this makes clearer the critical process of measurement at the screen S. The heuristic presentations also use the wave picture in the problem of the interaction of the object with the screen S, taking elementary diffraction theory as a basis. Here again, Beck and Nussenzveig (1958) give a more rigorous treatment; they are concerned principally with the difference between a very narrow gap and a wide gap, which then becomes relevant. In the purely wave-theory treatment, the function of the screen S as a measuring apparatus for determining position is not prominent. If this aspect is to be brought to the fore, Bohr, for example, uses the particle picture immediately (cf. sections I.2 and I.3), and regards the interaction concerned as being an exchange of momentum and energy between the particle and the screen. In the above presentation, the passage of the particle through the slit in S is taken to be a measurement of position with the qualification (6), and the mechanism of the reduction of states is consistently applied.

As regards the Heisenberg relations, the heuristic presentations define the position uncertainty ΔY by the slit width, while the momentum uncertainty Δp_y is obtained from the angle of the first diffraction minimum, using the de Broglie relation. It is not made clear what is the spatial significance of Δp_y. Beck and Nussenzveig (1958) in their stationary treatment take as ΔY and Δp_y neither the usual dispersions nor the heuristic values just mentioned, but the half-widths proportional to the latter. They show rigorously that Δp_y is independent of x for a wide slit but not for a narrow slit (cf. again Fig. 1 in section I.3). On the other hand, for any slit we have

$$(\Delta Y)_{x=0}(\Delta p_y)_{x\to+\infty} \leqslant 2p_x^0 d,$$

where p_x^0 is the momentum to the left of S, and d the slit width. Thus, the elementary heuristic treatment is found to be valid only for a wide slit, since otherwise it would be possible to violate the Heisenberg relations (cf. also Blokhintsev 1964, §16). Such problems do not occur in our time-dependent treatment, where ΔY and Δp_y are always unambiguously determined by the measurement or by the relevant description of state. Then the time to which they relate is also determined, and their mutual relationship is decided by the general theory together with the selection of two particular quantities. Their relationship to the parameters of the experimental arrangement may be undecided, however. For the four unsharpness values defined above, $\Delta_{me}A$, $\Delta_{\dot{me}}A$, $\Delta_W^* A$ and $\Delta_W A$, we had the general formula (10):

$$\Delta_{me}A \leqslant \Delta_{\dot{me}}A;$$

formula (6) in section 2:

$$\Delta_W A \leqslant \Delta_W^* A;$$

and the further relation (α_2):

$$\Delta_W^* A \leqslant \Delta_{me}A$$

on the assumption that the result of the measurement *me* in W is known with certainty. The Heisenberg relations for Y and p_y are originally related to the Δ_W. Then the second of the above three formulae gives them for the Δ_W^*, the third gives them for the Δ_{me}, and the first and third give them for the $\Delta_{\dot{me}}$, since in the two latter cases there are always descriptions of state which satisfy the assumption in (α_2). The dimensions of the screen S, however, also come into play unambiguously in the description of position immediately after the passage of a particle through the slit: by (5) for $\Delta_{\dot{me}}\ldots$, by (6) for $\Delta_{me}\ldots$, and by (7) for $\Delta_{\phi+}\ldots$ and $\Delta_{\phi+}^*\ldots$.

The sudden increase in the momentum uncertainty in consequence of the position measurement by S is shown directly by the separation of the reduced state function (8) on the way to the screen S_1, as represented by (22). Care must be taken, however, not to ascribe a definite momentum to an individual particle after it passes through the slit. On the orthodox view this would be meaningful only if the momentum could also be determined while maintaining the previous measurement of position and deriving from these a *prediction* of the point where the individual particles would strike S_1. The arguments against this have been discussed in sections I.2 and I.3 without using the quantum-mechanical formalism. In the context of this formalism, the subject will be further discussed in Chapters VI and VII. The experimental arrangement discussed here offers only the possibility of what might be called a *subsequent* determination of momentum by a second measurement of position at the screen S_1. For a particle appearing at (x_1, y), the momentum components would be

$$p_x = mx_1/t_1, \quad p_y = my/t_1. \tag{24}$$

Before this second measurement of position, however, the momentum is undetermined, and afterwards it cannot be used for a further prediction, since there has meanwhile been a fresh measurement of position (and in any case the particle is absorbed by S_1). Formulae (24) are strictly significant only in the probabilistic context. The position distribution on the screen S_1 gives for the point (x_1, y) a fraction of particles coming from the slit in S to this point which is correct according to (24) and the momentum distribution after passing through S. For free particles and long times t we in fact have approximately

$$\psi_t(x, y) \approx -\frac{m}{ht} e^{\pi i m(x^2+y^2)/ht} g(mx/ht, my/ht), \tag{25}$$

where g is the Fourier transform and therefore the momentum representation of ψ_0 (see, for example, Kemble 1937, section 15). With our choice (17) of the time $\overline{t_1}$, which according to (18) is sufficiently late, we find

$$|\psi_{\overline{t_1}}(x_1, y)|^2 \approx \frac{m^2}{h^2\overline{t_1}^2}|g(\overline{p_x}/h, my/h\overline{t_1})|^2, \tag{26}$$

from which the above statement easily follows.

Let us now turn to the two-hole experiment (cf. Fig. 2, section I.3). Here again the description can be divided into the same four stages as in the one-hole experiment.

(1) We make the same assumptions as for stage (1) of the one-hole experiment.

(2) Let S again serve as an apparatus for the measurement of position; this measurement must be differently described, however, because of the different form of S. The extent of S in the x direction is the same: $a = [-x_0, +x_0]$, but in the y direction the two slits must be described by the set (with y_1 in place of the y_0' of Fig. 3, section I.3)

$$\left.\begin{aligned} b &= [y_1-y_0, y_1+y_0] \cup [-y_1-y_0, -y_1+y_0], \\ &y_1 \gg y_0 > 0. \end{aligned}\right\} \tag{27}$$

The measurement $\dot{m}e$ is again to be made as in (5), using these a and b, and distinguishes the following possibilities:

$P_a^X P_b^Y$	passage through the pair of slits in S
$P_a^X P_{\mathbf{R}-b}^Y$	absorption by S
$P_{\mathbf{R}-a}^X$	no reaction with S

The arrangement again distinguishes the first possibility, and this leads to the selective measurement me defined as in (6), which reduces the incoming ϕ as in (7), but, because of the different significance of $P_a^X P_b^Y$, ϕ^+ has a different form from that in (8). If ϕ is symmetrical about the x axis, then

$$\left.\begin{aligned} \phi^+ &= \frac{1}{\sqrt{2}}(\chi_1^+ + \chi_2^+), \\ \chi_i^+ &= P_a^X P_{b_i}^Y \phi / \|P_a^X P_{b_i}^Y \phi\|, \end{aligned}\right\} \tag{28}$$

where b_i are the two parts of b in (27), and the form in (28) is obtained since

$$\left.\begin{aligned} P_b^Y &= P_{b_1}^Y + P_{b_2}^Y, \; P_{b_1}^Y P_{b_2}^Y = 0, \\ \|P_a^X P_{b_1}^Y \phi\| &= \|P_a^X P_{b_2}^Y \phi\|, \end{aligned}\right\} \tag{29}$$

the latter equation being valid only when ϕ is symmetrical about the x axis. Thus ϕ^+ is a superposition of two functions χ_i^+ having the form (8). The $\Delta_{\dot{m}e}$ are the same as in (9), but instead of (11) we have

$$\Delta_{me}X = 2x_0, \quad \Delta_{me}Y = 2(y_1+y_0) \tag{30}$$

and instead of (12)

$$\Delta_{\phi+}X \leqslant \Delta^*_{\phi+}X \leqslant 2x_0, \Delta_{\phi+}Y \leqslant \Delta^*_{\phi+}Y \leqslant 2(y_1+y_0). \tag{31}$$

(3) In this stage occurs the decisive interference between the components of ϕ^+ which result from the Schrödinger equation (16). To simplify the calculation we can, as in (13), take the following approximations for the χ_i^+:

$$\left.\begin{aligned}\chi_1^+(x, y) &\approx \frac{1}{\delta\sqrt{\pi}} \exp\left(-\frac{x^2+(y-y_1)^2}{2\delta^2}\right) \exp\left(\frac{i}{\hbar}\bar{p}_x x\right), \\ \chi_2^+(x, y) &\approx \frac{1}{\delta\sqrt{\pi}} \exp\left(-\frac{x^2+(y+y_1)^2}{2\delta^2}\right) \exp\left(\frac{i}{\hbar}\bar{p}_x x\right);\end{aligned}\right\} \tag{32}$$

δ is now related to the dispersions of the χ_i^+ as it is in (14) to those of ϕ^+ itself. Since $y_1 \gg y_0$, and therefore $y_1 \gg \delta$, (32) also provides a good approximation. On the same assumptions (17) and (18) as in the one-hole experiment, we then find on the screen S_1

$$|\phi_1(x_1, y)|^2 \approx 2\gamma \exp(-y^2/2\eta^2)\left[1+\cos\left(\frac{\bar{p}_x}{\hbar} \cdot \frac{2y_1}{x_1} y\right)\right], \tag{33}$$

where γ is an undetermined normalization factor. The cosine term shows that interference fringes appear at distances of $hx_1/2\bar{p}_x y_1$ (cf. Tomonaga 1966, §61).

(4) This is a final measurement of position on S_1, exactly as in the one-hole experiment.

The two-hole experiment may suggest that the passage of a particle through the screen S implies a decision as to which of the two slits it has passed through. This is admittedly nowhere recorded; but on the one hand the fact that a particle has passed through S at all is also not recorded, and, on the other hand, one is very much accustomed to the classical idea that a particle passing through the screen must pass through *either* one *or* the other slit. Arguments against this view have been presented in section I.3 to illustrate Bohr's standpoint. It is also prohibited in the present treatment using the quantum-mechanical formalism. The experiments previously mentioned, whose aim is to decide which path the particle has taken, amount to omitting the measurement $\dot{m}e$ described by (5) and (27) and performing instead the measurement $\dot{m}e'$ described by

$$\{P_a^X P_{b_1}^Y, P_a^X P_{b_2}^Y, P_a^X P_{\mathbf{R}-b}^Y, P_{\mathbf{R}-a}^X\}, \tag{34}$$

for which the possibilities are

$P_a^X P_{b_1}^Y$	passage through the upper slit
$P_a^X P_{b_2}^Y$	passage through the lower slit
$P_a^X P_{\mathbf{R}-b}^Y$	absorption
$P_{\mathbf{R}-a}^X$	no reaction

In contrast to the possibilities in $\dot{m}e$, the passages through the upper and lower slits appear explicitly here. The strong reduction of states accordingly gives for $\dot{m}e$ an assemblage of three components, but for $\dot{m}e'$ one of four components. If we consider only the particles which pass through, *one* component in the first assemblage corresponds to this case, namely ϕ^+ from (28), with the probability $\|P_a^X P_b^Y \phi\|^2$. In the assemblage corresponding to $\dot{m}e'$, however, the assemblage for this case has *two* components χ_1^+ and χ_2^+ from (28), with the (equal) probabilities $\|P_a^X P_{b_1}^Y \phi\|^2$ and $\|P_a^X P_{b_2}^Y \phi\|^2$. This difference, certainly, would not affect the recording of a position distribution immediately after the measurement $\dot{m}e$ or $\dot{m}e'$, since the calculation is based (after renormalization) firstly on ϕ^+ from (28), secondly on

$$W^+ = \tfrac{1}{2}(P_{\chi_1^+} + P_{\chi_2^+}). \tag{35}$$

Now we can prove immediately that in the y direction, for example,

$$\|P_c^Y\phi^+\|^2 = \mathrm{Tr}(W^+P_c^Y)$$
$$= \tfrac{1}{2}(\|P_c^Y\chi_1^+\|^2 + \|P_c^Y\chi_2^+\|^2) \tag{36}$$

for all sets of values c. The proof is based on the pairwise interchangeability of P_c^Y, P_a^X and P_b^Y. The situation is, however, quite different some time after the measurement $\dot{m}e$ or $\dot{m}e'$. In the further development of ϕ^+ the χ_i^+ interfere, but in that of W^+ they do not. If U_t is the solution of the Schrödinger equation (16), then we still have

$$\mathrm{Tr}(U_tW^+U_t^{-1}P_c^Y) = \tfrac{1}{2}(\|P_c^YU_t\chi_1^+\|^2 + \|P_c^YU_t\chi_2^+\|^2), \tag{37}$$

but

$$\|P_c^YU_t\phi^+\|^2 \neq \tfrac{1}{2}(\|P_c^YU_t\chi_1^+\|^2 + \|P_c^YU_t\chi_2^+\|^2). \tag{38}$$

In the calculation of $\|P_c^YU_t\phi^+\|^2$, the "interference terms" also would appear on the right.

From (37) we see in particular that the interference does not occur in the modified arrangement. This is a case of two mutually complementary phenomena as defined by Bohr (cf. section I.2). The two measurements $\dot{m}e$ and $\dot{m}e'$ in (5) and (34) not only can be made simultaneously; $\dot{m}e'$ is simply a refinement of $\dot{m}e$. According to the discussion at the end of section 3, therefore, the complementarity is here dependent on the initial description of state ϕ. For the description chosen above, the respective weak reductions of states, here restricted to particles which pass through, are different, and this is very obvious by the presence or absence of interference on the screen S_1. If we started from a ϕ with $P_a^XP_{b_1}^Y\phi = \phi$ or $P_a^XP_{b_2}^Y\phi = \phi$ (another screen with only an upper or a lower slit, directly to the left of S), there would be no perceptible difference between the two measurements $\dot{m}e$ and $\dot{m}e'$, but they would in fact be unnecessary, since the conditions which they could bring about would already exist.

CHAPTER V

Attempts at an Axiomatic Foundation of Orthodox Quantum Mechanics (The Von Neumann Programme)

It has already been indicated in Chapter II that there is another method of formulating a physical theory, besides the one used there. Whereas the method of Chapter II consists in a representation of physical concepts in a mathematical theory which is felt to be appropriate but is developed in a largely independent manner, in the new and much more profound method the physical concepts are directly axiomatized, and an attempt is made to arrive in this way at a complete physical characterization of the mathematical theory used previously, which was physically justified only by its results. This programme for quantum mechanics was first envisaged by von Neumann, who initiated it with some important contributions. Subsequently, the work was continued by other authors, but the programme has not yet been satisfactorily completed.

We do not intend to describe these developments. Instead, some especially clear cases of partial axiomatization will be used as examples to show what is involved in such an axiomatic foundation. After a brief introduction to the new method (section 1), we successively consider the characterizations of the quantum-mechanical propositions regarding expectation values of quantities as treated by von Neumann (section 2), the corresponding propositions regarding probabilities of properties as treated by Mackey and Gleason (section 3), and the quantum-mechanical reduction of states according to an idea due to Ludwig (section 4). All these treatments deal with relatively small parts of the standard formulation of quantum mechanics described in Chapters II and III. In particular, a Hilbert space is used *ad hoc* somewhere in each treatment. The general von Neumann programme, however, aims precisely at finding the physical conditions under which the classically unintelligible Hilbert space of quantum mechanics is necessary.

As regards this wider objective and its attainment, reference may at least be made here to the principal relevant literature. There are three broad types of attempted solution, using different procedures. The furthest advanced is the type based on lattice theory, using the lattice structure of the quantum-mechanical concept of a property. The classical work is that of Birkhoff and von Neumann (1936), leading to a result only in the finite-dimensional case. An unpublished paper by von Neumann (1937) leads to results for the infinite-dimensional case also, although these deviate somewhat from quantum mechanics proper, being based on continuous geometries. The most extensive results for quantum mechanics in its present form are those of Zierler (1961) and Piron (1964). The entire field has been described in a monograph by Varadarajan (1968), although this is entirely mathematical in treatment. (See also the recent discussion by Jauch and Piron (1969), Ochs (1972a, b), and the independent work of Gudder (1969, 1970b), who gives further references.)

The second type of solution makes use of various algebraic structures of the quantum-mechanical concept of a quantity. Definitive results for the finite-dimensional case were obtained by Jordan, von Neumann and Wigner (1934), and supplemented by Albert (1934). Von Neumann (1936) attempted a generalization, but without final success. This line was later continued by Segal (1947); Sherman (1956) related the results to the theory of C*-algebras. The connection between the lattice-theory and algebraic approaches is discussed by Mackay (1963) and Gudder and Boyce (1970).

A third type of solution is given by some combination of the first two. This admittedly vague description covers the recent work of Guenin (1966), Gunson (1967) and in particular Ludwig (1970).

V.1 Change of axiomatic standpoint

In section II.1 we outlined a method of formulating a physical theory whereby the concepts essential to the theory were presented in a purely mathematical theory given explicitly *a priori*. The same method has been consistently followed up to this point. It must now be replaced by another, since von Neumann's programme is dependent on such a change. The new method was expressly formulated as early as the first paper by von Neumann on quantum mechanics (Hilbert, von Neumann and Nordheim 1928). Because of the fundamental significance of the von Neumann programme, it is worth quoting this paper extensively.

In §1 it reads: "The fundamental physical idea of the whole theory is that the rigorous functional relations of ordinary mechanics are everywhere replaced by probability relations . . . The path which leads to this theory is as follows. Certain physical requirements are imposed on these probabilities, suggested by our experience and developments so far, and implying certain relations between the probabilities. Then we look for a simple analytical formalism involving quantities that satisfy just these relations. This formalism, and therefore the quantities that appear in it, can be given a physical interpretation based on the physical requirements. The aim is to formulate the physical requirements with just sufficient completeness to define precisely the analytical formalism. This is therefore an axiomatization, similar to that in geometry, for example. The axioms describe the relations between geometrical figures such as points, lines and planes, and it can then be shown that these relations are satisfied by an analytical formalism, namely the linear equations. In this way geometrical theorems can in turn be deduced from the properties of the linear equations. Likewise, in the new quantum mechanics, a particular procedure is used to assign formally to every mechanical quantity a mathematical quantity representing it; this is initially a purely theoretical quantity, but from it we can obtain statements about the representatives of other quantities, and so by retranslation statements about actual physical things . . . The axiomatization procedure indicated above is not usually employed in exactly that form in physics; the path to a new theory is as a rule, and in the present case also, as follows. The analytical formalism is usually guessed before the complete system of axioms has been set up, and the basic physical relations are derived only by the interpretation of the formalism. It is difficult to understand such a theory if these two, the formalism and its physical interpretation, are not kept clearly distinct. The distinction will be made as clear as possible here, although, in accordance with the present state of the theory, we do not desire to give a complete axiomatization."

This text could hardly be more clearly expressed, but it still merits some comments. In the passage quoted, the authors say that "the analytical formalism is usually guessed before the complete system of axioms has been set up". The sequence of operations thus indicated has also been used in our own presentation in Chapter II, and it corresponds entirely to historical facts. One partial exception is geometry, which is also mentioned in the text quoted above, and which can indeed be used to get a very clear idea of the position, despite this historical inversion. Since the time of Euclid there have been attempts to formulate geometry as a scientific theory by establishing a *direct* logical connection between certain of its concepts: for example, by requiring that just one straight line passes through two different points, that just one plane passes through three non-collinear points, and so on. This "axiomatic method" became, in the course of time, a model for the rigorous formulation of any science. Since the emergence of analytical geometry in modern times, there is another method available for geometry, namely, not to set up direct relations between geometrical concepts but to represent them in the algebraic-analytical theory of real numbers. Then, in the familiar manner, the points in space are represented by groups of three numbers, the straight lines and planes by sets of such triplets defined by systems of linear equations, the distance between two points by a certain function of their coordinates, and so on. In this respect it is immaterial whether the theory of numbers is presented axiomatically or constructively. The representation of the geometrical concepts by this theory provides again a logical connection between them, but *not a direct one*, since it involves the corresponding connection of the concepts which represent them. This is the method which has been used in Chapters II and III for the various formulations of classical mechanics and quantum mechanics. This preference over a direct axiomatization corresponds to the actual origin of physical theories, whose mathematical treatment was always developed from specific arithmetical models.

In fundamental research, however, this method is not sufficient, as it has the following disadvantages. The mathematical theory on which a physical theory is mapped by this method is in general stronger than the physical theory, in the sense that it contains a larger number of true theorems, in particular some which involve mathematical concepts having no physical interpretation in terms of the representation that gives the connection between the two theories. For example, the definition of a Hilbert space $\mathfrak{H}$ as a mathematical structure involves the basic relation $\phi+\psi = \chi$ between the vectors $\phi,\psi,\chi \in \mathfrak{H}$, and $\mathfrak{H}$ is stated to be an Abelian group in respect of this relation. At the time when quantum mechanics first appeared, this requirement was interpreted as a superposition principle, with reference to the wave picture of matter. However, in a rational quantum mechanics with a consistently probabilistic interpretation, this treatment has at most metaphorical significance and strictly should find no place. Since this point might lead to some dispute even today, another example may be given: the basic relation $\langle\phi|\chi\rangle = \alpha$ between two vectors ϕ, $\chi \in \mathfrak{H}$ and a number α. This again is an explicit constituent of the usual definition of the Hilbert space specified as linear in χ, antilinear in ϕ, positive definite, and so on. Yet this relation evidently does not appear directly in any of the requirements imposed on quantum mechanics in Chapters II and III. Instead, what is there directly physically interpreted are concepts *derived* from it together with the sum relation and other basic relations: for example, the concept of a self-adjoint linear operator, the relation $\mathrm{Tr}(WA) = \alpha$, and so on.

The demonstration of these circumstances leads on to the following entirely justifiable questions. What is the function of the former group of concepts in quantum mechanics? how can their appearance be vindicated (other than simply by results)? are they a necessary and indispensable constituent of the theory? if so, why are they not understandable? if not so, what would be an equivalent theory omitting them? There are other questions which can be directly related to the physical concepts actually represented. We may ask (and ought to do so, having regard to the peculiarities of quantum mechanics) how the quantities of a quantum-mechanical object and the states of statistical ensembles are to be represented as certain operators in a Hilbert space, why the probability functions defined by the states are to be represented by just the mathematical functions that do represent them in Hilbert space. In brief: how does it happen that quantum mechanics is what is given by the method of representation in Hilbert space? The method itself does not give the slightest indication. No-one can foresee the possible consequences of a formulation of a physical theory based on this method, and if unacceptable consequences appear no-one can point to any particular premises and say "This is the source of the trouble and must be changed in order to avoid such consequences". Instead, the premises themselves must first be determined, and this was the aim of von Neumann, at first evidently under the influence of Hilbert.

V.2 Characterization of expectation value functions (von Neumann's original proof)

The earliest contribution to the carrying out of the von Neumann programme is an axiomatic characterization of the expectation value functions and mean value functions defined in section IV.2 for the epistemic and statistical formulations of quantum mechanics respectively. This characterization was achieved by von Neumann himself (1927, §§II, III), who already coupled his result with the further one that quantum mechanics contains no dispersion-free descriptions of state (1927, §IV). In his book, published (in German) in 1932, he presented this topic afresh and extended it by asserting that quantum mechanics allows no classical theory of hidden parameters (1955, section IV.2). The complex of ideas thus outlined, and consisting of three parts, appears in the secondary literature somewhat lumped together under the name of the "von Neumann proof". Evidently only the first of these three parts belongs to what we here refer to as the von Neumann programme. Only the first part is concerned to assert and demonstrate the logical equivalence of two different formulations of quantum mechanics. Thus the content of such an assertion depends essentially on the possibility of different formulations for the same physical theory, whereas the content of either of the other two assertions, if these are physically meaningful, depends only on what is to be understood by quantum mechanics, regardless of whether such a statement occurs in one or another of a series of possible and logically equivalent formulations. Neglect of this distinction has led to errors in the evaluation of the "von Neumann proof". It must therefore be emphasized that only the particular characterization in question is at present under discussion. The other two assertions, especially the third-named, will come under consideration in section VII.1.

The von Neumann characterization relates to a part of the epistemic or statistical formulation of quantum mechanics, as given in section II.4, namely the part which defines the concepts of object or statistical ensemble, quantities of an object or a statistical ensemble, and states of knowledge or statistical states and the corresponding propositions regarding probabilities (subsections II.4(a), (c) and (d)(ii)). The actual formulation used as the starting-point of the characterization, however, is not exactly the one given previously. Though it is based on the older method of presentation (section II.1), it uses propositions regarding expectation values or mean values in place of propositions regarding probabilities as the contingent primitive propositions. This change leads to an extension of the fundamental concepts as regards quantities. The new formulation is as follows.

(a') *Objects* and *statistical ensembles* are represented as in subsection II.4(a);

(c') Several fundamental concepts occur, as already stated:

(c'_1) *Quantities* are represented as in subsection II.4(c);

(c'_2) *Bounded quantities* are represented by bounded self-adjoint linear operators in the Hilbert space $\mathfrak{H}$ of the object concerned;

(c'_3) A third concept is that of the *functional dependence of quantities*. The quantity B is dependent on the quantity A through the real Borel function f if

$$\bigwedge_b . P_b^B = P_{f^{-1}(b)}^A , \tag{1}$$

where the P are the spectral resolutions of the corresponding operators.

(d'_2) The *states of knowledge of an individual object* or the *states of statistical ensembles* are again represented by the statistical operators in the Hilbert space concerned, but the connection between these and the quantities is established only for bounded quantities, by the following propositional function:

$\mathbf{Exp}(W,A) = \alpha \rightleftharpoons \alpha$ *is the expectation value or mean value of the quantity* A *in the state of knowledge or statistical state* W, and the function **Exp** is mathematically represented by

$$\mathbf{Exp}(W,A) = \mathrm{Tr}(WA) \tag{2}$$

for bounded A. (For unbounded A, (2) would not always exist.)

This new formulation is found to be equivalent to the one in subsections II.4(a), (c) and (d)(ii), in precisely the sense described for another case in Chapter III. The equivalence proof is roughly as follows. From the previous formulation, the concept of a bounded quantity is obtained in the usual way by means of the spectrum of a quantity. The functional-dependence relation is obtained from

$$\bigwedge_b \bigwedge_W . \mathbf{pr}(W; B,b) = \mathbf{pr}(W; A,f^{-1}(b)). \tag{3}$$

The function **Exp**, finally, is described as in (2) in section IV.2. Conversely, if the new formulation is given, the probability function **pr** is all that is needed, and it is defined for *any* A by

$$\mathbf{pr}(W; A,a) = \mathbf{Exp}(W; \chi_a(A)) \tag{4}$$

with the characteristic function χ_a belonging to the Borel set a. Thus (c'_2) is used in (d'_2), and (c'_3) in (4).

Von Neumann constructed a formulation of quantum mechanics, equivalent to this expectation-value formulation, which as regards quantities adheres to the former method of presenting a physical theory but contains an abstract characterization of states of knowledge or statistical states and of the expectation-value or mean-value functions, i.e. states requirements which this theory does not connect *ad hoc* with certain entities in Hilbert space. An abstract characterization of the descriptions of state relative to the quantities which are naturally involved here is impossible without axiomatically emphasizing, and making available for the characterization of descriptions of state, a certain structure which the field of quantities acquires by the representation in Hilbert space. The new axiomatics therefore necessarily has the following form.

(α) Certain requirements are imposed in an *abstract* manner on the structure of the field of quantities.

(β) The resulting structure is identified in a *concrete* manner with a substructure in a Hilbert space.

(γ) Using only (α), certain requirements are imposed in an *abstract* manner on the structure of the field of states of knowledge or statistical states and of the expectation-value or mean-value functions.

Then, using (β) also, it is shown which states of knowledge or statistical states, and which expectation-value or mean-value functions, are obtained in a *concrete* manner from this. In this procedure, the method of mathematical presentation is evidently retained only in the form reduced to (β), while in (α) and (γ) a direct axiomatization is carried out.

The following explicit formulation of (α)$-$(γ) departs (for reasons of mathematical rigour) from that of von Neumann (1927, §II; 1955, sections IV.1, IV.2) in some inessential points.

(α_1) For an individual quantum-mechanical object Σ, distinguish the set $\mathfrak{G}$ of all measurable quantities of Σ.

(α_2) In $\mathfrak{G}$ distinguish a relation *fu*(A,B,f) of the functional dependence of the quantity B on the quantity A with the real Borel function f.

(α_3) In $\mathfrak{G}$ distinguish the set $\mathfrak{G}_b \subseteq \mathfrak{G}$ of bounded quantities of Σ.

(α_4) In $\mathfrak{G}_b$ distinguish an algebraic structure so that $\mathfrak{G}_b$ becomes a real vector space.

(α_5) In $\mathfrak{G}_b$ distinguish an ordering $\geqslant$. Let $\mathfrak{G}^+$ be the subset of bounded quantities $\geqslant 0$; 0 exists by (α_4). These are called positive quantities.

(α_6) In $\mathfrak{G}^+$ distinguish a constant quantity **1**.

(β) There exist a complex Hilbert space $\mathfrak{H}$ of countably infinite dimension, and a representation of the structure defined by (α_1)$-$(α_6) in $\mathfrak{H}$, so that:

(β_1) $\mathfrak{G}$ is isomorphically represented (as a set) by the set $\mathfrak{L}_{sa}$ of self-adjoint linear operators in $\mathfrak{H}$.

(β_2) *fu* becomes the relation (1).

(β_3) $\mathfrak{G}_b$ becomes the set $\mathfrak{L}_{b,sa}$ of the bounded self-adjoint linear operators.

(β_4) The real vector space structure of $\mathfrak{G}_b$ becomes the usual structure of $\mathfrak{L}_{b,sa}$.

(β_5) The ordering $\geqslant$ in $\mathfrak{G}_b$ becomes the canonical ordering in $\mathfrak{L}_{b,sa}$ ($A \geqslant B$ if $\langle A\phi|\phi\rangle \geqslant \langle B\phi|\phi\rangle$ for all $\phi \in \mathfrak{H}$).

(β_6) The quantity **1** becomes the unit operator **1** in $\mathfrak{H}$.

(γ) For an object or a statistical ensemble, distinguish the set $\mathfrak{S}$ of the possible states of knowledge or statistical states. These states of knowledge or statistical states W are related to the bounded quantities A distinguished by (α_3), through a real-valued function **Exp**. **Exp**(W, A) is the expectation value or mean value of A for the state of knowledge or statistical state W, and must satisfy the following conditions.

(γ_1) For every W, **Exp**(W, A) as a function of $A \in \mathfrak{G}_b$ is a linear form as regards the structure required by (α_4). It is positive as regards the set of positive quantities distinguished by (α_5), i.e. it does not take negative values in $\mathfrak{G}^+$, and it is normalized as regards the quantity **1** distinguished by (α_6), i.e. **Exp**(W, **1**) $= 1$. Finally, it is normal, in the following sense (cf. Dixmier 1957, section I.4.2): if $A \in \mathfrak{G}^+$ is the upper bound of a set $\mathfrak{F} \subseteq \mathfrak{G}^+$ filtered above as regards $\geqslant$ in (α_5), then **Exp**(W, A) is the upper bound of all numbers **Exp**(W, B) with $B \in \mathfrak{F}$. (A subset $\mathfrak{F}$ of a set ordered by $\geqslant$ is said to be filtered above if, for A and $B \in \mathfrak{F}$, there is $C \in \mathfrak{F}$ with $C \geqslant A$ and $C \geqslant B$.)

(γ_2) If **Exp**(W, A) $=$ **Exp**(W_1, A) for all $A \in \mathfrak{G}_b$, then $W = W_1$.

(γ_3) For every linear form L in $\mathfrak{G}_b$ which is positive, normal and normalized to 1, as defined in (γ_1), there is a $W \in \mathfrak{S}$ such that $L(A) = \mathbf{Exp}(W, A)$ for all $A \in \mathfrak{G}_b$.

This new formulation of a part of quantum mechanics can be proved, rigorously, to be logically equivalent to the expectation-value formulation, given above and based on the older method of presentation; the latter formulation is in turn, as already stated, logically equivalent to the formulation in subsections II.2 (a), (c) and (d) (ii). A first proof was given by von Neumann (1927, §III; 1955, section IV.2), and to this we need here add only two comments. The proof that the old formulation follows from the new one is based essentially on the purely mathematical theorem that every positive, normal and normalized linear form $L_r(A)$ has the form (2) with a von Neumann operator W. (For the proof of this, see Dixmier (1957, sections I.4 and I.6). The connection between his theory of the positive linear form L_c in the *complex* vector space $\mathfrak{L}_b$ of all bounded linear operators and our positive linear forms L_r in the *real* vector space $\mathfrak{L}_{b,sa}$ is given by the canonical embedding

$$L_c(A) = L_r\left(\frac{A+A^*}{2}\right)+iL_r\left(\frac{A-A^*}{2i}\right),$$

where A^* is the adjoint operator.) As regards the opposite derivation, the main point is how the further basic concepts (α_1)–(α_6) can be obtained from the old formulation. This is not entirely obvious for (α_4) and (α_5). The scalar multiplication in $\mathfrak{G}_b$ is easily found from the relation of functional dependence, which is also available in the old formulation. The sum relation $A = B+C$ in $\mathfrak{G}_b$ takes the form

$$A = B+C \leftrightarrow \bigwedge_W. \mathbf{Exp}(W, A) = \mathbf{Exp}(W, B)+\mathbf{Exp}(W, C). \tag{5}$$

The proofs that this gives altogether a vector space are easily found by means of (2). The ordering in (α_5) is also available in terms of **Exp**, thus:

$$A \geqslant B \leftrightarrow \bigwedge_W. \mathbf{Exp}(W, A) \geqslant \mathbf{Exp}(W, B). \tag{6}$$

Let us now see what has been gained by the foregoing characterization. We have asserted and proved the logical equivalence for two formulations of quantum mechanics, one of them proceeding entirely according to the method laid down in section II.1, the other at least in part, and especially in (γ), by the new method of direct axiomatization. An assertion of equivalence consists of two parts, of which the less trivial one here, and the one which evidently represents the real object of the exercise, is the one which makes the first formulation a logical consequence of the second. Thus an assessment of this result (apart from the general question of what constitutes a logical consequence) must consist in a discussion of the new hypotheses (α) and (γ). Of these, (α) by itself cannot be discussed, because the statements in it are transferred by (β) in their entirety to Hilbert space by the old method of presentation. They can be discussed only in combination with (γ). No problems arise from (α_1)–(α_3) or (α_6). For the ordering $\geqslant$ in (α_5), (6) is an equivalence, and the significance of this relation is thus clarified. For the sum relation in (α_4), (5) is an equivalence, and here of course the question arises as to the significance of $B+C$ when B and C are not at least simultaneously measurable with arbitrary accuracy (cf. section IV.2). This question arises because in that case the significance of $\mathbf{Exp}(W, B)+\mathbf{Exp}(W, C)$ is also not obvious. An answer has been found so far only in the sense that the requirement of additivity of **Exp** in (γ_1) (through the requirement of linearity) is dispensable if B and C are not simultaneously measurable with arbitrary accuracy, and so in this case the sum relation need not be made explicit, as will be seen in the next section. Of the other conditions on **Exp** imposed in (γ_1), only the normality causes any problem: this is a kind of continuity condition. It can in fact be shown that it is equivalent to the continuity of **Exp** as a function in $\mathfrak{L}_{b,sa}$ with ultra-weak topology (Dixmier 1957, section I.4, Theorem 1). We have given preference to the requirement of normality, since otherwise a topology would have to be distinguished in $\mathfrak{G}_b$ also, and this could not be obtained easily, if at all, from our older formulation (section II.4). The physical significance of the requirement is now being discussed in relativistic quantum theory (theory of local rings), and is not yet finally clarified. The significance of (γ_2) is obvious: the difference between two states of knowledge or statistical states must be expressed in **Exp**. Finally, (γ_3) is a routine requirement which merely completes what has been stated in (γ_1). Without (γ_3), the set of descriptions of state might be empty; with (γ_3), it covers all possibilities contained in (γ_1). Thus the latter is where the actual decision occurs.

V.3 Characterization of probability functions (Gleason and Mackey's proof)

In this section we shall discuss a refinement of von Neumann's proof which is based not on the formulation of quantum mechanics used by von Neumann himself and involving the concept of quantities and expectation values, but on the formulation given in section III.3 and involving properties and probabilities. It will be seen later that this also simplifies the formulation of the requirements. The refinement concerns in particular the relation (5) at the end of section 2, and is the result of a mathematical analysis by Gleason (1957). Mackey (1963, §2-2) has elegantly incorporated Gleason's result into a proof similar to that of von Neumann, and his treatment will be followed here; cf. also the purely mathematical formulation by Varadarajan (1968, §VII.2).

The basic idea of the formulation in relation to the more general programme stated in section 1 is again, as in section 2, the abstract characterization of the quantum-mechanical descriptions of state, now as probability functions over the domain of properties of an object. Thus we have first to give an abstract characterization of this domain of properties to the extent that is necessary in order to characterize the descriptions of state. This can be done as follows (cf. Mackey 1963, p. 67f.; Varadarajan 1962, p. 195ff.):

(α_1) For an individual object Σ, distinguish the set $\mathfrak{E}$ of all (contingent) properties of Σ.

(α_2) In $\mathfrak{E}$ distinguish a structure which makes $\mathfrak{E}$ an ordered orthocomplementary set: specifically, distinguish an element $\mathsf{U} \in \mathfrak{E}$, a mapping $^{\perp}$ of $\mathfrak{E}$ on itself and a two-place relation $\prec$ in $\mathfrak{E}$, such that:

(α_{21}) $\prec$ is an ordering and U its largest element.

(α_{22}) $^{\perp}$ is an involution which inverts the ordering $\prec$.

(α_{23}) Every sequence $E_1, E_2, \ldots$ of properties in $\mathfrak{E}$ which satisfies in pairs $E_\mu \prec E_\nu^{\perp}$ for $\mu \neq \nu$ has an upper bound in respect of the ordering $\prec$, denoted by $\cup_\mu E_\mu$ or $E_1 \cup E_2 \cup \ldots$.

(α_{24}) $E \cup E^{\perp} = \mathsf{U}$.

(α_{25}) For E and F in $\mathfrak{E}$ with $E \prec F$ there exists $E_1 \in \mathfrak{E}$ with $E_1 \prec E^{\perp}$ and $E \cup E_1 = F$.

In addition to these very general conditions on $\mathfrak{E}$, whose significance will be discussed shortly, there is again a fairly direct identification of $\mathfrak{E}$ with a substructure of a Hilbert space:

(β) There exists a complex separable Hilbert space $\mathfrak{H}$ of infinite dimension and an isomorphism of $\mathfrak{E}$ as an ordered orthocomplementary set with the set of all closed subspaces in $\mathfrak{H}$, with the usual inclusion and orthocomplementarity.

Finally, we have again the real objective of this formulation, namely the characterization of the descriptions of state:

(γ) For an object or a statistical ensemble, distinguish the set $\mathfrak{S}$ of possible states of knowledge or statistical states, W. These are related to the probabilities E by the probability $\mathbf{pr}_*$ (W, E) that the property E is found by a suitable measurement when the state of knowledge or the statistical state W is present. The conditions on $\mathbf{pr}_*$ are:

(γ_1) For every $W \in \mathfrak{S}$, $\mathbf{pr}_*(W, E)$ is a measure of probability in $\mathfrak{E}$, i.e.

(γ_{11}) $0 \leqslant \mathbf{pr}_*(W, E) \leqslant 1$,

(γ_{12}) $\mathbf{pr}_*(W, \cup_\mu E_\mu) = \Sigma_\mu \mathbf{pr}_*(W, E_\mu)$ for every sequence $E_1, E_2, \ldots$, such that $E_\mu \perp E_\nu$ (i.e. $E_\mu \prec E_\nu^\perp$) for $\mu \neq \nu$,

(γ_{13}) $\mathbf{pr}_*(W, \mathsf{U}) = 1$.

(γ_2) If $\mathbf{pr}_*(W, E) = \mathbf{pr}_*(W_1, E)$ for all $E \in \mathfrak{E}$, then $W = W_1$.

(γ_3) For every probability pr_* in $\mathfrak{E}$ (as defined by (γ_{11})$-$(γ_{13})), there is a $W \in \mathfrak{S}$ such that $pr_*(E) = \mathbf{pr}_*(W, E)$ for all $E \in \mathfrak{E}$.

The theory defined by these conditions can be shown to be logically equivalent to that part of the quantum mechanicsformulated in section III.3 by the old method of presentation which relates to the description of objects, their properties, and possible states of knowledge or statistical states. The derivation of the old formulation from the new one is based essentially on Gleason's theorem (1957) that every measure of probability pr_* in Hilbert space $\mathfrak{H}$ (with dimension > 2) has the form

$$pr_*(E) = \mathrm{Tr}(WP_E) \tag{1}$$

where E is any subspace of $\mathfrak{H}$ and P_E the projector on E, and W is a von Neumann operator. To prove the converse, we again need only refer to the derivation of the concepts used in (α_2) from the old formulation, as follows:

$$E \prec F \rightleftharpoons \wedge_W \mathbf{pr}_*(W, E) \leqslant \mathbf{pr}_*(W, F); \tag{2}$$

$$F = E^\perp \rightleftharpoons \wedge_W \mathbf{pr}_*(W, E) + \mathbf{pr}_*(W, F) = 1; \tag{3}$$

$$E = \mathsf{U} \rightleftharpoons \wedge_W \mathbf{pr}_*(W, E) = 1. \tag{4}$$

The relations (2) and (4), and a weaker form of (3), namely $F \prec E^\perp$, have been obtained in section III.4 by means of propositions regarding measurements. Here, we have an interpretation in terms of propositions regarding probabilities. (See Kakutani and Mackey (1946) for a generalization of Mackey and Gleason's result to Banach spaces by weakening (β).)

In general, the same comments can be made regarding the new formulation as have previously been made on the original procedure of von Neumann. The significance of the constants $\prec$, $^\perp$ and U for properties, used in (α), is sufficiently clear from (2)–(4), which arise as equivalences if we start from the new formulation. Instead of (2) we can also write

$$E \prec F \leftrightarrow \wedge_W \mathbf{pr}_*(W, E) = 1 \rightarrow \mathbf{pr}_*(W, F) = 1, \tag{5}$$

and this perhaps makes the relationship even clearer. Evidently $\prec$ is the quantum-mechanical analogue of the classical relation between two properties such that one always occurs when the other occurs. In the epistemic interpretation of quantum mechanics, if we use (5) for instance, this occurrence is replaced by the certainty of finding it by a suitable measurement (in an individual case), and $E \prec F$ signifies that this certainty always (i.e. for any state of knowledge) exists for F if it exists for E. In the statistical interpretation, the situation is similar. If a suitable statistical measurement would, on being performed, give E for every object in the ensemble, then it would also give F for every object.

$^{\perp}$ is the quantum-mechanical analogue of the classical relation between two properties whereby just one of the two occurs in any state. In quantum mechanics the corresponding situation is that in a suitable simultaneous measurement of E and $E^{\perp}$ the result is always either E or $E^{\perp}$.

Lastly, U is the property which trivially occurs always.

The conditions imposed on the probability functions $\mathbf{pr}_*$ in (γ_1) on the basis of the relations just discussed between properties are the simplest conceivable ones. In particular, no two properties appear which are not simultaneously measurable, and this is in fact the decisive advance over the original von Neumann axiomatics given in the previous section. See, however, the critique by Bell (1966).

V.4 Characterization of the reduction of states

As a final example of an abstract characterization, with some use of the older method of presentation, we shall take a characterization of the reduction of states. For this we first take as basis the quantum mechanics without individual states, in the property formulation given in section III.3. Here, three types of reduction of states have been distinguished: that which depends on a particular result of measurement, and the strong and weak reductions with the result of measurement undetermined. Of these, the first is contained in the strong reduction to the extent that the assemblage which furnishes the strong reduction consists, apart from the probabilities, simply of all the descriptions of state which are the result of the first-mentioned reduction for a particular result of measurement. It is just this transition

$$W_i^+ = \frac{P_i W P_i}{\mathrm{Tr}(WP_i)} \tag{1}$$

which is to be applied here for characterization. We regard it as a reduction of states including the result of measurement, i.e. as belonging to a measurement $(\{P_j\}_j, P_i)$. But the characterization concerned will clearly be independent of the selected P_i, and therefore we have at the same time a characterization of the strong reduction of states. The time suffix can be ignored, since only a single reduction will be involved.

In the property formulation (section III.3), an immediate consequence of (1) is the assertion

$$\mathbf{pr}_*(W_i^+; P_i) = 1. \tag{2}$$

This evidently signifies that

(N) a further measurement with the possible result P_i, made immediately after obtaining the result P_i in a measurement, will certainly have this same result.

In fact (N) *follows* from (β_2) in section III.3 if this condition may also be applied to a case me^+ and W^+ such as occurs in (N). This application has been excluded by (α_2) in section

III.3, whereby the next non-trivial measurement cannot take place until a finite time has elapsed. It is, however, usual to interpret (2) also in the form (N). From (β_2) in section III.3 we have

$$\mathbf{pr}_*(W_t; P) = 1\wedge\mathbf{M}^1_*(me_t; P) \rightarrow \mathbf{M}^2_*(me_t; P), \tag{3}$$

and the application to (2) for a measurement me^+ simultaneous with W_i^+ gives $\mathbf{M}^1_*(me^+; P_i) \rightarrow \mathbf{M}^2_*(me^+; P_i)$, which is ($N$). Von Neumann sought to provide a basis for the reduction of states in the form (N) (1927, §VI, notes 30 and 36; 1955, sections III.3 and IV.3). Using (N) or (2) as a basis for (1) is possible only for a maximum property $P_i = P_\phi$, since only then is W_i^+ uniquely defined by (2): $W_i^+ = P_\phi$. Von Neumann was not in possession of the reduction of states (1): as already mentioned in section II.4(e), this is due to Lüders (1951). However, von Neumann (1955, section III.3) did note that in the measurement of a multiple eigenvalue of a quantity the state after the measurement is not uniquely determined even if the result of the measurement is known. He nowhere indicated what description of state must occur in this case; instead, he seems to have assumed that the "measuring apparatus itself" would somehow by itself always perform a maximum measurement, but that we sometimes do not perceive this (1955, section V.1). This assumption was criticized by Lüders on the grounds that any measuring apparatus has a certain resolving power, sometimes very slight. (Cf. also Süssmann 1958, section III.C.) The criticism corresponds to the concept used here (sections II.4(d)(i), III.3) of a measurement whose description is not given by maximum systems $\{P_i\}_i$ alone.

Thus, more than (2) is needed for the characterization of (1). Following a suggestion by Ludwig (1954, §II.3), we impose the following condition. Assuming (α)–(γ) in section 3, let there be an initially present description of state W and a measurement me. The latter has a uniquely determined, (relatively) maximum actual result P_i. Let this be such that $\mathbf{pr}_*(W; P_i) > 0$. We seek a description of state W_i^+ for which

$$\mathbf{pr}_*(W_i^+; P) = \frac{\mathbf{pr}_*(W; P)}{\mathbf{pr}_*(W; P_i)} \qquad (P \prec P_i) \tag{4}$$

for all properties P such that $P \prec P_i$. Then it can easily be proved that there is just one such W_i^+ and that (1) is valid for the operators belonging to this W_i^+, W and P_i. The proof is based initially on the use of the result of section 3, which leads to

$$\mathrm{Tr}(W_i^+ P) = \frac{\mathrm{Tr}(WP)}{\mathrm{Tr}(WP_i)} \qquad (PP_i = P) \tag{5}$$

for the corresponding operators, instead of (4). The remainder of the proof, leading from (5) to (1), is simple; it is given by Ludwig (1954, §II.3).

We now have to consider the assessment of the condition (4). Let us imagine, in the statistical interpretation say, that the measurement $me = (\{P_j\}_j, P_i)$ is actually performed on an ensemble $\{\Sigma_k\}_{k\in K}$ in the state W. The ensemble then separates into as many parts as there are P_j with $\mathbf{pr}_*(W; P_j) > 0$, and these certainly include some which belong to P_i: the part of $\{\Sigma_k\}_{k\in K_i}$ which belongs to P_i consists of all Σ_k which gave the result P_i. If there is also a property P with $P \prec P_i$, the properties $P_j \neq P_i$ exclude this P. The measurement therefore provides regarding P at least the information that P was not the actual result for one of the $\{\Sigma_k\}_{k\in K_i}$ with $j \neq i$. If now not only $P \prec P_i$ but $P = P_i$, this justifies the special case (2) of (4): this P occurs throughout $\{\Sigma_k\}_{k\in K_i}$. If, on the other hand, $P \prec P_i$ and $P \neq P_i$, P is not a possible result of me, but from this argument we can say that, if a more accurate measurement me_1 having P as a possible result had been made, then P could have occurred only in the ensemble $\{\Sigma_k\}_{k \in K_i}$, and of course with the fraction stated by (4).

Let us finally glance briefly at the ontic formulation of quantum mechanics. In a purely formal manner, the condition (4) here corresponds to

$$\mathbf{pr}_*(\phi_i^+; P) = \frac{\mathbf{pr}_*(\phi; P)}{\mathbf{pr}_*(\phi; P_i)} \qquad (P \prec P_i) \tag{6}$$

for all properties P such that $P \prec P_i$, and for the same reasons as make (4) equivalent to (δ) in subsection II.4(e), this is equivalent to (δ) in section II.5. A further justification of (6), in this case without formal certainty, is, however, possible only if we use the epistemic or statistical extension of the ontic formulation. Then, from (OE) and (OE') or (OS) and (OS') in section II.5, we can argue as in the previous paragraph. Let there be initially an object Σ in a state ϕ, and let this object be regarded as a member of an ensemble $\{\Sigma_k\}_k$ where all the Σ_k are in the state ϕ. Then, from (OS) in section II.5, the ontically interpreted probabilities $\|P_a^A \phi\|^2$ are numerically equal to the statistically interpreted probabilities of obtaining the result a in a suitable measurement of A. If then the measurement $\dot{me} = \{P_j\}_j$ is made, the quotient on the right of (6) for $P \prec P_i$ has, for the part of the Σ_k which gives the result P_i, the same significance as was used in the previous paragraph; it can therefore be justified in the same way. Accordingly, the left-hand side of (6) must also be initially interpreted statistically, i.e. ϕ_i^+ denotes a purely statistical state of the specified part of $\{\Sigma_k\}_k$. Then using (OS') in section II.5, we again obtain from this an ontically interpreted probability for the individual object which belongs to this part and is in the individual state ϕ_i^+.

CHAPTER VI

Reduction of States and the Measurement Process

In the epistemic and statistical versions of classical mechanics (section II.3) and in the corresponding versions of the two principal formulations of orthodox quantum mechanics (sections II.4 and II.5) we find, as well as the continuous variation with time of an object or a statistical ensemble resulting from the postulated dynamics, also the abrupt reduction of states after acceptance or selection of the result of a measurement. In the classical case (section IV.3) this was an entirely intelligible process unaffected by the mere fact of making a measurement, and dependent only on the acceptance or selection accompanying it. That is to say, the reduction of states does not affect the object or ensemble itself, but only, in some way, the relation between it and the experimenter. In contrast, in quantum mechanics we find, in particular from the weak reduction of states ignoring the result (Section IV.3), that the mere making of a measurement influences the description of state immediately after the measurement; that the object or ensemble itself is affected by the measurement, and not only the experimenter's relation with it. In this situation, the axiomatization of the reduction of states in section V.4, embodying the result of the measurement, does not help much. It shows new assumptions under which the reduction of states is what it was assumed to be in the original formulation of quantum mechanics; but it does not do away with the phenomenon itself.

In section IV.3 we have given the analysis of the measuring process in quantum mechanics by Heisenberg (1930, section IV.2), who resolved it into the strong reduction of states without consideration of the result of the measurement (a specifically quantum-mechanical transition without a classical analogue or, in Heisenberg's later view, the transition from possibility to actuality), followed by the reduction to the result of the measurement, which in itself does not differ from the corresponding classical reduction, i.e. it *is* the classical reduction. Heisenberg calls the first of these two parts of the measuring process a "real process". This view suggests the desirability of further analysing the first part by asking to what extent this "real process" can be understood as a quantum-mechanical interaction between the object being measured and a measuring apparatus. The orthodox reply to this question is that such a replacement is not completely possible. In Heisenberg's words (1930, p. 58): "It must also be emphasized that the statistical character of the relation depends on the fact that the influence of the measuring device is treated in a different manner than the interaction of the various parts of the system on one another. This last interaction also causes changes in the direction of the vector representing the system in the Hilbert space, but these are completely determined. If one were to treat the measuring device as part of the system—which would necessitate an extension of the Hilbert space—then the changes considered above as indeterminate would appear determinate. But no use could be made of this determinateness unless our observation of the measuring device were free of indeterminateness.

For these observations, however, the same considerations are valid as those given above, and we should be forced, for example, to include our own eyes as part of the system, and so on. The chain of cause and effect could be quantitatively verified only if the whole universe were considered as a single system—but then physics has vanished, and only a mathematical scheme remains. The partition of the world into observing and observed system prevents a sharp formulation of the law of cause and effect. (The observing system need not always be a human being; it may also be an inanimate apparatus, such as a photographic plate.)" In fact, the Hilbert-space formulation of quantum mechanics shows more precisely that the weak reduction of states, but not the strong one, is reproducible in the sense of quantum mechanics by an interaction of an object with a measuring apparatus. As soon as we include explicitly in the analysis, in order to attain our objective, a measuring apparatus, which *itself* is, as it must be, treated quantum-mechanically, it necessarily ceases to have the properties of a measuring apparatus. Its interaction with the object then provides the weak reduction of states, which is indeed necessary, but not the strong reduction, which alone would be sufficient.

This negative result does yield the following positive point—although practically only as a necessary condition. The impossibility of making a measuring apparatus *as such* completely objective implies that we must always recognize something as completing the measurement process which is outside the part of the world placed with the object and treated quantum-mechanically together with the object. Whether we follow Bohr (as Heisenberg and Pauli did on this point) and regard this thing as a "final" measuring apparatus to be treated classically, or regard it as a "final" observer who effectively acquires knowledge of something (a view to which von Neumann seemed to incline), there is in any case the necessity of making a "cut" or division between these two parts. Whereas the *existence* of this division is unavoidable on the orthodox view, its *position* is found to be largely arbitrary; that is, when a measuring apparatus is incorporated in the system to be observed, the reduction of states can always be so arranged in respect of the resulting combined system that it agrees with the corresponding reduction for the original object. Thus the reduction of states can be moved to the composite system, in such a way that here it appears as a consequence of a measurement of the now included measuring apparatus—and this is a genuine gain.

The technical details of the relationships outlined above, which will now be presented more fully, were worked out in 1932 by von Neumann (1955, Chapter VI). Somewhat later, London and Bauer (1939) gave an elementary but very clear account. More recently, Süssmann (1957, 1958) in particular has given a full orthodox analysis of the process of measurement. On the other side, several papers have been published in the last few years which seek to give some explanation of the reduction of states, if only in the negative sense of showing that it can be practically eliminated at the macroscopic level. These papers cannot be discussed in detail here, but are asterisked in the bibliography, together with other contributions to the topic.

VI.1 Reproduction of the weak reduction of states without taking account of the result of measurement, by a dynamic process

The problem to be discussed in this section will first be considered for the epistemic or statistical formulation of quantum mechanics, without involving the states of individual objects. In the epistemic formulation we take an individual object, denoted by I (as an abbreviation of Σ^I) to distinguish it from another object II (for Σ^{II}) which will shortly be brought in. The concept of measurement for such an object I has been defined in subsection II.4(d)(i). If *me* is a measurement of I, the corresponding propositions regarding measurement $\mathbf{M}(me, A, \mathfrak{a}, a)$ in the epistemic interpretation state that the measurement *me* determines the quantity A of I with accuracy $\mathfrak{a}$ and result a. If at a time t there is a state of knowledge V regarding I and if the measurement *me*, described by $(\{P_i\}_i, P_j)$, is then carried out, the immediately subsequent state of knowledge is (see subsection II.4(e))

$$V_j^+ = \frac{P_j V P_j}{\mathrm{Tr}(VP_j)} \qquad (\mathrm{Tr}(VP_j) > 0); \tag{1}$$

we must have $\mathrm{Tr}(VP_j) > 0$, since otherwise the measurement *me*, which already embodies a result, could not be the one it is for the given V. The change from V to V_j^+ in accordance with (1) has been called the reduction of states taking account of the the result of measurement. From the measurements *me* which embody results, we derived by abstraction the measurements $\dot{me}$ in which the result is ignored (subsection II.4(d)(i)). The corresponding propositions regarding measurements were $\dot{\mathbf{M}}(\dot{me}, A, \mathfrak{a})$, stating that the measurement $\dot{me}$ of I measures the quantity A with accuracy $\mathfrak{a}$. No result is mentioned. Correspondingly, the mathematical description of $\dot{me}$ involved only a system $\{P_i\}_i$. A measurement $\dot{me}$ changes an initial state of knowledge V first into the assemblage of possible states of knowledge

$$\dot{V}^+ = \{V_j^+, \mathrm{Tr}(VP_j)\}_{j:\,\mathrm{Tr}(VP_j)>0} \tag{2}$$

in which V_j^+ is given by (1); there will remain just one of these states, with probability $\mathrm{Tr}(VP_j)$, if the actual result of the measurement, expected with this same probability, is taken into account. If the result is definitely ignored, the state of knowledge available for further predictions is only

$$V^+ = \sum_{j:\,\mathrm{Tr}(VP_j)>0} \mathrm{Tr}(VP_j) V_j^+. \tag{3}$$

The change from V to $\dot{V}^+$ or V^+ was termed respectively the strong or weak reduction of states without taking account of the result of measurement.

In the statistical formulation of quantum mechanics also, the formalism just recalled, and in particular formulae (1)–(3), were available by using a different interpretation. Here the basis is a statistical ensemble $\{I_k\}_k$ of objects I_k all having the same structure. We consider statistical measurements of such an ensemble, each such measurement being an ensemble of objectively interpreted individual measurements $me = (\{P_i\}_i, P_j)$ with fixed $\{P_i\}_i$ but variable results P_j. As such, the statistical measurement $\dot{me} = \{P_i\}_i$ is formally identical with an individual measurement in which the result is ignored. If the ensemble is initially in the (statistical) state V, a statistical measurement $\dot{me} = \{P_i\}_i$ changes it first into the assemblage $\dot{V}^+$ given by (2), which is now an assemblage of ensembles in the statistical states V_j^+ occurring with the probabilities (proportional to the relative frequencies) $\mathrm{Tr}(VP_j)$. The V_j^+ ensemble consists of just the I_k for which the result P_j occurs. If no selection of one of these V_j^+ ensembles is made, the new state is given by (3) when the old ensemble is V^+. One can also make a measurement with selection, however, and this is formally identical with an individual measurement $me = (\{P_i\}_i, P_j)$ in which the result is considered. The selected ensemble with the result P_j is then in the state V_j^+ given by (1).

In section IV.3 we have already examined in more detail the transition $V \rightarrowtail V_j^+$ in accordance with (1), and the transitions $V \rightarrowtail \dot{V}^+$ and $V \rightarrowtail V^+$ in accordance with (2) and (3) respectively, and found that the actual quantum-mechanical peculiarity resides in the two latter transitions: in contrast to the classical case, in general $V^+ \neq V$, and just this result (that the state of knowledge or the statistical state is changed by a measurement without acceptance or selection) made necessary a distinction between states of knowledge or statistical states and assemblages thereof, and in particular a distinction between the strong reduction of states (2) and the weak reduction of states (3). The reduction of states (1) was tentatively resolved into the initial transition (2) and a subsequent acceptance or selection by

$$\dot{V}^+ \rightarrowtail V_j^+, \tag{4}$$

the latter being entirely classical. The question now arises whether the classically unintelligible reduction of states without taking account of the result of measurement, i.e. (2) or (3), can be regarded as a process dependent on a normal interaction of the object I, or of each of the objects I_k, with a measuring apparatus that is used to perform the measurement $\dot{m}e$ with which (2) or (3) is correlated. By a "normal" interaction we here mean one which can be described *in quantum-mechanical terms* by a Hamiltonian operator for the object+measuring apparatus system. This question will be treated in the present section first of all in respect of the weak reduction of states (3).

In order to pursue this question, which was first attacked by von Neumann (1955, Chapter VI), we must explicitly define, as well as the fundamental object I or the I_k of an ensemble, a measuring apparatus II or an ensemble II_k of these. Like I or I_k, II or II_k and the composite object $I \otimes II$ or $I_k \otimes II_k$ is to be treated quantum-mechanically in relation to the question under discussion. Thus this leads to the Hilbert spaces $\mathfrak{H}^I$ of I or I_k, $\mathfrak{H}^{II}$ of II or II_k, and $\mathfrak{H}^I \otimes \mathfrak{H}^{II}$ of $I \otimes II$ or $I_k \otimes II_k$. The choice of $\mathfrak{H}^I \otimes \mathfrak{H}^{II}$ to describe the composite object in accordance with subsection II.4(b) is appropriate here because of the structural difference between object and apparatus. As we have idealized the transition (3) as occurring at a point in time, in order to obtain it we must assume a pulsed interaction between the object and the apparatus. This is most simply done by means of a unitary operator U in $\mathfrak{H}^I \otimes \mathfrak{H}^{II}$ which effects the instantaneous transition from an initial state of knowledge or statistical state $W^{\otimes}$ of $I \otimes II$ or $\{I_k \otimes II_k\}_k$ to a description of state $U \circ W^{\otimes}$ after the measurement, in the usual form

$$U \circ W^{\otimes} = UW^{\otimes}U^{-1}; \tag{5}$$

cf. subsection II.4(e). We shall further assume that $W^{\otimes}$ as a description of state before the measurement always has the special form

$$W^{\otimes} = V \otimes W_0. \tag{6}$$

This means that before the measurement the object and the apparatus are independently known, or uncorrelated. The subscript zero written with W on the right, but not with V, indicates that W_0 is involved in the desired statement of the problem as a neutral description of the state of the measuring apparatus; this is not true of V, since the dynamic recovery of (3) must be possible for *any* initial condition V of the object: every interaction in measurement thus begins with W_0 as regards the apparatus, whereas the V of the object can vary in any manner. After the interaction, the description of state (i.e. state of knowledge or statistical state) of object $\otimes$ apparatus is $U \circ (V \otimes W_0)$. In arriving at the transition (3), we need consider only the resulting description of state of the object. This is in general determined as follows. A description of state $W^{\otimes}$ of the product $I \otimes II$ yields for I the probabilities $\mathbf{pr}(W; A \otimes 1, a)$ with a quantity A of I. It can be shown that there is just one description of state $W_I^{\otimes}$ of I such that

$$\mathbf{pr}(W_I^{\otimes}; A, a) = \mathbf{pr}(W^{\otimes}; A \otimes 1, a) \tag{7}$$

for all A and a. This is because, for $W^{\otimes}$ as a von Neumann operator in $\mathfrak{H}^I \otimes \mathfrak{H}^{II}$, there is just one $W_I^{\otimes}$ in $\mathfrak{H}^I$ such that identically

$$\mathrm{Tr}(W_I^{\otimes} P_a^A) = \mathrm{Tr}(W^{\otimes} P_a^{A \otimes 1}) \tag{8}$$

(see, e.g., von Neumann 1955, section VI.2). The above description of state $U \circ (V \otimes W_0)$ therefore gives for I or $\{I_k\}_k$ unambiguously the description of state $(U \circ (V \otimes W_0))_I$. If now $\dot{m}e = \{P_i\}_i$ is the measurement for which the transition (3) is to be obtained, we have the reproduction of (3) if for U, W_0 and $\{P_i\}_i$

$$(U \circ (V \otimes W_0))_I = \Sigma_i P_i V P_i \tag{9}$$

for all V of I or $\{I_k\}_k$.

So far, we have indicated only a relationship between a (pulsed) dynamics U on $I \otimes II$ or all the $I_k \otimes II_k$, an initial description of state W_0 for II or $\{II_k\}_k$, and an individual measurement (ignoring the result) of I or a statistical measurement of $\{I_k\}_k$, viz. $\dot{m}e = \{P_i\}_i$: the relationship exists between the three specified entities if and only if (9) is valid for all initial descriptions of state V of I or $\{I_k\}_k$. In this case we shall say that *the dynamics U and the initial description of state W_0 of the apparatus replace the weak reduction of states* (3) *of the object corresponding to the measurement $\dot{m}e$*. In section 3 we shall show, under much more stringent conditions, that for a given measurement $\dot{m}e = \{P_i\}_i$ there are always a U and a W_0 such that the relationship just discussed is satisfied. This solves the replacement problem initially raised in the present section. We shall, however, still seek a deeper understanding of the replacement relation, and for this purpose use two other conditions in place of the defining condition (9).

It is immediately evident that there are two consequences of (9) for all V of I:

(α) If ϕ is any state of maximum knowledge of I through which the value of every quantity measured by $\dot{m}e$ with absolute accuracy is already known, or a pure state of $\{I_k\}_k$ in which every quantity measured by $\dot{m}e$ with absolute accuracy is free from dispersion (cf. section IV.2), then

$$(U \circ (P_\phi \otimes W_0))_I = P_\phi.$$

(β) For all descriptions of state V of I or $\{I_k\}_k$, $(U \circ (V \otimes W_0))_I$ is invariant under a (further) weak reduction of states effected by $\dot{m}e$, i.e. for all V

$$(U \circ (V \otimes W_0))_I = \Sigma_i P_i (U \circ (V \otimes W_0))_I P_i.$$

To prove (α) from (9) we need only note that the assumption about ϕ and $\dot{m}e = \{P_i\}_i$ whose content is specified there is formally equivalent to

$$\|P_i \phi\|^2 = \delta_{i i_0} \tag{10}$$

with a certain i_0 for all i. Then, with

$$P_i P_\phi P_i = \|P_i \phi\|^2 P_{P_i \phi}, \tag{11}$$

we obtain (α), and (β) follows by simple calculation.

The implication of the condition (α) is easily understood. The dynamics U leaves unchanged in respect of I or $\{I_k\}_k$ any initial maximum or pure description of state which would make unnecessary the measurement whose weak reduction of states is to be replaced by U. Everything which the measurement could furnish is already known or free from dispersion. It also follows easily that the same is true of *any* description of state if it is true for a maximum or pure description of state, as specified by (α). Hence we might say that (α) requires ideal conditions in the classical sense (the description of state to be unchanged by the measurement process in so far as it can be described as a dynamic process) at least for the descriptions of state which are trivial in relation to the measurement. The measurement involves only changes which are unavoidable in the quantum-mechanical sense.

Measurements which embody no unnecessary classical changes of state (even if these are capable of surveillance) were called measurements of the first kind by Pauli (1933, section A.9). The condition (α) in fact demands the preservation of certain descriptions of state. Von Neumann (1955, section VI.3, para. 5) discusses somewhat cryptically a conservation of the probabilities of the P_i for any descriptions of state ϕ. His condition is

(α') For any maximum state of knowledge ϕ of I or any pure state of $\{I_k\}_k$,

$$\mathrm{Tr}((U \circ (P_\phi \otimes W_0))_I P_i) = \mathrm{Tr}(P_\phi P_i),$$

where $\dot{m}e = \{P_i\}_i$ is the relevant measurement.

This follows from (α). The converse, however, is valid only for maximum measurements (see Mathematical Appendix 1 to the present section). The implication of (α') also is easily understood. The interaction must not alter the probabilities for the possible results of the measurement whose weak reduction of states is brought about by that interaction.

The following two-part condition on U, W_0 and $\{P_i\}_i$ is exactly equivalent to (α):

(α''_1) For all V of I or $\{I_k\}_k$,
$\Sigma_i P_i (U \circ (V \otimes W_0))_I P_i = \Sigma_i P_i V P_i$.

(α''_2) For all V of I or $\{I_k\}_k$ and all i, if
$\mathrm{Tr}((U \circ (V \otimes W_0))_I P_i) > 0$, then $\mathrm{Tr}(VP_i) > 0$ and

$$\frac{P_i (U \circ (V \otimes W_0))_I P_i}{\mathrm{Tr}((U \circ (V \otimes W_0))_I P_i)} = \frac{P_i V P_i}{\mathrm{Tr}(VP_i)}.$$

(α''_1) asserts that it does not matter whether the weak reduction of states (3) is carried out on V or only on $(U \circ (V \otimes W_0))_I$. ($\alpha''_2$) asserts the same for each of the reductions of states (1). Since, from (α''_1) and (α''_2),

$$\mathrm{Tr}((U \circ (V \otimes W_0))_I P_i) = \mathrm{Tr}(VP_i) \tag{12}$$

for all i, the same assertion is made also for the strong reduction of states (2). Thus all possible reductions of states are preserved by the interaction in measurement. This once again shows very clearly that the condition (α) for the conversion of V into $(U \circ (V \otimes W_0))_I$ prevents something negative rather than achieving something positive. It prevents the involvement of changes of state which ought specifically to be taken into account in the reduction of states. The quantum-mechanically indispensable change of V is, however, not present in (α), the latter being compatible even with the solution $U = 1$; it appears only in (β). We shall use the characterization of (α) by (α'') in proving the equivalence of (9) to (α) and (β), and also in section 3. It is also concerned in the proof of (α') from (α). The somewhat complicated proof of equivalence is in Mathematical Appendix 2 to the present section.

Passing to the consideration of (β), we may mention, as a final characterization of (α),

(α''') For all V of I or $\{I_k\}_k$ such that $\Sigma_i P_i V P_i = V$,

$$(U \circ (V \otimes W_0))_I = V.$$

Here the premise for the description of state V *before* the measurement includes the (restrictive) condition that is also required without restriction for the description of state $(U \circ (V \otimes W_0))_I$ *after* the measurement, according to (β). And for all such V it is then asserted that they are unaltered by the measurement. Whatever the implication of the condition presupposed for V (this will have to be elucidated later), the role of (α''') relative to (β) is clear. Since (β) has to be satisfied for the description of state after the measurement, (α''') formulates the idea that a V which already satisfies this condition before the measurement is unaltered by the latter; this is consistent, since what is to be achieved is already true. The equivalence of (α''') and (α) is proved in Mathematical Appendix 3 to the present section.

Our understanding of the distinctive features of quantum mechanics itself is dependent on a correct understanding of the condition (β). This will become clear in section 2, and especially in Chapter VII. To prepare for the further discussion here, we shall merely refer to the problems related to this condition. For any description of state V, and not only for those of the form $(U \circ (V \otimes W_0))_I$, and any measurement $\dot{m}e = \{P_i\}_i$, (β) becomes

$$\Sigma_i P_i V P_i = V, \tag{13}$$

and we know that it is a necessary condition for V^+ given by (3), and so also for $(U \circ (V \otimes W_0))_I$ given by (9). Moreover, it is evidently true for the V_j^+ given by (1): all these descriptions of state after a measurement satisfy (13). The only question is whether *conversely*, for an existing description of state related to a measurement by (13), we can say that the measurement can be regarded as having been performed, at least in the sense that the possible results are already objectively in existence and are at most unknown or subject to dispersion. It can be seen immediately that this view is unobjectionable if, as in the case of the V_j^+ given by (1), the last proviso is not invoked, i.e. in general if $\mathrm{Tr}(VP_i)$ is a δ-function distribution. Then, for every quantity measured with absolute accuracy by $\{P_i\}_i$ (cf. section IV.2), we can say that it *has* a well-defined value in V: for the individual object I, because this value is in fact known; for the ensemble $\{I_k\}_k$, because it is not subject to dispersion in that ensemble. There are, however, also cases of the relation (13) in which the results of measurement P_i have genuine probabilities, and for these the problem is a serious one. On the orthodox view there can be no affirmative answer without abandoning the completeness of quantum mechanics, as we shall see in Chapter VII. The difference between the strong and weak reduction of states already foreshadows this negative result.

Hitherto, (α) and (β) have been considered and discussed only as conditions necessary for (9), but they are also sufficient for (9). For, if (α) is valid, so is (α_1''), and with (β) this proves (9). This quick proof of course conceals the somewhat more complicated derivation of (α_1'') from (α), given in the Mathematical Appendix.

Finally, something must be said about the form of these problems of replacing the weak reduction of states when put in the ontic formulation of quantum mechanics, i.e. using states for individual objects. As we know from section II.5, in this version of quantum mechanics only the reduction of states *with* allowance for the result of measurement (here to be regarded purely as a record) can be formulated, since both the probabilities which occur explicitly in the strong reduction of states (without taking account of the result of measurement) and the new description of state which occurs in the weak reduction of states have no possible expression in the ontic version. Such a possibility occurs only in the epistemic or statistical extension of the ontic formulation of quantum mechanics. But there the position is essentially the same as was described above for the two corresponding formulations of quantum mechanics without individual states. This may be seen as follows.

Before the start of the interaction, V and W_0 would each have to be replaced by an assemblage of states, $\{\phi_m, v_m\}_m$ and $\{\psi_n^0, w_n^0\}_n$ respectively, and the entire object by the assemblage

$$\{(\phi_m \otimes \psi_n^0), v_m \cdot w_n^0\}_{m,n} \tag{14}$$

by analogy with (6). Then U converts this into the assemblage

$$\{U \circ (\phi_m \otimes \psi_n^0), v_m \cdot w_n^0\}_{m,n} \tag{15}$$

by analogy with (5). The weak description of state regarding I or I_k would then be given by

$$\begin{aligned} &\Sigma_{m,n} v_m w_n^0 U \circ (P_{\phi_m} \otimes P_{\psi_n^0}))_I \\ &\quad = (U \circ ((\Sigma_m v_m P_{\phi_m}) \otimes (\Sigma_n w_n^0 P_{\psi_n^0})))_I. \end{aligned} \tag{16}$$

Here we have immediately expressed the equality with the right-hand side, from which it now follows that the initial description by (14) is involved only to the extent that the weak description of state

$$(\Sigma_m v_m P_{\phi_m}) \otimes (\Sigma_n w_n^0 P_{\psi_n^0}) \tag{17}$$

is uniquely related to this strong description; (16) depends only on (17). Thus the left-hand side of the condition (9) is to be written exactly as previously. The same applies to the right-hand side of (9); as we already know from section II.5, the weak reduction of states in the ontic formulation depends only on the weak description of state before the measurement. The problem for the ontic formulation is thus reduced to the corresponding one, already discussed, for the formulation without individual states.

Mathematical Appendix to section VI. 1

1. The relationship between (α) and (α')

It will be shown in part 2 of this Appendix, formula (21), that (α') follows from (α). The converse derivation of (α) from (α') for a *maximum* $\{P_i\}_i = \{P_{\phi_i}\}_i$ is easy. If $\mathrm{Tr}(P_\phi \cdot P_{\phi_i})$ is a delta function of i, ϕ must be equal to one of the ϕ_i. Thus the premise of (α) reduces to $\phi = \phi_{i_0}$. Then, for this ϕ, (α') gives

$$\mathrm{Tr}((U \circ (P_\phi \otimes W_0))_I P_{\phi_i}) = \delta_{ii_0};$$

hence we have in general

$$(U \circ (P_\phi \otimes W_0))_I \phi_i = \delta_{ii_0} \phi_i$$

and so, since $\phi = \phi_{i_0}$,

$$(U \circ (P_\phi \otimes W_0))_I = P_\phi,$$

which is the conclusion of (α).

2. Proof that (α) and (α'') are equivalent

We shall first show that (α) follows from (α''). From (α_1''),

$$\sum_{j:\,\mathrm{Tr}((U\circ(V\otimes W_0))_I P_j)>0} P_j(U\circ(V\otimes W_0))_I P_j = \Sigma_j P_j V P_j. \tag{1}$$

Substitution from (α_2'') on the left of (1) gives

$$\sum_{j:\,\mathrm{Tr}((U\circ(V\otimes W_0))_I P_j)>0} P_j V P_j \cdot \frac{\mathrm{Tr}((U\circ(V\otimes W_0))_I P_j)}{\mathrm{Tr}(VP_j)} = \Sigma_j P_j V P_j. \tag{2}$$

Hence, for all j,

$$\mathrm{Tr}((U\circ(V\otimes W_0))_I P_j) = \mathrm{Tr}(VP_j); \tag{3}$$

for, on multiplying both sides of (2) on the right by P_i, if $\mathrm{Tr}((U\circ(V\otimes W_0))_I P_i) = 0$, we have $0 = P_i V P_i$ and hence $\mathrm{Tr}(VP_i) = 0$, which is (3) for this case. If $\mathrm{Tr}((U\circ(V\otimes W_0))_I P_i) > 0$, then

$$P_i V P_i \cdot \frac{\mathrm{Tr}((U\circ(V\otimes W_0))_I P_i)}{\mathrm{Tr}(VP_i)} = P_i V P_i$$

and hence (3) again follows, since now $P_i V P_i \neq 0$.

Now let $\mathrm{Tr}(P_\phi \cdot P_i)$ be a δ-function distribution. Then, from (3), the same is true of $\mathrm{Tr}((U\circ(P_\phi\otimes W_0))_I P_i)$ with the same i as argument to give the value 1 as for $\mathrm{Tr}(P_\phi P_i)$. Thus, for the same i,

$$\mathrm{Tr}((U\circ(P_\phi\otimes W_0))_I P_i) = 1 \text{ and } \mathrm{Tr}(P_\phi\cdot P_i) = 1. \tag{4}$$

From this we have generally

$$P_i(U\circ(P_\phi\otimes W_0))_I P_i = (U\circ(P_\phi\otimes W_0))_I,\ P_i P_\phi P_i = P_\phi. \tag{5}$$

By (4), the denominators of the equation in (α_2'') are equal. Since the numerators are then equal, the equation in (α) follows from (5).

Conversely, let (α) be valid. We first deduce (α'') only for $V = P_\phi$, i.e.

$$\Sigma_i P_i(U\circ(P_\phi\otimes W_0))_I P_i = \Sigma_i P_i P_\phi P_i, \tag{6}$$

and if $\mathrm{Tr}((U\circ(P_\phi\otimes W_0))_I P_i) > 0$

$$\frac{P_i(U\circ(P_\phi\otimes W_0))_I P_i}{\mathrm{Tr}((U\circ(P_\phi\otimes W_0))_I P_i)} = \frac{P_i P_\phi P_i}{\mathrm{Tr}(P_\phi P_i)}. \tag{7}$$

To calculate the left-hand side of (7) we must now derive the most explicit possible form of $U\circ(P_\phi\otimes W_0)$.

For this purpose we select a basis $\{\phi_\mu\}_\mu$ for $\mathfrak{H}^I$ through the P_i, i.e.

$$P_i\phi_\mu = \delta_{ig(\mu)}\phi_\mu \tag{8}$$

with a mapping g which brings together the μ belonging to the same i. Then, for every μ, $\mathrm{Tr}(P_{\phi_\mu}P_i)$ is a δ-function distribution, and hence, from (α),

$$(U\circ(P_{\phi_\mu}\otimes W_0))_I = P_{\phi_\mu} \tag{9}$$

for all μ. From (9) it follows that

$$U \circ (P_{\phi_\mu} \otimes W_0) = P_{\phi_\mu} \otimes W^{(\mu)} \tag{10}$$

with certain von Neumann operators $W^{(\mu)}$. For W_0, there are numbers w_m^0 such that

$$w_m^0 > 0, \ \Sigma_m w_m^0 = 1, \tag{11}$$

and a normalized orthogonal system $\{\psi_m^0\}_m$ in $\mathfrak{H}^{II}$ such that

$$W_0 = \Sigma_m w_m^0 P_{\psi_m^0}. \tag{12}$$

For every $\phi \in \mathfrak{H}^I$, it follows that

$$U \circ (P_\phi \otimes W_0) = \Sigma_m w_m^0 P_{U(\phi_\mu \otimes \psi_m^0)}. \tag{13}$$

For a fixed μ, the $\phi_\mu \otimes \psi_m^0$ are eigenvectors of $P_{\phi_\mu} \otimes W_0$ with eigenvalues $w_m^0 > 0$; hence the $U(\phi_\mu \otimes \psi_m^0)$ are eigenvectors of $U \circ (P_{\phi_\mu} \otimes W_0)$ with the same positive eigenvalues. From (10), however, $U \circ (P_{\phi_\mu} \otimes W_0)$ has the form $P_{\phi_\mu} \otimes W^{(\mu)}$. The $U(\phi_\mu \otimes \psi_m^0)$ must therefore have the form

$$U(\phi_\mu \otimes \psi_m^0) = \phi_\mu \otimes \psi_m^{(\mu)}. \tag{14}$$

Since U is unitary,

$$\langle \psi_m^{(\mu)} | \psi_n^{(\mu)} \rangle = \delta_{mn}, \tag{15}$$

and it may be mentioned in passing that (10), (12) and (14) give

$$W^{(\mu)} = \Sigma_m w_m^0 P_{\psi_m(\mu)}. \tag{16}$$

The principal result from (14) is, however,

$$U(\phi \otimes \psi_m^0) = \Sigma_\mu \langle \phi_\mu | \phi \rangle (\phi_\mu \otimes \psi_m^{(\mu)}) \tag{17}$$

and hence, from (13),

$$U \circ (P_\phi \otimes W_0) = \Sigma_m w_m^0 P_{\Sigma_\mu < \phi_\mu | \phi > (\phi_\mu \otimes \psi_m(\mu))}. \tag{18}$$

Using (18), we can now easily calculate for $\phi \neq 0$

$$(P_i \otimes 1) U \circ (P_\phi \otimes W_0)(P_i \otimes 1) = \frac{\|P_i \phi\|^2}{\|\phi\|^2}, \tag{19}$$

using (8) and $\|\psi_m^{(\mu)}\|^2 = 1$ from (15). For $\phi \neq 0$,

$$\mathrm{Tr}(P_\phi P_i) = \frac{\|P_i \phi\|^2}{\|\phi\|^2}, \tag{20}$$

and so from (19)

$$\mathrm{Tr}((U \circ (P_\phi \otimes W_0))_I P_i) = \mathrm{Tr}(P_\phi P_i) \tag{21}$$

for all ϕ and i. For $\mathrm{Tr}((U \circ (P_\phi \otimes W_0))_I P_i) > 0$ we therefore have also

$$\frac{(P_i \otimes 1) U \circ (P_\phi \otimes W_0)(P_i \otimes 1)}{\mathrm{Tr}((U \circ (P_\phi \otimes W_0))(P_i \otimes 1))}$$
$$= \Sigma_m w_m^0 P_{\Sigma_\mu < \phi_\mu | \phi > (\phi_\mu \otimes \psi_m(\mu))}. \tag{22}$$

When (18) is written for $P_i\phi$ instead of ϕ, we have on the right-hand side just the right-hand side of (22). Hence

$$\frac{(P_i \otimes 1)U \circ (P_\phi \otimes W_0)(P_i \otimes 1)}{\mathrm{Tr}(U \circ (P_\phi \otimes W_0)(P_i \otimes 1))} = U \circ (P_{P_i\phi} \otimes W_0). \tag{23}$$

When ϕ is replaced by $P_i\phi$, (α) gives

$$(U \circ (P_{P_i\phi} \otimes W_0))_I = P_{P_i\phi} \tag{24}$$

for all ϕ and i, since $\mathrm{Tr}(P_{P_i\phi}P_j)$ is always a δ-function distribution and the premise of (α) is therefore fulfilled. From (23) and (24), when $\mathrm{Tr}((U \circ (P_\phi \otimes W_0))_I P_i) > 0$, we have

$$\left(\frac{(P_i \otimes 1)U \circ (P_\phi \otimes W_0)(P_i \otimes 1)}{\mathrm{Tr}(U \circ (P_\phi \otimes W_0)(P_i \otimes 1))}\right)_I = P_{P_i\phi}. \tag{25}$$

In general, as is easily proved,

$$\frac{P_i(U \circ (P_\phi \otimes W_0))_I P_i}{\mathrm{Tr}(U \circ (P_\phi \otimes W_0))_I P_i} = \left(\frac{(P_i \otimes 1)U \circ (P_\phi \otimes W_0)(P_i \otimes 1)}{\mathrm{Tr}(U \circ (P_\phi \otimes W_0)(P_i \otimes 1))}\right)_I. \tag{26}$$

If $\mathrm{Tr}(P_\phi P_i) > 0$,

$$\frac{P_i P_\phi P_i}{\mathrm{Tr}(P_\phi P_i)} = P_{P_i\phi}. \tag{27}$$

Then, from (25)–(27), we have (7), which with (21) gives (6).

We now have to generalize (6) and (7) to all statistical operators. With the representation $V = \Sigma_n v_n P_{\chi_n}$,

$$(U \circ (V \otimes W_0))_I = \Sigma_n v_n (U \circ (P_{\chi_n} \otimes W_0))_I. \tag{28}$$

Multiplying by P_i, taking the trace, and using (21), we have

$$\mathrm{Tr}((U \circ (V \otimes W_0))_I P_i) = \mathrm{Tr}(VP_i). \tag{29}$$

Multiplication of (28) by P_i on the right and on the left gives

$$P_i(U \circ (V \otimes W_0))_I P_i = \Sigma_n v_n P_i (U \circ (P_{\chi_n} \otimes W_0))_I P_i. \tag{30}$$

The numerators in (7) are equal, according to (21). Thus substitution on the right of (30) gives immediately

$$P_i(U \circ (V \otimes W_0))_I P_i = P_i V P_i. \tag{31}$$

From this we have (α_1''), and (29) then gives (α_2'').

3. Proof that (α) and (α''') are equivalent

The derivation of (α) from (α''') is trivial. Conversely, let (α) be valid. The premise of (α''') is, by general theorems, equivalent to the existence of an orthonormal basis $\{\phi_k\}_k$ for $\mathfrak{H}^I$ which (1) passes through all the P_i, i.e. satisfies (8) in Mathematical Appendix 2, and (2) forms a complete set of eigenvectors for V: $V = \Sigma_k v_k P_{\phi_k}$. From this it follows generally that

$$(U \circ (V \otimes W_0))_I = \Sigma_k v_k (U \circ (P_{\phi_k} \otimes W_0))_I$$

and the conclusion of (α''') follows from (α) and (8) in Mathematical Appendix 2.

VI.2 Impossibility of reproduction of the strong reduction of states without taking account of the result of measurement, by a dynamic process

Having shown in section 1 that the weak reduction of states, equation (3), can be effected for an individual object I or ensemble $\{I_k\}_k$ of such objects by an interaction of I or $\{I_k\}_k$ with a measuring apparatus II or an ensemble $\{II_k\}_k$ of these, and how this can be done, we now ask whether, and if so how, the same can be achieved in respect of the strong reduction of states, (2) in section 1, and hence in respect of the complete effect of measurement on I or on the I_k. For the quantum mechanics without states of individual objects, it is easily seen that such a reproduction is impossible for a purely formal reason. If the problem is again stated in the same way as was done for the reproduction of the weak reduction of states in section 1, then before the interaction in measurement there would be a description of state $V \otimes W_0$ of the whole system. This $V \otimes W_0$ would then be converted by U into $U \circ (V \otimes W_0)$, and we should again have to ask what is known about I in the new situation or what kind of statistics is valid for $\{I_k\}_k$. The only available answer is that the $(U \circ (V \otimes W_0))_I$ defined already by (5) in section 1 describes the new situation. Thus the usual formal treatment of the interaction by a Hamiltonian operator for the whole system does not lead to a multiplicity of descriptions of state for I or $\{I_k\}_k$, accompanied by a multiplicity of probabilities, such as would be necessary for deriving the V_j^+ and $\mathrm{Tr}(VP_j)$ of the strong reduction of states in (2), section 1. Such a multiplicity could be obtained only by taking as an assemblage the initial description of state V or W_0 or both. The question, however, is whether such a treatment can possibly be proved *necessary*, since only then might it be feasible to arrive at a reproduction of the strong reduction of states in *every* case.

One could argue that the *apparatus* must be a macroscopic body, and therefore is never the subject of a state of maximum knowledge or—from the statistical viewpoint—never belongs to a pure ensemble. But this does not immediately tell us any more than that the description of the apparatus cannot be maximum or pure, as is indeed necessary in order to be able to take a true assemblage for the apparatus. It is also sufficient to make such a treatment possible. But the necessity of this treatment is not yet evident, especially as the fact that the apparatus is largely unknown rather suggests that it should be described by a W_0 and not by a corresponding assemblage. Thus it seems that the proof is not attainable for quantum mechanics without states of individual objects.

In this respect the situation is different in the ontic version with states for the individual object. Here we start from the assumption that the individual object *is* in a certain state. If the state is unknown, the epistemic extension of the theory says that the true state is just one of a series $\{\psi_n^0\}_n$ of possible states ψ_n^0 which occur with the epistemically interpreted probabilities w_n^0. If, on the other hand, there is an ensemble, each of its objects is said to be in a certain state ψ_n^0, which occurs in the ensemble with a certain relative frequency w_n^0. In both cases we necessarily obtain an assemblage $\{\psi_n^0, w_n^0\}_n$. However, there are some problems resulting from the view that a quantum-mechanical object *is* in a certain state. For, if an object treated quantum-mechanically, as in this case the measuring apparatus, is coupled with another quantum-mechanical object in such a way that the state of the combined object is not a $\otimes$ product, the components of this state do not distinguish an assemblage of states as regards the constituent objects; cf. (7) in section 1. It is therefore not clear how we may say in such a case that the constituent objects *are* in certain states. This important point will be the subject of detailed discussion in Chapter VII; in the present context it is mentioned simply to show that only with the special hypothesis that the apparatus is completely "uncoupled" can we argue that an assemblage of states has necessarily to be assumed for it. In respect of the object, this hypothesis is in any case made for the instant preceding the interaction in measurement; cf. section 1, (6) and (14). Thus, to be sure that it is satisfied for the apparatus also, we need only ensure that no other object is concerned.

If this is done, it then follows in fact, therefore, that we have an assemblage of states in the form $\{\psi_m^0, w_m^0\}_m$ for the apparatus. If we now further assume that this is an assemblage of high order, because the measuring apparatus is macroscopic, we might hope that the assemblage which contains the object after the measurement can be generated by an interaction process. In fact, however, this is impossible, as is most clearly seen by assuming that the state ϕ of the object I before the interaction is known or (in the statistical case) that all the I_k in the ensemble $\{I_k\}_k$ are in the same state. We then start with the assemblage $\{\phi \otimes \psi_n^0, w_n^0\}_n$ and finish (after the interaction) with the assemblage $\{U \circ (\phi \otimes \psi_n^0), w_n^0\}_n$. Let us try to make things easy by assuming that this assemblage has the same multiplicity as the assemblage to be reproduced

$$\left\{\frac{P_i\phi}{\|P_i\phi\|}, \|P_i\phi\|^2\right\}_{i:\|P_i\phi\|^2>0} \tag{1}$$

from (2) in section 1, in which now $V = P_\phi$. Then we have $\{U \circ (\phi \otimes \psi_i^0), w_i^0\}_i$. From the remark just made, we know that in general this assemblage does not uniquely determine an assemblage of individual states with respect to I or $\{I_k\}_k$. On the other hand, of course, the possible states of (still maximum) knowledge or the possible (and still pure) statistical states $U \circ (\phi \otimes \psi_i^0)$ will uniquely give the possible states of knowledge or statistical states $(U \circ (\phi \otimes \psi_i^0))_I$. Even if these are in general, as already stated, not *inherently* maximum or pure states, we might now *impose* the equation

$$(U \circ (\phi \otimes \psi_i^0))_I = \frac{P_i\phi}{\|P_i\phi\|} \tag{2}$$

and regard it as at least part of the reproduction condition corresponding to (9) in section 1. Indeed there is no other possible choice. The other part of the reproduction condition must relate to the probabilities $\|P_i\phi\|^2$ in (1). Since the reproduction of (1) must certainly reproduce the weak reduction of state belonging to (1), (9) in section 1 must itself be valid, i.e. in this case

$$(U \circ (P_\phi \otimes \Sigma_i w_i^0 P_{\psi_i^0}))_I = \Sigma_i \|P_i\phi\|^2 P_{P_i\phi}. \tag{3}$$

From (2) and (3) we have immediately

$$\Sigma_i w_i^0 P_{P_i\phi} = \Sigma_i \|P_i\phi\|^2 P_{P_i\phi}$$

for all ϕ, and hence

$$w_i^0 = \|P_i\phi\|^2$$

for all ϕ and i such that $P_i\phi \neq 0$. Since the w_i^0 are specified independently of ϕ, this is clearly impossible.

This negative result, first deduced by von Neumann (1955, section VI.3) using a similar argument, can be explained as inherent in quantum mechanics. In section 1 it has been shown how the reproduction of the weak reduction of states (3) can be attained by a dynamic process of interaction between the object I and the apparatus II. The condition to be satisfied by the measurement $\dot{m}e = \{P_i\}_i$, the interaction U and a neutral initial description W_0 of the apparatus was (9) in section 1, for all descriptions of state V of I. This was in turn represented by (α) and (β), the latter condition giving rise to the question whether (β) is sufficient, as well as necessary, to ensure that $(U \circ (V \otimes W_0))_I$ can be replaced by the assemblage

$$\left\{\frac{P_i(U \circ (V \otimes W_0))_I P_i}{\mathrm{Tr}((U \circ (V \otimes W_0))_I P_i)}, \mathrm{Tr}((U \circ (V \otimes W_0))_I P_i\right\}_i. \tag{4}$$

Together with (α) in section 1 (cf. the representation of (α) by (α_1'') and (α_2'')), this would mean, however, that $(U \circ (V \otimes W_0))_I$ could be replaced also by the assemblage (2) in section 1. We have seen that the latter is impossible, and it therefore follows that the replacement of $(U \circ (V \otimes W_0))_I$ by (4) is not ensured only by (9) in section 1. The process of measurement is not yet completed by the transition from V to $(U \circ (V \otimes W_0))_I$.

Setting aside the objections to this view which will be discussed in Chapter VII, according to the analysis so far all that can be achieved is a shift of the problem of the measurement $\dot{m}e = \{P_i\}_i$ of I to a measurement $\dot{m}e^{\otimes} = \{P_i \otimes 1\}_i$ of $I \otimes II$, in which an interaction applied can again lead only to (9) in section 1; this must be followed by a reduction of the description of state $U \circ (V \otimes W_0)$ for the *whole* system, which would lead as a weak reduction to a new description of state

$$\Sigma_i(P_i \otimes 1)U \circ (V \otimes W_0)(P_i \otimes 1), \tag{5}$$

but as a strong reduction to the assemblage

$$\left\{\frac{(P_i \otimes 1)U \circ (V \otimes W_0)(P_i \otimes 1)}{\mathrm{Tr}(U \circ (V \otimes W_0)(P_i \otimes 1))}, \mathrm{Tr}(U \circ (V \otimes W_0)(P_i \otimes 1))\right\}_i. \tag{6}$$

Since in general

$$(\Sigma_i(P_i \otimes 1)U \circ (V \otimes W_0)(P_i \otimes 1))_I = \Sigma_i P_i(U \circ (V \otimes W_0))_I P_i, \tag{7}$$

and for all i

$$\left(\frac{(P_i \otimes 1)U \circ (V \otimes W_0)(P_i \otimes 1)}{\mathrm{Tr}(U \circ (V \otimes W_0)(P_i \otimes 1))}\right)_I = \frac{P_i(U \circ (V \otimes W_0))_I P_i}{\mathrm{Tr}((U \circ (V \otimes W_0))_I P_i)}, \tag{8}$$

we have from (α) (section 1) alone

$$(\Sigma_i(P_i \otimes 1)U \circ (V \otimes W_0)(P_i \otimes 1))_I = \Sigma_i P_i V P_i \tag{9}$$

and

$$\left(\frac{(P_i \otimes 1)U \circ (V \otimes W_0)(P_i \otimes 1)}{\mathrm{Tr}\,(U \circ (V \otimes W_0)(P_i \otimes 1))}\right)_I = \frac{P_i V P_i}{\mathrm{Tr}(VP_i)}, \tag{10}$$

as can be seen immediately from the representation (α_1'') and (α_2'') of (α) in section 1. Conversely, (9) and (10) lead to (α), because of (7) and (8) and the equivalence of (α) and (α_1'')+(α_2''). Thus (α) not only preserves the various reductions of states of system I in the interaction U, as has been shown in section 1, but also ensures that this reduction can be shifted from I to the whole system $I \otimes II$.

If (β) in section 1 is valid as well as (α), U generates the weak reduction of states for the measurement $\dot{me} = \{P_i\}_i$ of I. It is *not* true, however, that the corresponding weak reduction

$$U \circ (V \otimes W_0) = \Sigma_i (P_i \otimes 1) U \circ (V \otimes W_0)(P_i \otimes 1) \tag{11}$$

for the whole system $I \otimes II$ is generated; from (11) for a particular V, (β) follows for the same V, but not conversely. Thus, as regards the whole system, the interaction U does not effect the measurement $\dot{me}^{\otimes} = \{P_i \otimes 1\}_i$. We cannot say that the value of any of the quantities of I measurable with absolute accuracy by $\dot{me} = \{P_i\}_i$ has been measured, in respect of the description of state $U \circ (V \otimes W_0)$ of the whole system $I \otimes II$ which is valid after the interaction: (11) is a necessary (not even sufficient) condition for this, and is not satisfied. This is one reason for the failure of the attempt to achieve, by means of the interaction U, not only the weak but also the strong reduction for I. The fact that (11) does not follow from (9) in section 1 is of course not dependent on the particular form of the description of state $W^{\otimes} = U \circ (V \otimes W_0)$. But in the present special context one may ask for which V (11) is valid. For this we can prove:

(α) If the condition (α) in section 1 is valid for $\{P_i\}_i$, U and W_0, then (11) is valid for a V of the system I if and only if

$$V = \Sigma_i P_i V P_i. \tag{12}$$

From the equivalence of (α''') and (α) in section 1, we know that, if the latter is assumed, the interaction in measurement U does not affect a V which satisfies (12): the weak reduction has already been performed, and does not need to be effected by U. We obtain (11) only in this entirely trivial case. The result (α) above was first derived by Süssmann (1958, section III.B) for maximum measurements. The proof of the foregoing generalization to non-maximum measurements is given in the Mathematical Appendix to this section.

The negative result of this section confirms Bohr's view, described in Chapter I and repeated at the beginning of this chapter, that a purely quantum-mechanical analysis of what physically occurs in a process of measurement is impossible. From some point onwards, where the "cut" is made, the measuring equipment has to be treated *classically*. We only seem here to be deriving as a conclusion what Bohr regarded as a postulate, in that this postulate has its equivalent in the treatment of quantum mechanics according to sections II.4 and II.5, and we have here shown that, once this equivalence has been put into the theory, it cannot be got rid of. The equivalent concerned is the treatment of the strong reduction of states, which completes a measurement process without acceptance or selection, as a *classical* assemblage. It can be shown that a quantum-dynamical process strictly cannot yield such an assemblage, but may do so approximately (in a quantitative sense) whenever a measurement is possible, since the equation (11) is approximately valid (in a sense to be defined). This cannot be further considered here, but the view in question depends essentially on stressing the necessary occurrence of both a quantum-mechanical and a classical part in the complete analysis of a quantum phenomenon. On the other hand, von Neumann (1955, section VI.1) appears to have sought to except quantum mechanics alone from the process of (subjective) acceptance of the result of a measurement. This view has a disadvantage of asserting that only the acceptance of the result decides what has been measured. Before continuing the study of these topics in Chapter VII, we must first consider a sharper form of the positive conclusion of section 1, namely the at least partial analysis of the measurement by a dynamic process.

Mathematical Appendix to section VI.2

Equation (11) above is equivalent to

$$(P_i \otimes 1)U \circ (V \otimes W_0)(P_j \otimes 1) = 0 \tag{1}$$

for $i \neq j$. With a basis $\{\phi_\mu\}$ for $\mathfrak{H}^I$ through the P_i (cf. Mathematical Appendix to section VI.1, part 2, equation (8)), (1) is equivalent to

$$\langle U \circ (V \otimes W_0)(\phi_\mu \otimes \psi)|(\phi_\lambda \otimes \chi)\rangle = 0 \tag{2}$$

for $g(\mu) \neq g(\lambda)$ and identically in $\psi, \chi \in \mathfrak{H}^{II}$. With a representation

$$V = \Sigma_n v_n P_{\zeta n} \tag{3}$$

for V,

$$U \circ (V \otimes W_0) = \Sigma_n v_n U \circ (P_{\zeta_n} \otimes W_0). \tag{4}$$

With the representation of $U \circ (P_{\zeta_n} \otimes W_0)$ given by equation (18) in the Mathematical Appendix to section VI.1, part 2, which follows from (α), we have

$$U \circ (V \otimes W_0) = \Sigma_n v_n \Sigma_m w_m^0 P_{\Sigma_\mu <\phi_\mu|\zeta_n> (\phi_\mu \otimes \psi_m(\mu))}. \tag{5}$$

Substitution of (5) in (2) gives the new condition

$$(\Sigma_n v_n \langle \phi_\mu|\zeta_n\rangle\langle\zeta_n|\phi_\lambda\rangle)(\Sigma_m w_m^0 \langle\psi|\psi_m^{(\mu)}\rangle\langle\psi_m^{(\lambda)}|\chi\rangle) = 0. \tag{6}$$

Since this must be identically valid in ψ and χ, it is equivalent to

$$\Sigma_n v_n \langle\phi_\mu|\zeta_n\rangle\langle\zeta_n|\phi_\lambda\rangle = 0 \tag{7}$$

for $g(\mu) \neq g(\lambda)$. From (3), this is simply

$$\langle V\phi_\mu|\phi_\lambda\rangle = 0 \tag{8}$$

for $g(\mu) \neq g(\lambda)$, and (8) is equivalent to (12) in section 2, in the same way as (2) is equivalent to (11) there.

VI.3 Inclusion of a measurement on the apparatus

In the last two sections we have seen the position as regards the reproduction of the reduction of states without taking account of the result of measurement of an object I or an ensemble $\{I_r\}_r$, by the interaction of these with another object II or ensemble $\{II_r\}_r$ of objects regarded as a measuring apparatus. Let us now consider the same problem of reproduction by incorporating a further component besides the interaction. Since the following discussion, like that in sections 1 and 2, provides no new aspect as regards the ontic formulation of quantum mechanics, it will be given only for the weaker formulation, without including individual states. We then have to deal with a measurement $\{P_j\}_{j \in J}$ of I or of an ensemble $\{I_r\}_r$, an interaction U between I or the I_r and II or the II_r, and an initial description of state W_0 for II or $\{II_r\}_r$, and as our new elements a measurement $\{Q_k\}_{k \in K}$ of II or of $\{II_r\}_r$ (again without taking account of the result of measurement) and a one-to-one subscript mapping $f: I \rightarrowtail K$. For this total of five entities we shall consider a certain relationship, to the effect that *for the initial description of state W_0 the interaction U, together with the* *measurement* $\{Q_k\}_k$ *which is connected with* $\{P_i\}_i$ *by* f, *replaces the latter measurement.* That is to say, not only will U reproduce the weak reduction of states connected with $\{P_i\}_i$, but at the same time the measurement $\{P_i\}_i$ of the object concerned will be replaced *explicitly* by a measurement $\{Q_k\}_k$ of the apparatus concerned. Such an explicit inclusion of a measurement on the apparatus is the more necessary in that, as we have seen, an interaction between the object and the apparatus in which the latter also is treated quantum-mechanically cannot constitute the measurement of the object. On the other hand, it must be emphasized that the (probabilistic) correlation which allows the replacement of $\{P_i\}_i$ by $\{Q_k\}_k$ is not a typically quantum-mechanical correlation; it has an exact counterpart in classical statistical mechanics. In both theories there must first be an interaction which correlates the descriptions of state (assumed initially independent) for the object and apparatus in such a way that information obtained from the apparatus allows definite conclusions as to the behaviour of the object. In quantum mechanics, this interaction must also produce the weak reduction of states of the object. We shall see that it necessarily does so. In classical mechanics, the interaction has only to provide the correlation; as we know, in classical mechanics the weak reduction of states leaves the description of state invariant, so that no additional problem arises. After completion of the interaction in measurement, both theories must again ensure that the reduction of states for the apparatus, taking account of the result, leads to the same information as would have been obtained by the corresponding direct reduction of the object. In particular, the strong reduction of states is thereby reproduced also. As has been shown in section 2, this process is in general again a true quantum-mechanical reduction, whereas in the classical case we only accept or select what is already in existence.

By way of technical detail, let us subject the replacement relation between $\{P_i\}_i$, U, W_0, $\{Q_k\}_k$ and f, which so far has been described only as regards content, to the following two conditions:

(α) A condition identical with (α) in section 1, involving only $\{P_i\}_i$, U and W_0.

(β) For all $j \in J$ and $k \in K$ with $k \neq f(j)$ and any V,

$$\mathrm{Tr}(U \circ (V \otimes W_0)(P_j \otimes Q_k)) = 0.$$

Here (α), the same condition as in section 1, requires that the interaction in measurement should be conservative (measurements of the first kind as defined by Pauli (1933, section A.9)). In the classical case, where there are no maximum or pure descriptions of state, (α) would have to be imposed for every description of state of I or $\{I_r\}_r$. The condition (β), in this form, was stated by von Neumann (1955, section VI.3, para. 7), together with the condition (α′) in section 1, which is a sufficient replacement for (α) only for maximum measurements $\{P_i\}_i$. It will be shown in Mathematical Appendix 1 to this section that (β) in section 1 follows from (β) above. Consequently, the reproduction condition (9) discussed in section 1 follows, as it should, from the corresponding conditions (α) and (β) in this section.

The question to what extent, conversely, (α) and (β) in section 1 as conditions on $\{P_j\}_{j \in J}$, U and W_0 are sufficient for the existence of $\{Q_k\}_{k \in K}$ and f, so that all five together satisfy (α) and (β) above, seems to be still unresolved. They are probably not sufficient, and (β) is therefore a genuine additional condition on the dynamic measurement process U also. This can be seen from the following conditions relating to only the first three entities, which are necessary and sufficient for the existence of $\{Q_k\}_{k \in K}$ and f with the above (α) and (β):

(α') The same as (α), and therefore the same as (α) in section 1.

(β') It has been shown in part 2 of the Mathematical Appendix to section 1 (formula (10)) that, for a ϕ such that $P_i\phi = \phi$ with some $i \in J$,

$$U \circ (P_\phi \otimes W_0) = P_\phi \otimes W^{(\phi)}, \tag{1}$$

where $W^{(\phi)}$ is a von Neumann operator. Let $P_{(W^{(\phi)})}$ be the carrier of $W^{(\phi)}$. The condition imposed is that

$$P_{(W^{(\phi)})}\, P_{(W^{(\phi_1)})} = 0 \tag{2}$$

for ϕ such that $P_i\phi = \phi$ and ϕ_1 such that $P_j\phi_1 = \phi_1$, with $i \neq j$.

The significance of (β') is as follows (in the epistemic formulation). If there is a maximum state of knowledge ϕ of the object I at the beginning of the measuring process, so that a definite result of the measurement is already known, then $W^{(\phi)}$ is the state of knowledge of the apparatus II after the interaction, and $P_{(W^{(\phi)})}$ is the property of II that implies all other properties known with certainty after the interaction with II. The condition (2) requires that such properties exclude each other if the initial states of knowledge of I which give rise to them belong to mutually exclusive P_j and P_i; cf. Süssmann (1958, section III.A.2.b). Only in this way can the apparatus separate the properties P_i and P_j in such a manner that the result of the measurement as shown by the apparatus gives unambiguous conclusions about the object. The proof that (α') and (β') are necessary and sufficient for the existence of $\{Q_k\}_k$ and f with (α) and (β) is given in Mathematical Appendix 2 to the present section.

The condition (α) has been sufficiently discussed in section 1. We can therefore proceed at once to a corresponding discussion of (β). First, it should perhaps be pointed out, since the point will be disregarded in the subsequent analysis, that (β) gives a particularly strong connection between the five entities mentioned in it, in that it is a universal assertion concerning the description of state V of I or $\{I_r\}_r$. But, apart from this, even when V is given, the equation in (β) is important, as being still valid for all $j \in J$ and $k \in K$ with $k \neq f(j)$. For given V, the equation represents the situation existing after completion of the interaction, and in order to assess it we need only replace $U \circ (V \otimes W_0)$ by an unspecified $W^{\otimes}$. Then the relation obtained between $W^{\otimes}$, $\{P_j\}_{j \in J}$, $\{Q_k\}_{k \in K}$ and f is

(β_1) For $j \in J$ and $k \in K$ with $k \neq f(j)$,

$$\mathrm{Tr}(W^{\otimes}(P_j \otimes Q_k)) = 0.$$

This asserts directly that, if the measurements $\{P_j\}_j$ of I or $\{I_r\}_r$ and $\{Q_k\}_k$ of II or $\{II_r\}_r$ are made simultaneously (cf. section IV.2) or (which amounts to the same thing) if the measurement $\{P_j \otimes Q_k\}_{j \in J,\, k \in K}$ of $I \otimes II$ or $\{I_r \otimes II_r\}_r$ is made, properties P_j and Q_k with $f(j) \neq k$ cannot occur simultaneously as results, or, equivalently, that the properties $P_j \otimes Q_k$ with $f(j) \neq k$ are not possible results. This already foreshadows the correlation envisaged between the possible results P_j and Q_k: results are possible only in the combination P_j and $Q_{f(j)}$, or $P_j \otimes Q_{f(j)}$. It must be understood that this connection arises only *contingently*, from the particular $W^{\otimes}$ which happens to occur. This is simply an extreme case of the quantum-mechanical theory of probabilistic correlations—a theory whose classical analogue is familiar from the ordinary theory of probability. The theories are indeed identical so long as only properties in a Boolean sublattice (cf. section V.1) are concerned, as is true here. The counterpart of the correlation (1) is the case where the two measurements are independent in $W^{\otimes}$; this is, of course, defined by

$$\mathrm{Tr}(W^{\otimes}P_j \otimes Q_k) = \mathrm{Tr}(W_I^{\otimes}P_j)\,\mathrm{Tr}(W_{II}^{\otimes}Q_k). \tag{3}$$

This occurs, in particular, when $W^{\otimes} = W_I^{\otimes} \otimes W_{II}^{\otimes}$. From this we see that $W^{\otimes}$ in (β_1) cannot be a product, i.e. going back to (β) itself for a moment, that U must have converted the initial product $V \otimes W_0$ into a non-product. This is the only possible way in which the information originally distributed entirely between the descriptions of state V of I and W_0 of II can become available also for relations between the systems; for this and the following analysis, compare the discussion by Schrödinger (1935a, §10; 1935b; 1936).

Further insights into (β_1) and hence into (β) are obtained by considering formulations equivalent to (β_1), of which three will be given here. The first is

(β_{21}) $\mathrm{Tr}(W^{\otimes}(1 \otimes Q_k)) = 0$ for all k not in $f(J)$;

(β_{22}) $\mathrm{Tr}(W^{\otimes}(P_j \otimes Q_{f(j)})) = \mathrm{Tr}(W^{\otimes}(1 \otimes Q_{f(j)}))$ for all $j \in J$.

Thus (β_{21}) asserts that results for II which are not related by f to results for I cannot occur. This subsidiary requirement, used to ensure the generality of $\{Q_k\}_{k \in K}$, is combined with (β_{22}), in which it is formally asserted that, in a simultaneous measurement of I and II, the results P_i and $Q_{f(i)}$ related by f are to be expected with the same frequency as the result $Q_{f(i)}$ when a measurement of II only is made. A third form is

(β_{31}) same as (β_{21});

(β_{32}) for all $j \in J$, if $\mathrm{Tr}(W^{\otimes}(1 \otimes Q_{f(j)})) > 0$, then

$$\mathrm{Tr}\left(\frac{(1 \otimes Q_{f(j)})\, W^{\otimes}(1 \otimes Q_{f(j)})}{\mathrm{Tr}(W^{\otimes}(1 \otimes Q_{f(j)}))}(P_j \otimes 1)\right) = 1.$$

(β_{32}) can be interpreted as follows. If a measurement of II is to give a result corresponding to $Q_{f(j)}$, then a check measurement on I will certainly give the result corresponding to P_j. This makes the connection of P_j with $Q_{f(j)}$ particularly clear. After an actual measurement of II, the result obtained can be used to deduce unambiguously a property which certainly exists in I. A fourth version, the last to be given here, goes still further in this direction:

(β_{41}) same as (β_{21});

(β_{42}) $\Sigma_k(1 \otimes Q_k)\, W^{\otimes}(1 \otimes Q_k)$
$= \Sigma_j(P_j \otimes 1)\, W^{\otimes}(P_j \otimes 1)$;

(β_{43}) for all $j \in J$, if $\mathrm{Tr}(W^{\otimes}(1 \otimes Q_{f(j)})) > 0$ then also $\mathrm{Tr}(W^{\otimes}(P_j \otimes 1)) > 0$ and

$$\frac{(1 \otimes Q_{f(j)})\, W^{\otimes}(1 \otimes Q_{f(j)})}{\mathrm{Tr}(W^{\otimes}(1 \otimes Q_{f(j)}))} = \frac{(P_j \otimes 1)\, W^{\otimes}(P_j \otimes 1)}{\mathrm{Tr}(W^{\otimes}(P_j \otimes 1))}.$$

In (β_4) it is directly asserted that the reduction of states (of any kind) can be undertaken after the measurement of II (as a part of $I \otimes II$) as if the measurement of I (likewise as a part of $I \otimes II$) had taken place. Unlike the situation in (β_1), (β_2) and (β_3), where the component assertions are independent, (β_{42}) is a consequence of (β_{41}) and (β_{43}). Here, for reasons of an immediate economy of proof and to facilitate insight into the logical structure of the relationships to be discussed, it is included as a component of (β_4).

The equivalence of the four versions is proved in the Mathematical Appendix to this section, parts 3 to 5.

The treatment so far shows quite easily that the conditions (α) and (β) are necessary and sufficient for the interaction U together with the measurement $\{Q_k\}_k$, related to the measurement $\{P_i\}_i$ by f, to replace the latter for the initial description of state W_0 or (using an abridged wording in place of the original one) for the direct measurement of I or $\{I_r\}_r$ to be replaced by a measurement of II or $\{II_r\}_r$; in the latter, an interposed interaction between I or the I_r and II or the II_r can and does effect this replacement. The proof is as follows: (α) and (β) are equivalent to

(γ) For every V in I or $\{I_r\}_r$,

(γ_1) $\mathrm{Tr}(U \circ (V \otimes W_0)(1 \otimes Q_k)) = 0$ for $k \notin f(J)$,

(γ_2) $(\Sigma_k(1 \otimes Q_k)\, U \circ (V \otimes W_0)(1 \otimes Q_k))_I = \Sigma_i P_i V P_i$,

(γ_3) for every $i \in J$ such that $\mathrm{Tr}(U \circ (V \otimes W_0)(1 \otimes Q_{f(i)})) > 0$, we have also $\mathrm{Tr}(VP_i) > 0$ and

$$\left(\frac{(1 \otimes Q_{f(i)})\, U \circ (V \otimes W_0)(1 \otimes Q_{f(i)})}{\mathrm{Tr}(U \circ (V \otimes W_0)(1 \otimes Q_{f(i)}))}\right)_I = \frac{P_i V P_i}{\mathrm{Tr}(VP_i)}.$$

This shows immediately how the three types of reduction of states (1), (2), (3) in section 1, belonging to the measurement $\{P_i\}_i$ for I or $\{I_r\}_r$, are replaced by the corresponding reductions belonging to the measurement $\{Q_k\}_k$ after the interaction U for II or $\{II_r\}_r$. For the reduction taking account of the result, the answer is in (γ_3), for the weak reduction in (γ_2), and for the strong reduction in (γ_3) again. The proof that (α) and (β) are equivalent to (γ) is given in Mathematical Appendix 6 to this section.

In sections 1 and 3 we have only expressed and discussed in various forms the essential relation between $\{P_i\}_{i \in J}$, U and W_0 or between these, $\{Q_k\}_{k \in K}$ and $f\colon J \rightarrowtail K$; a proof of existence must finally be given, to ensure that these relations can in fact be satisfied. The least that has to be shown, as part of the general plan of this chapter, is that for given $\{P_i\}_i$ there exist U and W_0 for the relation discussed in section 1, or that for given $\{P_i\}_i$ there exist U, W_0, $\{Q_k\}_k$ and f for the relation discussed in this section. Both cases concern the replacement of an originally specified measurement $\{P_i\}_i$ of the object. Since (9) in section 1 follows from (α) and (β) in the present section, the second existence theorem would suffice. Actually, we can prove a much more far-reaching assertion. It is the interaction U whose existence certainly has to be proved, and we may therefore ask to what extent the other entities can be specified arbitrarily. Some caution is necessary here: for example, there are quite strong relationships between W_0 and $\{Q_k\}_k$; it can be shown that W_0 cannot be faithful if $\{Q_k\}_k$ is not trivial, i.e. there must then be at least one property of the apparatus that is non-trivial and known with certainty in W_0. It can also be shown that, for maximum $\{Q_k\}_k$, W_0 must likewise be a maximum. Details will not be given here; the following is taken as a sufficient case:

(E) Let $\mathfrak{H}^{II}$ be of infinite dimension. (We assume that the dimension is always countable.) For any $\{P_i\}_{i \in J}$, maximum $W_0 = P_{\psi_0}$, maximum $\{Q_k\}_{k \in K}$ and any (one-to-one) $f\colon J \rightarrowtail K$, there exists a U such that (α) and (β) are satisfied.

The proof of (E) is given in Mathematical Appendix 7 to the present section.

Mathematical Appendix to section VI.3

1. Proof of (β) in section 1 from (β) in section 3

Let V be chosen in any manner from $\mathfrak{H}^I$. Since, in this proof, V occurs only as $U \circ (V \otimes W_0)$, we use the abbreviation

$$W^{\otimes} = U \circ (V \otimes W_0). \tag{1}$$

To prove the equation in (β), section 1, we first note the general result (not yet using (β) in section 3)

$$W_I^{\otimes} = \Sigma_k((1 \otimes Q_k)\, W^{\otimes}(1 \otimes Q_k))_I. \tag{2}$$

For

$$\begin{aligned}
&\mathrm{Tr}((\Sigma_k((1 \otimes Q_k)\, W^{\otimes}(1 \otimes Q_k))_I)P)\\
&= \Sigma_k\, \mathrm{Tr}(((1 \otimes Q_k)\, W^{\otimes}(1 \otimes Q_k))_I P)\\
&= \Sigma_k\, \mathrm{Tr}((1 \otimes Q_k)\, W^{\otimes}(1 \otimes Q_k)\,(P \otimes 1))\\
&= \Sigma_k\, \mathrm{Tr}((P \otimes Q_k)\, W^{\otimes}(P \otimes Q_k))\\
&= \Sigma_k\, \mathrm{Tr}(W^{\otimes}(P \otimes Q_k))\\
&= \mathrm{Tr}(W^{\otimes}(P \otimes \Sigma_k Q_k))\\
&= \mathrm{Tr}(W^{\otimes}(P \otimes 1))\\
&= \mathrm{Tr}(W_I^{\otimes} P),
\end{aligned}$$

from which (2) follows immediately. We now use (β) in section 3. As will be shown in parts 3 to 5 of this Appendix, this gives

$$\Sigma_k(1 \otimes Q_k)\, W^{\otimes}(1 \otimes Q_k) = \Sigma_j(P_j \otimes 1)\, W^{\otimes}(P_j \otimes 1). \tag{3}$$

Hence we have at once

$$\Sigma_k((1 \otimes Q_k)\, W^{\otimes}(1 \otimes Q_k))_I = \Sigma_j P_j W_I^{\otimes} P_j. \tag{4}$$

The statement to be proved follows from (2) and (4).

2. The relationship between (α), (β) and (α'), (β')

Let $\{P_i\}_{i \in J}$, U and W_0 be given. If then $\{Q_k\}_{k \in K}$ and $f\colon J \rightarrowtail K$ exist with the properties (α) and (β), then, as mentioned in connection with (β'), equation (1) follows from (α) for $P_i\phi = \phi$. From (β), for $V = P_\phi$, we have $U \circ (P_\phi \otimes W_0)\,(P_i \otimes Q_k) = 0$ for $k \neq f(i)$. Altogether, for $P_i\phi = \phi$ and $k \neq f(i)$,

$$(P_\phi \otimes W^{(\phi)})\,(P_i \otimes Q_k) = P_\phi \otimes W^{(\phi)}Q_k = 0,$$

i.e.

$$W^{(\phi)}Q_k = 0. \tag{1}$$

Since $\Sigma_k Q_k = 1$,

$$W^{(\phi)} = W^{(\phi)} \cdot \Sigma_k Q_k = W^{(\phi)}Q_{f(i)} + \Sigma_{k \neq f(i)} W^{(\phi)}Q_k,$$

and hence from (1), with $P_i\phi = \phi$,

$$W^{(\phi)} = W^{(\phi)}Q_{f(i)}; \tag{2}$$

hence for the carrier $P_{(W^{(\phi)})}$ (see Dixmier 1957, section I.4.6)

$$P_{(W^{(\phi)})}Q_{f(i)} = P_{(W^{(\phi)})}. \tag{3}$$

From this and (β'), (2) follows immediately, since $Q_k Q_l = 0$ for $k \neq l$ and f is one-to-one.

Conversely, let (α') and (β') be satisfied for $\{P_i\}_{i \in J}$, U and W_0. For a given P_i, we have the set of $W^{(\phi)}$ with $P_i\phi = \phi$ and (1) from (β'), and therefore the set of corresponding carriers $P_{(W^{(\phi)})}$ for $P_i\phi = \phi$. Let Q_i be the upper bound of these. Then, from (2), $Q_i Q_j = 0$ for $i \neq j$. If the ensemble of Q_i thus obtained is not yet complete ($\Sigma_i Q_i = 1$), it can be made complete. This leads to a new set K of subscripts for the Q_k, and a canonical one-to-one mapping f of J on K. These are what is required: for the proof of (β) we need only take

$V = P_\phi$ and replace W_0 by one of its eigenvectors ψ_m^0. From here onwards we use the argument which leads from (8) to (14) in part 2 of the Mathematical Appendix to section VI.1, and which is based on the condition (α') (or, in that argument, (α) in section 1). Using the result (14), we have by simple calculation

$$(P_i \otimes Q_k)\, U \circ (\phi \otimes \psi_m^0)$$
$$= \Sigma_{\mu:\, g(\mu)=i} \langle \phi_\mu | \phi \rangle (\phi_\mu \otimes Q_k \psi_m^{(\mu)}). \tag{4}$$

When the Q_k have been constructed we have, again as in (3),

$$P_{(W^{(\mu)})} Q_{f(i)} = P_{(W^{(\mu)})}$$

for $g(\mu) = i$. Since

$$P_{W^{(\mu)}} = \Sigma_m P_{\psi_m^{(\mu)}},$$

it follows that

$$Q_k \psi_m^{(\mu)} = 0$$

for $k \neq f(g(\mu))$. The right-hand side of (4) is therefore zero, and this completes the essentials of the proof.

3. The equivalence of (β_1) and (β_2)

Let (β_1) be satisfied. If $k \notin f(J)$, then $\mathrm{Tr}(W^\otimes(P_i \otimes Q_k)) = 0$ for *all* $i \in J$, and so

$$\Sigma_i \mathrm{Tr}(W^\otimes(P_i \otimes Q_k)) = \mathrm{Tr}(W^\otimes((\Sigma_i P_i) \otimes Q_k))$$
$$= \mathrm{Tr}(W^\otimes(1 \otimes Q_k)) = 0,$$

which gives (β_{21}). To prove (β_{22}), we use the identity

$$\mathrm{Tr}(W^\otimes(1 \otimes Q_{f(i)})) = \mathrm{Tr}(W^\otimes(P_i \otimes Q_{f(i)}))$$
$$+ \Sigma_{j:\, j \neq i} \mathrm{Tr}(W^\otimes(P_j \otimes Q_{f(i)})). \tag{1}$$

From (β_1), $\mathrm{Tr}(W^\otimes(P_j \otimes Q_{f(i)})) = 0$ for $f(i) \neq f(j)$, and therefore for $i \neq j$. Hence (β_{22}) follows immediately from (1).

Conversely, let (β_2) be satisfied. From (1) and (β_2) we have at once $\mathrm{Tr}(W^\otimes(P_j \otimes Q_{f(i)})) = 0$ for $j \neq i$, and hence (β_1) for the case where k has the form $f(i)$. If not, (β_{21}) is used to show that $\mathrm{Tr}(W^\otimes(1 \otimes Q_k)) = 0$, and from

$$\mathrm{Tr}(W^\otimes(P \otimes Q_k)) \leqslant \mathrm{Tr}(W^\otimes(1 \otimes Q_k)) \tag{2}$$

we have (β_1) in this case also.

4. The equivalence of (β_2) and (β_3)

Let (β_2) be satisfied. Then (β_{31}) is trivial, and (β_{32}) is obtained by direct calculation from (β_{22}) for $\mathrm{Tr}(W^\otimes(1 \otimes Q_{f(i)})) > 0$. Conversely, let ($\beta_3$) be satisfied. Then ($\beta_{21}$) is trivial, and ($\beta_{22}$) is again obtained by direct calculation from (β_{32}), if $\mathrm{Tr}(W^\otimes(1 \otimes Q_{f(i)})) > 0$. If, however, $\mathrm{Tr}(W^\otimes(1 \otimes Q_{f(i)})) = 0$, then from (2) in part 3 we have $\mathrm{Tr}(W^\otimes(P_i \otimes Q_{f(i)})) = 0$ also, again leading to (β_{22}).

5. The equivalence of (β_3) and (β_4)

Let (β_3) be satisfied. Then (β_{41}) is trivial. To prove the other two parts, we first seek to prove that

$$(1 \otimes Q_{f(i)})\, W^{\otimes}(1 \otimes Q_{f(i)}) = (P_i \otimes 1)\, W^{\otimes}(P_i \otimes 1) \tag{1}$$

for all $i \in J$. We begin from the trivial identity

$$W^{\otimes}(1 \otimes Q_{f(i)}) = W^{\otimes}(P_i \otimes Q_{f(i)}) + \Sigma_{j:\, j \neq i} W^{\otimes}(P_j \otimes Q_{f(i)}). \tag{2}$$

From the equivalence of (β_3) and (β_1), already proved, it follows that $\mathrm{Tr}(W^{\otimes}(P_j \otimes Q_{f(i)})) = 0$ for $j \neq i$ if (β_3) is satisfied. Hence also $W^{\otimes}(P_j \otimes Q_{f(i)}) = 0$ for $j \neq i$ and so, from (2),

$$W^{\otimes}(1 \otimes Q_{f(i)}) = W^{\otimes}(P_i \otimes Q_{f(i)}). \tag{3}$$

Taking the adjoint, we have

$$(1 \otimes Q_{f(i)})W^{\otimes} = (P_i \otimes Q_{f(i)})W^{\otimes}, \tag{4}$$

and from (3) and (4) together

$$(1 \otimes Q_{f(i)})W^{\otimes}(1 \otimes Q_{f(i)}) = (P_i \otimes Q_{f(i)})W^{\otimes}(P_i \otimes Q_{f(i)}). \tag{5}$$

In exactly the same way, we can prove that

$$(P_i \otimes 1)W^{\otimes}(P_i \otimes 1) = (P_i \otimes Q_{f(i)})W^{\otimes}(P_i \otimes Q_{f(i)}), \tag{6}$$

in this case expanding the 1 in $W^{\otimes}(P_i \otimes 1)$ in terms of the Q_k (cf. (2)) and also using (β_{31}). From (5) and (6) we have (1). Now summing (1) over i and again using (β_{31}) on the left-hand side, we get (β_{42}). From (1),

$$\mathrm{Tr}(W^{\otimes}(1 \otimes Q_{f(i)})) = \mathrm{Tr}(W^{\otimes}(P_i \otimes 1)), \tag{7}$$

and from (1) and (7) we have immediately (β_{43}).

Conversely, let (β_4) be satisfied. Then (β_{31}) is trivial, and (β_{32}) follows immediately from (β_{43}). (Note that, as stated in section 3, (β_{42}) is a consequence of (β_{41}) and (β_{43}). The proof has just been given, by showing that (β_3) follows from (β_{41}) and (β_{43}) only, and (β_{42}) from (β_3).)

6. Proof that (α) and (β) are equivalent to (γ)

Let us first assume (α) and (β). (α) is equivalent to (α_1'') and (α_2'') in section 1. For (β) we use the representation (β_4) of (β_1), replacing $W^{\otimes}$ by $U \circ (V \otimes W_0)$ and generalizing over V. From (α_1'') and (α_2'') in section 1 and this form of (β_4), (γ) easily follows.

Conversely, let (γ) be satisfied. Then (β) follows from (γ_1) and (γ_3); this is most easily seen by using (β_1) in the form (β_3) and converting this (like (β_4) in the previous paragraph) back to the present context by replacing $W^{\otimes}$ by $U \circ (V \otimes W_0)$ and generalizing over V.

(α) follows easily in the form of (α'') in section 1. (β) has already been proved. We now take it in the (restored) form of (β_4). This leads with (γ_2) immediately to (α_1'') in section 1. Finally, (α_2'') follows from the (restored) (β_{43}) with (γ_3), using (7) in part 4 of this Appendix as a consequence of (β_4).

7. Proof of (E)

We take a basis $\{\phi_\mu\}_\mu$ of $\mathfrak{H}^I$ passing through the P_i:

$$P_i\phi_\mu = \delta_{ig(\mu)}\phi_\mu. \tag{1}$$

Let the vectors $\chi_{k'} \in \mathfrak{H}^{II}$ together with the given vector $\psi^0 \in \mathfrak{H}^{II}$ form a basis of $\mathfrak{H}^{II}$. Then

$$\{\phi_\mu \otimes \psi^0\}_\mu \text{ and } \{\phi_\mu \otimes \chi_{k'}\}_{\mu,k'} \tag{2}$$

form one basis of $\mathfrak{H}^I \otimes \mathfrak{H}^{II}$. By hypothesis, $\{Q_k\}_k$ is maximum, and therefore provides a further basis of $\mathfrak{H}^{II}$, say $\{\psi_k\}_k$, so that the $\phi_\mu \otimes \psi_k$ form a second basis of $\mathfrak{H}^I \otimes \mathfrak{H}^{II}$. These can be divided as follows:

$$\{\phi_\mu \otimes \psi_{fg(\mu)}\}_\mu \text{ and } \{\phi_\mu \otimes \psi_k\}_{\mu,\, k \neq fg(\mu)}. \tag{3}$$

The first two parts in (2) and (3) already correspond element-to-element. Since $\mathfrak{H}^{II}$ is assumed to be of countably infinite dimension, the other two parts in (2) and (3) are certainly of equal power. There is therefore a unitary mapping U of $\mathfrak{H}^I \otimes \mathfrak{H}^{II}$ on itself, so that

$$U(\phi_\mu \otimes \psi^0) = \phi_\mu \otimes \psi_{fg(\mu)} \tag{4}$$

for all μ. We now have to prove (α) and (β) in section 3 for this U. From (1), a ϕ which satisfies the premise of (α) has the form

$$\phi = \Sigma_{\mu:\, g(\mu) = i}\, c_\mu \phi_\mu.$$

Then the conclusion of (α) follows at once from (4); (β) is calculated first for $V = P_\phi$; calculation using (4) gives

$$\|(P_j \otimes Q_{\psi_k})U(\phi \otimes \psi^0)\| = \|P_j\phi \otimes \delta_{kf(j)}\psi_k\|^2.$$

This is zero if $k \neq f(j)$. The generalization to arbitrary V is trivial.

CHAPTER VII

Completeness and Reality

ONE of the fundamental problems raised immediately by quantum mechanics is whether it is a *complete* theory, in the sense of providing a complete description of the state of an object. Doubts on this point arise in the first place because the quantum-mechanical description of state is a probabilistic one. Independently of the various possible ways of interpreting the propositions regarding probabilities, this implies an incompleteness in comparison with the probability-free descriptions of state in classical physics, taken to be an in principle attainable ideal. This initially very superficial comparison leads naturally to the question whether the probabilistic description of state in quantum mechanics is unavoidable or whether it can in some way be reduced to a description not involving probabilities, which would, however, have to be empirically accessible. The question is made more acute by the fact that classical phenomenological thermodynamics can be replaced by a theory involving probabilistic descriptions of state, based on the probability-free descriptions from exact mechanics. This would represent a model in which the probabilistic nature of a theory can be explained by the use of *hidden parameters.*

The completeness problem in quantum mechanics has in fact been mainly viewed and discussed from this aspect. Theories of hidden parameters have been proposed for quantum mechanics, and proofs have been given that certain other theories of this kind cannot exist. It has already been mentioned in the Introduction to this book that the former proposals will not be discussed here. The first of the impossibility proofs was given by von Neumann, whose proof was subsequently refined, generalized and modified. This line is discussed in section 1 of the present chapter. We shall try, to the extent that is necessary and appropriate from the purely logical aspect, to specify the hypotheses underlying the proofs in question, and this will lead spontaneously to a classification of the proofs. The main emphasis is, of course, on what is required of a theory of hidden parameters, and how it is to be related to quantum mechanics in particular. On the latter point, von Neumann's hypotheses are so restricted that, for comparison, classical mechanics would not be a theory of hidden parameters in relation to classical statistical mechanics, but his successors have achieved some progress in just this respect. In the classification given below, all the immediately relevant publications will be cited, but only some of them will be discussed, partly with references to sections V.2 and V.3. Other publications which approach the von Neumann proof from other standpoints, and are mainly critical of it, are marked with a dagger (†) in the bibliography. The relation between von Neumann-type impossibility proofs and proposals for hidden-variable theories is discussed by Capasso, Fortunato and Selleri (1970) and Gudder (1970a).

In sections 2 and 3, the attempts to ensure the completeness of quantum mechanics by proving the impossibility of using hidden parameters will be contrasted with an attempt to question this completeness, due to Einstein (first published in his celebrated paper with Podolsky and Rosen, 1935). It does not seek an explicit theory of hidden parameters:

although this seems at first sight to be the only alternative, Einstein succeeded in finding a middle way which possesses a conviction of its own. It consists in weighing against the claim of completeness a certain idea of what is meant by physical reality. More precisely, Einstein constructed a sufficient condition for a physical quantity of an object to have a definite value, and then sought to show that this criterion of reality excludes the possibility that quantum mechanics is complete. The content and logical structure of the arguments used are by no means clear at a glance. In the secondary literature, Einstein's comparison of his criterion of reality with the facts of quantum mechanics is usually represented as something that appears paradoxical in the light of physical intuition, and an attempt is made to mitigate the paradox in some way. In contrast to this, we shall try to show by a thorough analysis that Einstein's criterion, when precisely stated, is logically incompatible in the strict sense with the precisely stated assumptions regarding completeness in quantum mechanics. Some of the secondary literature is marked with a double dagger (‡) in the bibliography, but will not be discussed here.

VII.1 The problem of hidden parameters and the "von Neumann proof"

In sections V.2 and V.3 we have dealt with the problem of a physically plausible, abstract characterization of the probabilistic descriptions of state used *de facto* in orthodox quantum mechanics, with emphasis on the separation of this problem from two others, those of the existence of classical descriptions of state in quantum mechanics and the existence of a theory of hidden parameters in quantum mechanics. This was done in order to avoid misunderstandings such as can occur (and have occurred) in connection with this group of problems. It is fairly easy to see why caution is necessary and what are the dangers. In the problem of characterization, we do not go outside quantum mechanics: it is accepted in a particular formulation given in section II.4, and we then seek a new formulation equivalent to this, but subject to certain general conditions given in section V.1, and therefore having certain *epistemological* advantages over the other, despite their *logical* equivalence. In particular, the new formulation has the advantage of being (at least partially) based on axioms which are physically more intelligible than are the axioms of the standard formulation. The solutions of the characterization problem are definitive and can differ at most by being better or less good as solutions of the general epistemological problem of arriving at satisfiable axiomatizations of a physical theory. There can be no doubt concerning what is asserted and proved by von Neumann according to section V.2 and by Gleason and Mackey according to section V.3, and the respective advantages of von Neumann's formulation over the standard formulation and of Mackey's formulation over von Neumann's formulation are likewise evident.

Now let us compare with this problem of characterization the two other problems mentioned above, beginning with the second of them: the existence of a classical theory of hidden parameters for quantum mechanics. It is clear from the outset that here a completely new problem of *meaning* enters, namely the problem of the meaning of the concept of hidden parameters or of the whole conception of a theory of hidden parameters with respect to quantum mechanics. Thus, contrary to the situation met with in the characterization problem, we should have to go outside the conceptual framework of quantum mechanics: it is obviously impossible to answer the question whether there is a theory of hidden parameters for quantum mechanics without having defined the concept of such a theory. And such a definition cannot be given on the sole basis of concepts belonging to quantum mechanics itself, though it will be related to these concepts. It is here that some uncertainty justifiably arises about what von Neumann meant by asserting that quantum mechanics allows no theory of hidden parameters, whether he proved this assertion, and if so in what sense. Even if it were possible to achieve certainty on these points, the conclusions would be open to criticism as regards the definitions on which they depend—a criticism which could hardly be applied to a solution of the characterization problem.

Now, apart from some misleading formulations, the passages in von Neumann's book (1955, sections III.2 and IV.2) suggest fairly reliably that he finally adopts the following form for the problem of hidden parameters. He asks (1) whether the expectation or mean value functions allowed by purely epistemic or purely statistical quantum mechanics (section II.4) and characterized by him through specified properties (cf. section V.2) include any which are free from dispersion; (2) whether the expectation or mean value functions allowed by quantum mechanics can be represented as mixtures of those among them (if any) which are free from dispersion.

The expectation or mean value functions sought in answer to question (1) would then be possible *classical descriptions of state.* Having as their defining property $(\text{Exp}(W, A))^2 = \text{Exp}(W, A^2)$, the states W could be reinterpreted as classical states in such a way that $\text{Exp}(W, A)$ would be the value which the quantity A of the (individual) object *actually has* (whereas in the general case it is only the expectation value with respect to the knowledge W about the object or the mean value in the statistical state W of an ensemble). Then quantum mechanics, confined to these classical descriptions of state for an individual object, *would be the desired theory of hidden parameters,* and would be such a theory with respect to the original unlimited quantum mechanics in the sense stated in question (2), namely that every expectation or mean value function of quantum mechanics would appear as a mixture of the dispersion-free classical descriptions of state.

Von Neumann put forward two arguments against the possibility of such a theory of hidden parameters for quantum mechanics. If we leave aside for a moment the definition of a theory of hidden parameters for quantum mechanics just given, the two arguments are simply refutations of a positive solution of the questions (1) and (2) respectively. As regards (2), *quantum mechanics has descriptions of state,* namely the maximum expectation value functions or the pure mean value functions, *which demonstrably cannot be represented as mixtures of others among the descriptions of state allowed by quantum mechanics*; cf. subsection II.4(d)(ii). Thus, *a fortiori,* they cannot be represented as mixtures of dispersion-free descriptions of state allowed by quantum mechanics, whether there are any such descriptions or not. The second argument relates to question (1), and shows that the possibility referred to in question (2) cannot arise, *there being no dispersion-free descriptions of state in quantum mechanics.* Both these arguments, being purely logical refutations within the conceptual framework of (1) and (2), are as unobjectionable as were the characterizations in Chapter V. Objections can arise only in the second step, when these negative results are combined with the definitions of a classical description of state and a classical theory of hidden parameters for quantum mechanics to yield the result that such a theory cannot exist. But even then it is not the inference drawn which is objectionable: a theory of hidden parameters *in the sense defined* really does not exist. The crucial question is *whether our intuitions of what a theory of hidden parameters should look like are adequately expressed in von Neumann's definition.*

Before continuing the discussion about this question, let us have a look at von Neumann's result in the light of his characterization given in section V.2. The negative answer to the question (1) being the core of the matter, one should, firstly, realize the fact that this result is a logical consequence of the relevant part of the standard formulation of quantum mechanics (cf. section II.4) as well as of the corresponding formulation of that part given in section V.2. The reason is that the two formulations are logically equivalent. Seen from an epistemological standpoint, however, the non-existence of dispersion-free descriptions of state appears differently in the two cases. As a consequence of the standard formulation it is a fairly trivial result. As a consequence of the partial axiomatization (α)–(γ) of section V.2 it appears much less trivial. The somewhat unfortunate circumstance, mentioned at the beginning of section V.2, that von Neumann placed his characterization in immediate juxtaposition to the other two assertions (of the non-existence of dispersion-free descriptions of state and of the non-existence of hidden parameters), discredited the merits of the characterization, particularly with respect to the assertion in question *as a consequence of the new formulation* (α)–(γ) of section V.2. There has even been speculation whether von Neumann in fact assumed what he was seeking to prove. De Broglie (1953, p. 18), for example, summarizes his otherwise very important analysis of the von Neumann proof in the remark: "He merely showed that, if one assumes the ideas underlying the purely probabilistic interpretation, that interpretation necessarily follows. Thus there is a kind of vicious circle." And elsewhere (de Broglie 1957, p. 27), without any reference to the characterization considered here, the same author writes "In reality, von Neumann's ingenious, if somewhat laborious, proof tells us nothing really new. If the Heisenberg relations are known . . . we already know that . . . every quantity is subject to dispersion." Such statements are at best misleading, and glide too facilely over the subtleties of axiomatic analyses. A physical theory can be stated in various forms, and it is only this difference in formulation which determines the assessment of assertions of equivalence and of further consequences using this equivalence. There is no question that von Neumann proposed the partial axiomatization (α)–(γ) of section V.2 in order to ascertain whether the explicit descriptions of state of the standard formulation are the only physically plausible ones. It is possible, though rare, that important possibilities are overlooked, and these might even have included some states without dispersion. The perception that the old and new formulations are equivalent shows that no such thing has occurred. From this standpoint, the question of dispersion-free descriptions of state is thus seen to form part of the question of the equivalence of two formulations of quantum mechanics. We do not assume what is to be proved, but assume the definite formulation (α)–(γ) of section V.2 and seek to prove that it leads back to the standard formulation, and in particular to the non-existence of dispersion-free descriptions of state. The entire point here is the difference between the two formulations.

Thus, as regards the treatment of the problem of hidden parameters by von Neumann himself, i.e. his celebrated proof, we can say that he (a) made precise and proved his assertion that there exists no theory of hidden parameters for orthodox quantum mechanics, in the sense of the discussion following (1) and (2) above; (b) used his characterization (section V.2) of quantum-mechanical descriptions of state to show that there is a non-trivial formulation among the mutually equivalent formulations of the assertion in (a).

All further contributions to the problem of hidden parameters in quantum mechanics can be classified, on the basis of these results of von Neumann's, into those which retain the basic von Neumann idea in (a), specified previously, but refine the characterization mentioned in (b), and those which regard as in some way too narrow or as altogether misconceived von Neumann's conception of what a theory of hidden parameters should look like, and appropriately extend or amend it. Using this classification, some other contributions will be discussed in the present section which have the same tendency as von Neumann's, viz. to demonstrate the impossibility of hidden parameters in quantum mechanics.

A typical result which *leaves unaffected* von Neumann's conception of a theory of hidden parameters but refines the characterization of the descriptions of state is evidently obtained from Gleason and Mackey's characterization given in section V.3. From that section and the results of Chapter III it is seen that the conditions here imposed on a description of state relate to only a subset of quantities, namely those which are described by projection operators. Since this subset is considerably smaller than the set of all quantities, von Neumann's result is strengthened by using Gleason and Mackey's characterization.

As we know from section V.2, however, this characterization relates primarily to the formulation of quantum mechanics in terms of properties. According to (γ_1) in section V.3, the dispersion-free descriptions of state would be given by the following conditions:

(α) $pr_*(E) = 0$ or 1,
(β) $pr_*(\cup_\mu E_\mu) = \Sigma_\mu pr_*(E_\mu)$ for every countable set $\{E_\mu\}_\mu$ with each pair of E_μ orthogonal,
(γ) $pr_*(\mathsf{U}) = 1$.

Here the $pr_*(E)$ are the probabilities of the properties E, to be interpreted at first either epistemically or statistically. (The dependence on W shown previously is here, for simplicity, included in the (variable) pr_*.)

Let the structure $\mathfrak{E}$, defined by the conditions (α_2) of section V.3, with (α_{23}) relating only to pairs of properties, be called a *Mackey structure*, and let $\mathfrak{E}$ defined by (α_2) of section V.3 as it stands be called a *σ-complete* Mackey structure. The conditions (α)–(γ) are, assuming only a σ-complete Mackey structure, equivalent to

(α_1) same as (α),
(β_1) for every countable set $\{E_\mu\}_\mu$ with each pair of E_μ orthogonal, $pr_*(\cup_\mu E_\mu) = 1 \rightarrow \mathsf{V}_\mu . pr_*(E_\mu) = 1$,
(γ_1) $pr_*(E) = 1 \wedge E \prec F \rightarrow pr_*(F) = 1$,
(δ_1) $pr_*(E) = 1 \rightarrow pr_*(E^\perp) = 0$,
(ε_1) $pr_*(\mathsf{U}) = 1$.

The proof of (α_1)–(ε_1) from (α)–(γ) is trivial throughout. Conversely, if (α_1)–(ε_1) are valid, (β) can be derived as follows. If $pr_*(\cup_\mu E_\mu) = 1$, then from (β_1) there is some μ such that $pr_*(E_\mu) = 1$. For all other E_ν, we must then have $pr_*(E_\nu) = 0$. This follows, since $E_\mu \perp E_\nu$ implies that $E_\mu \prec E_\nu^\perp$; from $pr_*(E_\mu) = 1$ and (γ_1) we therefore find $pr_*(E_\nu^\perp) = 1$, and (δ_1) then gives $pr_*(E_\nu) = 0$. Altogether, then, $\Sigma_\nu pr_*(E_\nu) = 1$, which is ($\beta$) in this case. If on the other hand $pr_*(\cup_\mu E_\mu) = 0$, then from (γ_1) and (α_1) $pr_*(E_\mu) = 0$ for all μ, since $E_\mu \prec \cup_\mu E_\mu$; hence $\Sigma_\mu pr_*(E_\mu) = 0$, which is ($\beta$) in this case.

The new characterization of the dispersion-free descriptions of state pr_* affords the possibility of replacing these by probability-free classical descriptions of state on_*, for which we have in correspondence with (β_1)–(ε_1)

(β_2) for every countable set $\{E_\mu\}_\mu$ of E_μ orthogonal in pairs, $on_*(\cup_\mu E_\mu) \rightarrow \mathsf{V}_\mu\, on_*(E_\mu)$,
(γ_2) $on_*(E) \wedge E \prec F \rightarrow on_*(F)$,
(δ_2) $on_*(E) \rightarrow \neg\, on_*(E^\perp)$,
(ε_2) $on_*(\mathsf{U})$.

Starting from pr_*, we must use the definition

$$on_*(E) \rightleftharpoons pr_*(E) = 1; \tag{1}$$

starting from on_*,

$$\left.\begin{aligned} pr_*(E) &= 1 \text{ for } on_*(E), \\ &= 0 \text{ for } \neg\, on_*(E). \end{aligned}\right\} \tag{2}$$

From (β_2)–(ε_2) it is very clear what cannot occur in quantum mechanics. The states in a theory of hidden parameters of the von Neumann type would be states which would have to satisfy *only* the conditions (β_2)–(ε_2); according to Gleason and Mackey's result, such states do not exist. At the same time, (β_2) can be seen to be the particular condition which makes impossible a von Neumann theory of hidden parameters; states which satisfy only (γ_2)–(ε_2) are readily found, for example the states of the ontic formulation of quantum mechanics (section II.5), with

$$on_*^{\phi}(E) \rightleftharpoons \mathbf{pr}_*(\phi; E, 1). \tag{3}$$

Then (γ_2)–(ε_2) follow immediately, but, of course, (β_2) is not valid.

In accordance with our programme, we now consider contributions which *modify* von Neumann's conception of a theory of hidden parameters for quantum mechanics, but do so only to an extent which leaves the results negative. Such modifications are of two kinds.

Firstly, the parts of quantum mechanics relevant to the problem, in their purely epistemic or purely statistical form (sections II.4, III.3) are simply generalized. This involves no modification, in the sense that the new generalized part is not in contradiction with the standard formulation of section II.4 or III.3; in fact, it contains the corresponding part of this as a special case. Thus, for such a generalization in the strictly logical sense, the advantage of a related proof of the non-existence of hidden parameters is precisely that certain assumptions are discovered which are weaker than those in the standard formulation but nevertheless sufficient to obtain the proof in question. Such a treatment becomes relevant when doubts arise as to some hypothesis in the standard formulation, which do not affect the weaker assumptions. For instance, some of the generalizations considered here are capable of embracing the case of a quantum mechanics with superselection rules. Modifications of this first kind are discussed by Jauch and Piron (1963), Jauch (1968, Chapter 7), Gudder (1968a, b) and Ochs (1972c).

Whereas such generalizations refer to quantum mechanics itself, there are, secondly, also modifications which affect the concept of a theory of hidden parameters. Von Neumann evidently made this concept a very narrow one: the range of quantities or properties in his theory is the same as in quantum mechanics, and the desired "hidden" descriptions of state are introduced as a subset of the set of quantum-mechanically specified descriptions of state. Here, therefore, interest resides in considering a wider range of possibilities. Contributions to this second kind of modification (often including the first kind) are to be found in the papers by Kamber (1964, 1965), Dombrowski and Horneffer (1964), Kochen and Specker (1967), Misra (1967) and Ochs (1970, 1971). Here we shall give a brief account of the relatively little-known results of Kamber (1964), with a few unimportant changes.

The concepts used by Kamber are the most general ones so far considered in the problem of a theory of hidden parameters in quantum mechanics. With regard to the range of properties of an object, the lattice condition is weakened so as to avoid any undesirable occurrence of incommensurable properties. Kamber postulates the range $\mathfrak{E}$ of properties as a "partial Boolean algebra" or, somewhat more strongly, as a "semi-Boolean algebra". We shall assume that $\mathfrak{E}$ is a Mackey structure as defined above. Every semi-Boolean algebra as defined by Kamber is a Mackey structure, and the converse is probably also true. Let a probability function in a Mackey structure be described by the conditions (γ_1) in section V.3, with (γ_{12}) relating only to pairs of properties. If the Mackey structure is σ-complete, i.e. if (α_{23}) in section V.3 is valid without restriction, then we shall consider only σ-additive probability functions in it, i.e. functions for which (γ_{12}) in section V.3 is valid without restriction. As a generalization of classical statistical mechanics or of quantum mechanics in the formulation in terms of properties (Chapter III), we shall consider pairs $(\mathfrak{E}, M)$, where $\mathfrak{E}$ is a Mackey structure and M a set of probability functions in $\mathfrak{E}$. The elements of $\mathfrak{E}$ represent possible properties of an object, and those of M represent possible states of knowledge or statistical states.

We correlate these pairs $(\mathfrak{E}, M)$ with pairs $(\mathfrak{E}', H)$, in which $\mathfrak{E}'$ is a quasi-modular orthocomplementary lattice (see Kamber 1965, §§1 and 2), i.e. a somewhat more specialized structure than has been assumed for $\mathfrak{E}$. Likewise, let H be more specialized than M: a set of *two*-valued probability functions in $\mathfrak{E}'$. The individual $h \in H$ is thus assumed to satisfy (γ_1) in section V.3, but (γ_{12}) again only for pairs of properties (Kamber 1965, §5, (W1) and (W2) but not (W3)). Further, let h have only the values 1 and 0. The elements of $\mathfrak{E}'$ are again interpreted as properties, and those of H, because of their two-valuedness, are classical states: an $h \in H$ decides, for each property $E' \in \mathfrak{E}'$, whether it exists or not. $(\mathfrak{E}', H)$ generalizes a classical-mechanical theory in the sense of section III.1, and is contrasted with the generalizations $(\mathfrak{E}, M)$ of the probabilistic theories as a candidate for being a theory of hidden parameters for the latter. To $(\mathfrak{E}', H)$ there corresponds uniquely the homomorphism

$$\beta(E') = \{h(E') = 1\}_{h \in H} \tag{4}$$

between the *Mackey* structure in $\mathfrak{E}'$ and the corresponding structure supported by the (Boolean) power set lattice $\mathfrak{N}(H)$. The Boolean lattice generated by $\{\beta(E')\}_{E' \in \mathfrak{E}'}$ in $\mathfrak{N}(H)$ will be denoted by $\mathfrak{B}'$.

The theories $(\mathfrak{E}, M)$ and $(\mathfrak{E}', H)$ are now to be related in a way which allows $(\mathfrak{E}', H)$ to be regarded as a theory of hidden parameters for $(\mathfrak{E}, M)$. We shall say that this relationship exists if there is a relation R between the elements of $\mathfrak{E}'$ and those of $\mathfrak{E}$ which satisfies the following conditions (cf. Kamber 1964, §7):

(α_3) R is compatible with the Mackey structures of $\mathfrak{E}$ and $\mathfrak{E}'$. In particular:

(α_{31}) $R(\cup', \cup)$ and $R(\cap', \cap)$;

(α_{32}) if, for $E, F \in \mathfrak{E}$ with $E \perp F$, there exist elements $E', F' \in \mathfrak{E}'$ with $R(E', E)$ and $R(F', F)$, then there exist such elements with these properties and also $R(E' \cup F', E \cup F)$, and the same is assumed valid in reverse from $\mathfrak{E}'$ to $\mathfrak{E}$;

(α_{33}) if, for $E \in \mathfrak{E}$, there exists $E' \in \mathfrak{E}'$ with $R(E', E)$, then there exists E' with this property and also $R(E'^{\perp}, E^{\perp})$, and the same in reverse.

(β_3) For every $E \in \mathfrak{E}$ there exists $E' \in \mathfrak{E}'$ with $R(E', E)$.

(γ_3) For every probability function $\mu \in M$ there exists a probability function μ' in $\mathfrak{E}'$ such that

(γ_{31}) $\bigwedge_{E', F' \in \mathfrak{E}'} . \bigvee_{E \in \mathfrak{E}} R(E', E) \wedge R(F', E) \rightarrow \mu'(E' \Delta F') = 0$, where Δ is the symmetric difference of E' and F';

(γ_{32}) for any $E \in \mathfrak{E}$ and $E' \in \mathfrak{E}'$ with $R(E', E)$,

$$\mu'(E') = \mu(E);$$

(γ_{33}) μ' can be represented by H, i.e. there exists a probability function ν in $\mathfrak{B}'$ with $\mu'(E') = \nu(\beta(E'))$ for all $E' \in \mathfrak{E}'$.

These conditions evidently connect $(\mathfrak{E}, M)$ and $(\mathfrak{E}', H)$ in a certain manner. The connection is not very strong, but this is desirable in seeking to prove the non-existence of a theory $(\mathfrak{E}', H)$ as a theory of hidden parameters corresponding to a theory $(\mathfrak{E}, M)$.

Let us first see how the von Neumann proof in the Mackey-Gleason form responds to these new concepts. For $\mathfrak{E}$ we take here the (σ-complete) Mackey structure for the subspaces of a Hilbert space $\mathfrak{H}$, as already used in section V.3. M is the ensemble of quantum-mechanical (σ-additive) probability functions in this $\mathfrak{E}$ (cf. sections III.3 and V.3). This is the given theory $(\mathfrak{E}, M)$. Now the desired theory of hidden parameters $(\mathfrak{E}', H)$ is not just any theory permitted by the foregoing; it is subject to certain restrictions. $\mathfrak{E}'$ is taken to be the usual orthocomplementary (and quasi-modular) subspace lattice of $\mathfrak{H}$. In terms of sets, $\mathfrak{E} = \mathfrak{E}'$, and their Mackey structures are also identical. H is the ensemble of σ-additive two-valued probability functions in the Mackey structure of $\mathfrak{E}$. This is von Neumann's condition (1) (see above). The existence of a relation R in accordance with (α_3)–(γ_3), together with (γ_{33}), would correspond to von Neumann's condition (2). According to Gleason's theorem (see section V.3), however, H is empty if $\dim \mathfrak{H} > 2$. The $\mathfrak{B}'$ defined above is then degenerate, consisting of only one element. Since M is not empty, (γ_{33}) shows that, whatever the choice of R, $(\mathfrak{E}', H)$ is not a theory of hidden parameters corresponding to $(\mathfrak{E}, M)$. The closer connection between $(\mathfrak{E}, M)$ and $(\mathfrak{E}', H)$ is irrelevant here; it is of course such that H cannot be empty if M is not empty, and H is in fact empty here.

The class of hidden-parameter theories for ordinary quantum mechanics that are allowed by von Neumann is extremely small. This is especially clear from the observation that classical statistical mechanics has no hidden-parameter theory in this class, although it does have a hidden-parameter theory in the extended sense defined above, namely ordinary classical mechanics, which is usually assumed to be such a theory. Let $\mathfrak{M}$ be the phase space of a classical-mechanical object, $\mathfrak{E}'$ the set of Borel sets in $\mathfrak{M}$ as a (σ-complete) Boolean lattice. If $x \in \mathfrak{M}$, then $h_x(E') = 1$ or 0, according as $x \in E'$ or $x \notin E'$, is a (σ-additive) two-valued probability function in $\mathfrak{E}'$. Let H be the set of all such functions. Then $(\mathfrak{E}', H)$ is the part of ordinary classical mechanics which would rank as a hidden-parameter theory for classical statistical mechanics. The latter theory has been derived in the following manner (section III.2), as regards the part that is of interest here. $\mathfrak{E}$ is formed from $\mathfrak{E}'$ by constructing equivalence classes, E' and $F' \in \mathfrak{E}'$ being equivalent if their symmetric difference has Lebesgue measure zero. $\mathfrak{E}$ also is a σ-complete Boolean lattice, related to $\mathfrak{E}'$ by the canonical homomorphism $\pi: \mathfrak{E}' \rightarrowtail \mathfrak{E}$. Next, let M be the ensemble of (σ-additive) probability functions in $\mathfrak{E}$ allowed for classical statistical mechanics in section III.2. Then $(\mathfrak{E}, M)$ is the part of classical statistical mechanics relevant here. Just as in quantum mechanics, there exist in $\mathfrak{E}$ no σ-additive two-valued probability functions (cf. Kamber 1965, corollary to theorem XII). The construction of a hidden-parameter theory for $(\mathfrak{E}, M)$ described in the previous paragraph would thus fail in the classical case also. But, as just stated, $(\mathfrak{E}', H)$ *is* a hidden-parameter theory, in the extended sense, for $(\mathfrak{E}, M)$: we take $R(E', E) \rightharpoonup \pi(E') = E$, and ($\alpha_3$)–($\gamma_3$) are easily proved.

From this argument it is seen to be necessary to generalize the von Neumann proof towards an extended concept of a theory of hidden parameters. This can be done as follows.

For any given theory $(\mathfrak{E}, M)$, we restrict the desired theories of hidden parameters $(\mathfrak{E}', H)$ by the condition

$$\bigwedge_{E' \in \mathfrak{E}'} \cdot \bigwedge_{h \in H} \cdot h(E') = 0 \rightarrow E' = \cap'. \tag{5}$$

This limitation very plausibly requires the existence of a sufficient number of classical states. We also impose the condition on each $h \in H$ that

$$\bigwedge_{E', F' \in \mathfrak{E}'} \cdot h(E') = h(F') = 0 \rightarrow h(E' \cup F') = 0. \tag{6}$$

If $(\mathfrak{E}', H)$ satisfies the conditions (5) and (6), $\mathfrak{E}'$ is a Boolean lattice (Kamber 1965, corollary to theorem IV). Let R be a relation which, according to (α_3)–(γ_3), makes $(\mathfrak{E}', H)$ a hidden-parameter theory for $(\mathfrak{E}, M)$. If $I' \subseteq \mathfrak{E}'$ is the $\cup$-ideal generated by all elements $E' \Delta F'$ with $R(E', E)$ and $R(F', E)$ for a suitable $E \in \mathfrak{E}$, we use the quotient lattice $\mathfrak{E}'/I'$, which is also Boolean. If $\pi\colon \mathfrak{E}' \rightarrowtail \mathfrak{E}'/I'$ is the canonical homomorphism, then $\phi\colon \mathfrak{E} \rightarrowtail \mathfrak{E}'/I'$, $\phi(E) = \pi(E')$ with $R(E', E)$ uniquely defines a homomorphism of $\mathfrak{E}$ as a Mackey structure in $\mathfrak{E}'/I'$. There is no great difficulty in using (α_3)–(γ_3) to prove that, for every $\mu \in M$, there exists a probability function v in $\mathfrak{E}'/I'$ with $\mu(E) = v(\phi(E))$ for all $E \in \mathfrak{E}$. If now we assume $(\mathfrak{E}, M)$ such that

$$\bigwedge_{E,F \in \mathfrak{E}} \cdot \bigwedge_{\mu \in M} \cdot \mu(E) = \mu(F). \rightarrow E = F, \tag{7}$$

i.e. that, as in (6), M includes a sufficient number of probability functions, then it follows at once that ϕ is one-to-one. This completes in outline the proof of the result (Kamber, 1964, §4)

(A) If $(\mathfrak{E}, M)$ with the property (7) has a hidden-parameter theory $(\mathfrak{E}', H)$ with the properties (5) and (6), then the Mackey structure $\mathfrak{E}$ allows a one-to-one homomorphism ϕ in a Boolean lattice generated by $\phi(\mathfrak{E})$.

It can now be shown that this conclusion (A) is not satisfied by the Mackey structure of quantum mechanics. First of all, we have the general result (Kamber 1964, theorem 5.1)

(B) Let $\mathfrak{E}$ be a Mackey structure, $\mathfrak{B}$ a Boolean lattice and $\phi\colon \mathfrak{E} \rightarrowtail \mathfrak{B}$ a homomorphism such that $\{\phi(E)\}_{E \in \mathfrak{E}}$ generates $\mathfrak{B}$; let $\mathfrak{B}_{\mathfrak{E}}$ be a Boolean lattice defined through $\beta_{\mathfrak{E}}$ in (4), both with respect to $H_{\mathfrak{E}}$, the set of all two-valued probability functions in $\mathfrak{E}$; then there is a homomorphism $\pi\colon \mathfrak{B}_{\mathfrak{E}} \rightarrowtail \mathfrak{B}$ such that $\phi = \pi \circ \beta_{\mathfrak{E}}$.

Now let $(\mathfrak{E}, M)$ be restricted to quantum mechanics as above. Then (7) is satisfied. If there were a hidden-parameter theory $(\mathfrak{E}', H)$ with the properties (5) and (6), the conclusion in (A) would follow, and so (B) would make the canonical homomorphism $\beta_{\mathfrak{E}}$ one-to-one. If now the Hilbert space $\mathfrak{H}$ on which $(\mathfrak{E}, M)$ is based has a finite dimension > 2, Gleason's theorem (section V.3) shows that $H_{\mathfrak{E}}$ is empty. Thus $\beta_{\mathfrak{E}}$ cannot be one-to-one. If, on the other hand, $\mathfrak{H}$ has a countably infinite dimension, the proof goes as follows. Let $E \in \mathfrak{E}$ be a subspace of $\mathfrak{H}$, with $2 < \dim E < +\infty$. If, for such an E, there were some $\mu \in M$ with $\mu(E) = 1$, this would be a two-valued probability function in the Mackey structure $\mathfrak{E}_E = \{F \subseteq E\}_{F \in \mathfrak{E}}$, but according to Gleason's theorem this cannot happen. Thus there must be at least one $E \in \mathfrak{E}$, $E \neq \cap$, with $\beta_{\mathfrak{E}}(E) = \beta_{\mathfrak{E}}(\cap) = \emptyset$. Therefore $\beta_{\mathfrak{E}}$ is not one-to-one. Hence

(C) Ordinary quantum mechanics has no hidden-parameter theory $(\mathfrak{E}', H)$ with the properties (5) and (6).

Note that, unlike the von Neumann proof, the argument makes use of Gleason's theorem only for Hilbert spaces of finite dimension. In the exceptional case of a two-dimensional Hilbert space, theories of hidden parameters have been given, but have not so far achieved any physical significance (Specker 1960, Weidlich 1960, Bell 1966, Kochen and Specker 1967).

VII.2 Nine logical variations on a theme of Einstein

THEME

Among the critics of orthodox quantum mechanics who, as Heisenberg once put it (1959, p. 115), "express rather their general dissatisfaction with the quantum theory, without making definite counterproposals" is found the prominent figure of Einstein, who sought to show that quantum mechanics is an incomplete theory, both in many oral discussions with the supporters of orthodoxy (of which a lively account is given by Bohr (1949)) and also in a sequence of publications (Einstein 1936, 1948, 1949, 1953a, b; Einstein, Podolsky and Rosen 1935). The present section offers a logical analysis of the celebrated paper of 1935 by Einstein, Podolsky and Rosen.

Before proceeding to the analysis, a brief general description of Einstein's position will be useful. It can be called a classical realism, with certain qualifications. The basic conviction which underlies the whole of Einstein's critical comments regarding quantum mechanics is that it must make sense to speak of particular, objectively existing situations in nature, and that the aim of physics is "the complete description of any (individual) real situation (as it supposedly exists irrespective of any act of observation or substantiation)" (Einstein 1949, p. 667). The classical concept of state is naturally adduced to express this conviction, for example: "*There is such a thing as the 'real state'* of a physical system, which exists objectively and independently of any observation or measurement and can in principle be described in the language of physics" (Einstein 1953a). Similarly, Pauli in a letter of 1954 to Born describes discussions with Einstein: "Now from my conversations with Einstein I have seen that he takes exception to the assumption, essential to quantum mechanics, that *the state of a system is defined only by specification of an experimental arrangement. Einstein wants to know nothing of this*" (Born 1971, p. 218).

One must not suppose, however, that Einstein's view of reality, thus expressed, was a naive realism, and certainly not that a deliberately metaphysical approach was adopted. On the contrary, Einstein qualified the above type of assertion, necessarily expressed in an inherited ontological vocabulary, by saying that there are certain mental constructs which determine what we regard as "real": " 'Being' is always something which is mentally constructed by us, that is, something which we freely posit (in the logical sense). The justification of such constructs does not lie in their derivation from what is given by the senses. Such a type of derivation (in the sense of logical deducibility) is nowhere to be had, not even in the domain of pre-scientific thinking. The justification of the constructs, which represent 'reality' for us, lies alone in their quality of making intelligible what is sensorily given" (Einstein 1949, p. 669). Einstein's own ontology, if such can be said to exist, thus has conventional features, although the assertions that can be made with reason within a chosen system of concepts are not completely arbitrary. The choice that Einstein was inclined to make here was no doubt governed largely by his deep distaste for any kind of positivism, and the "serious and fundamental misgivings which I feel regarding the basis of the statistical quantum theory" (Einstein 1953a) probably arose from the fact that the quantum theory at least allows a positivist interpretation, if not actually supporting it. Einstein saw the conceptual systems of classical physics as just the opposite of such a view of physical reality, since these systems enable us to refer meaningfully to events or situations regarding individual objects without actually observing such objects. With this classical view as a basis,

Einstein's problem was to reconcile with it the existing form of quantum mechanics. He felt the need to achieve such a reconciliation because, to an extent not easily defined but certainly significant, he regarded quantum mechanics as a theory which functioned satisfactorily and was unusually well confirmed empirically: "Probably never before has a theory been evolved which has given a key to the interpretation and calculation of such a heterogeneous group of phenomena of experience as has the quantum theory" (Einstein 1936, p. 374). Later, he concedes that "the formal relations which are given in this theory—i.e. its entire mathematical formalism—will probably have to be contained, in the form of logical inferences, in every useful future theory" (Einstein 1949, p. 667). The latter quotation indicates that Einstein did not, however, regard quantum mechanics as a final theory, in the sense that he considered it to involve an *avoidable incompleteness* in the description of reality.

In support of this view, Einstein, Podolsky and Rosen (to be denoted by EPR), in their paper already quoted (1935) put forward an argument whose precise import is not obvious at first sight. First of all, note that the authors evidently accept quantum mechanics in some form, yet propose to prove (not merely to assert) that something is wrong with it. The question then is, in which particular form do they accept it and what exactly do they propose to prove on this basis? On the latter point, it has already been mentioned that incompleteness is to be demonstrated on the foundation of classical realism. Such a proof, it must be noticed from the start, is a relational proof: not only is incompleteness proved, but quantum mechanics, or a particular description of state, is shown to be incomplete in a certain sense *if* certain reality requirements are imposed. For such a proof, the theory must obviously be given some latitude for the introduction of new requirements. One cannot investigate quantum mechanics as a fully defined theory with no options remaining, and take this as a basis for argument using a new hypothesis about the same theory. In the present case it is also fairly clear where one has to draw the line between what is to be accepted as a firm basis of proof and what must be left temporarily undecided. The purely epistemic or purely statistical treatment (section II.4), in which states of individual objects do not occur, can be accepted, whereas the question of extending this treatment by bringing in the concept of the state of an individual object is still open for discussion: this concept can still be interpreted in various ways, and we have already seen how Einstein uses it in his realistic conception. On the other hand, concepts of completeness can be introduced as the decisive link between the concept of a state and the epistemic or statistical concepts of the treatment without states. It seems less important to decide whether the epistemic or the statistical formulation is to be the basis. We shall follow the language of EPR by choosing the epistemic formulation, since this relates exclusively to the individual object and can therefore incorporate immediately the concept of the state of an individual object.

The result is that the EPR arguments are presented as at least a proof of the incompatibility of certain reality requirements and certain completeness requirements pertaining to the epistemic formulation of quantum mechanics as extended by the concept of the state of an individual object. The necessary refinements having been made, it will be found that this incompatibility has an extremely restrictive effect, since such an extension can take place, while maintaining both the reality and the completeness requirements, only at the cost of *a contradiction in the strictest sense.* Although such a result is all too painfully familiar, its significance must not be overestimated. It does not compel a decision as to which group of postulates is to be accepted, but simply shows that one cannot have both at the same time. Thus any decision in this respect must come from elsewhere. The problem will be further discussed in section 3; here, we aim simply to make precise the completeness requirements and the reality requirements, and to show that they are incompatible.

VARIATION I

The following two new basic concepts are axiomatically added to the purely epistemic formulation of the quantum mechanics of an individual object Σ:

(α') the concept of the *state* of Σ,

(β') the relation between a state, a quantity and a value of a quantity of Σ, such that *the quantity A has the value α in the state s.*

The necessary comments on (α') have already been made. The relation given in (β') is to be compared with the universal relation S that has been used for the formulation of ordinary classical mechanics (section II.2). This is indicated in (β') by expressing the content of the relation in the same way as in Chapter II. It is of course quite another matter whether the content of the relation in (β') can in fact be the same as that of the classical S, since here we have a connection between the states, the *quantum-mechanical* quantities and their possible values. The purpose of using (β') must be a classical one, however: EPR have (β') in the form that they sometimes refer to a value that a quantity (actually) has as an "element of physical reality" (see their third and fourth paragraphs). They evidently regard the state of an object as the combination of all the "elements of physical reality" which can be present in the object at a particular time. This corresponds to the form used in (β'): for a given state s, all pairs (A, α) are "elements of physical reality" for which it is true that A has the value α in the state s.

We now proceed to the first derived relation:

(α'_d) A state of knowledge W of Σ is called *a knowledge of the state* **s** *of* Σ if, for every A and every value α of A, $\mathbf{pr}(W; A, \alpha) > 0$ when A has the value α in s and $\mathbf{pr}(W; A, \alpha) < 1$ when A does not have the value α in s.

This relation connects the states of Σ and the states of knowledge concerning Σ. It would be closer to the language of EPR to define a corresponding relation between von Neumann operators in the Hilbert space of Σ and the states of Σ, to the effect that the operator W describes the state s. But, on account of the one-to-one relation between states of knowledge and von Neumann operators (subsection II.4(d)(ii)), the difference would be only a linguistic one. EPR consider explicitly only the case of maximum knowledge of Σ, which can be directly represented by vectors in the Hilbert space of Σ. Their treatment, however, also implicates the non-maximum case, so that it is desirable to include this expressly from the start.

Our second derived relation is

(β'_d) A quantity A of Σ is said to be *real in the state s of* Σ if there is a value α such that A has the value α in the state s.

In the classical view, this relation is valid for any A and s, i.e. it is trivial and therefore unnecessary. It is important to appreciate from the start that EPR's argument does *not* exclude the possibility that a quantity in a state is not real. Einstein may well have been convinced that a classical theory of hidden parameters for quantum mechanics will in the end prove not to include this critical case, i.e. that every quantity is real in every state. But this is not assumed in the EPR argument, and we shall see (Variation IV) that the point is essential to the value of their argument.

Variation II

Next, we must deal with the concept of completeness. EPR have intermingled two questions here: the meaning of saying that individual knowledge of a state is complete knowledge of it, and the meaning of saying that the entire theory is complete (see their third, tenth, eleventh, third-from-last, and last paragraphs). In consequence, it is not clear which of two possible and independent requirements concerning the completeness of the theory is to be refuted. We shall define the concept of complete knowledge of a state as follows:

(γ'_d) A state of knowledge W of Σ is called *a complete knowledge of the state s of* Σ if, for every quantity A and every value α of Σ, when A has the value α in s, this value is known in W, i.e. $\mathbf{pr}(W; A, \alpha) = 1$, and when A does not have the value α in s, $\mathbf{pr}(W; A, \alpha) < 1$.

We must distinguish from the definition (γ'_d) two requirements concerning the completeness of the theory, which can be formulated in terms of the concept which this definition provides:

(Q_1) For every state s of Σ there is a state of knowledge W of Σ that is a complete knowledge of s.

(Q_2) Every maximum knowledge P_ϕ of Σ that is a knowledge of the state s of Σ is a complete knowledge of s.

The definition (γ'_d) is supported by its use in an argument in the tenth paragraph of EPR's paper, as will be seen later. Like (Q_1), it is supported by the ambiguous expression used in their third paragraph, where the "condition of completeness" is stated to be that "*every element of the physical reality must have a counterpart in the physical theory*". If this refers to the "elements of physical reality" that pertain to a single state, we get part of the definition (γ'_d): if W is a complete knowledge of s, then every pair (A, α) that is real in s (= an "element of physical reality") must be marked as known in the catalogue of information W concerning s (= a "counterpart in the physical theory"). If, however, this passage is meant to leave the state of the system unspecified, we have (Q_1). This is confirmed by the fact that EPR, just before stating their condition, refer to it as a "requirement for a complete *theory*" (italics E.S.). A third factor in this uncertainty is that in two places (their eleventh and third-from-last paragraphs) EPR act as if they wished to refute (Q_2), which is not the same as (Q_1). Here it is unnecessary to discuss the text in detail, since the situation has been sufficiently clarified, and a satisfactory basis obtained for subsequent proofs, by distinguishing (γ'_d) from (Q_1) and (Q_2) and stressing the difference between the latter pair. However, the refutation of (Q_1) and (Q_2) on the basis of EPR's requirements will be possible only by means of a further postulate, which causes no difficulty and therefore may be stated immediately.

(Q_3) Every state of knowledge of Σ is a knowledge of at least one state of Σ.

Variation III

The extension of the epistemic formulation of quantum mechanics given by (α'), (β') and the definitions (α'_d)–(γ'_d) is already fairly explicit. The full classical propositions regarding states are formally present in (β'). (α'_d)–(γ'_d) are explicit definitions; (α'_d) and (γ'_d) make explicit use of quantum-mechanical propositions regarding probabilities. This explicitness has the advantage that the concepts defined are made sufficiently clear, and they have been introduced for just this reason. On the other hand, such an extensive commitment as results in particular from the explicit definition of the most important concepts runs the risk of being more easily attacked. In making an explication, as is done here for the EPR argument, it is always advisable to bring in nothing more than is absolutely necessary. It is found that in fact a weaker form of (α'), (β') and (α'_d)–(γ'_d) is sufficient to rebuild EPR's argument. In that case not two but four basic concepts are used:

(α) the concept of the *state* of Σ,

(β) the relation between a quantity and a state of Σ, that *the quantity A is real in the state s*,

(γ) the relation between a state of knowledge and a state of Σ, that *the state of knowledge W is a knowledge of the state s*,

(δ) the relation between a state of knowledge and a state of Σ, that *the state of knowledge W is a complete knowledge of the state s*.

These basic concepts are connected by only two postulates:

(α_s) every complete knowledge of a state s of Σ is a knowledge of s,

(β_s) if W is a complete knowledge of the state s of Σ, then for every quantity A of Σ, if A is real in s, then A is known in W, i.e. $\mathbf{pr}(W;\, A,\, a)$ is a delta function of a.

The resulting theory is clearly obtainable in a very simple manner from that described by (α'), (β') and (α'_d)–(γ'_d). (α) is identical with (α'). (β) is defined in (β'_d), (γ) in (α'_d), and (δ) in (γ'_d). (α_s) is a trivial consequence of the definitions (α'_d) and (γ'_d); (β_s) is a trivial consequence of (β'_d) and (γ'_d). Finally, the conditions (Q_1)–(Q_3) to be considered involve only concepts used in (α)–(δ). We shall see that the limitation to (α)–(δ), (α_s) and (β_s) can be maintained for the whole of the discussion in this section. The theory thus defined will be denoted by $\mathfrak{t}$.

Variation IV

It has already been indicated in Variation I that EPR's objective becomes trivial if we assume from the start any over-strong classical hypotheses about the relation (β') or, in the subsequent form, the relation (δ), for example, that

(K) every quantity of Σ having at least one eigenvalue is real in every state of Σ.

For

(Γ) in the theory $\mathfrak{t}$, the assertions (K) and (Q_1) are incompatible, as are (K), (Q_2) and (Q_3).

To prove this, let (K) be valid. Then, if W is a complete knowledge of s, *all* the quantities referred to in (K) must be known in W, which is obviously impossible. Hence (K) implies that there are no W and s such that W is a complete knowledge of s. This disproves (Q_1). If, secondly, (K) and (Q_3) are valid, then there are states of maximum knowledge, and states described by these. Because of the conclusion just drawn from (K), however, complete knowledge cannot occur. Thus there must be states of maximum knowledge which are not states of complete knowledge, which contradicts (Q_2).

What now is the significance of this result? We shall see later that the postulate stated by EPR for refuting the completeness assertions (Q_1) and (Q_2) is satisfied if (K) is valid. Hence it follows that they are interested in a sharper result than (Γ), and the reason is not difficult to see. A crude requirement such as (K) does not penetrate into quantum mechanics at all.

The real obligation on anyone who regards (K) as important is therefore to prove a physically meaningful extension of quantum mechanics in which (K) is valid. EPR, however, do not specify any particular extension of quantum mechanics, either making (K) valid or satisfying their own postulate. For this very reason, however, they must be concerned to anchor their postulate as firmly as possible in quantum mechanics. This is their only chance of making the postulate credible, a point which is not correctly perceived in the analysis of the EPR paradox by Bohm (1951, §22.15), where the condition denoted by (3) is equivalent to (K) and is treated as an implicit assumption made by EPR. As already mentioned in Variation I, Einstein may have believed that (K) is true, but it is not *used* in their argument, which otherwise would be trivial and in need of no further treatment here. But our discussion is really only beginning, as we turn to EPR's criterion of reality.

VARIATION V

The relevant passage (in the fourth paragraph of EPR's paper) is: "*If, without in any way disturbing a system, we can predict with certainty* (*i.e. with probability equal to unity*) *the value of a physical quantity, then there exists an element of physical reality corresponding to this physical quantity*". At first sight one might suppose this to be only a fairly trivial remark. If we ask in which situation the value of a quantity A of the object Σ can be predicted with certainty, the most obvious answer is: if we have a knowledge W of Σ which includes a knowledge of the value of A. In such a case, the prediction of the value of A will evidently be possible "without disturbing Σ in any way": nothing need be done in order to find the value, for it is already known. Lastly, it is easy to agree that in such a case the quantity A actually has a definite value: what is known as something must exist as that thing. However, this trivial explanation immediately calls forth the objection that it deals only with cases where the added "without in any way disturbing a system" is a *consequence* of the assumed situation and could therefore be omitted. In fact, we now see that it must be an essential (not necessarily satisfied) condition in the premise of the criterion quoted above. In that case, the remainder of the premise, viz. the possibility of reliable prediction, cannot be meant to signify that the knowledge needed for such a prediction is already possessed, but rather that this knowledge is desired. The case where it is already possessed would be just a trivial special case of this. Now we know from the purely epistemic formulation of quantum mechanics (subsection II.4(e)) that, with an initial state of knowledge concerning Σ, a measurement of the quantity A always leads via the reduction of states to a new state of knowledge which allows the desired prediction of the measured value, although in general not until after the measurement. Moreover, some perturbation (in a certain sense) of the object is unavoidable in this process (sections IV.3 and VI.2). This is precisely the point at which EPR's criterion comes in. We are concerned not with the exceptional case where the possibility of prediction is based on an already existing state of (certain) knowledge, so that no perturbation of the object can be in question, nor with the possibility, which always exists, of simply measuring the quantity concerned and perhaps causing a perturbation, but with the intermediate possibility of obtaining the knowledge needed for a prediction (with certainty) of the value of a quantity A *without in any way perturbing the object.* The significance of the condition quoted is that if this possibility exists, then the quantity A has a well-defined value, and that this value is the one which can and will be known.

Thus, with the concept of state included, EPR's condition becomes

(EPR) Let s be a state and A a quantity of the object Σ. If it is possible, without in any way perturbing the state s, to acquire a knowledge of s such that the value of A can be predicted with certainty, then A is real in s.

The significance of this is, of course, that the value of A which can thereby be predicted was in fact already present in the state s, before the knowledge in question was acquired, since the state s is, by hypothesis, unchanged by such acquisition of knowledge. For the condition (EPR) to be effective, however, it is necessary to say more precisely what is meant by its premise. So far we have seen only that the premise certainly does not imply that the knowledge mentioned is already possessed. The wording now chosen, that a knowledge of s is acquired, without any perturbation of s, such that the value of A can be predicted, is still not within the vocabulary of the epistemic version of quantum mechanics extended by (α)–(δ), which is under consideration here. It must therefore be interpreted in the language of this version. We shall see that *everything depends on how this interpretation is effected.* It is here that EPR make a choice which is not acceptable on the orthodox view, since it makes the condition (EPR) incompatible with the assertions regarding completeness discussed in Variation II. Secondly, it will be shown in section 3 that there does exist an orthodox interpretation in which (EPR) is a straightforward conclusion from the epistemic extension of the ontic version of quantum mechanics.

First of all, we shall give a simplified form of EPR's interpretation of the premise of their condition, which avoids the need to invoke a second object as EPR do. Let there be a single object Σ. First, we have to obtain, in the purely epistemic formulation of quantum mechanics, a relation between a state of knowledge W of Σ and a quantity A of Σ, which holds if and only if (as we shall put it) *the value of A can be determined without perturbing W.* In order to be able to determine a (sharp) value of A, the latter must possess eigenvalues. Let $\{\alpha_i\}_i$ be the non-empty set of the eigenvalues of A. To determine these, a sufficiently accurate measurement is needed, for which we take

$$\{\{P^A_{\{\alpha i\}}\}_i, 1-\Sigma_i P^A_{\{\alpha i\}}\}. \tag{1}$$

This is a measurement *without* acceptance of the result. It must contain $1-\Sigma_i P^A_{\{\alpha i\}}$ because we do not assume that the set $\{P^A_{\{\alpha i\}}\}$ is complete. If the state of knowledge before the measurement is W, then the state of knowledge of Σ after the measurement (1) is given by the result of either the strong or the weak reduction of states (cf. section II.4). If we further assume that W can include only the eigenvalues of A as results of measurement, then after the measurement we therefore have either the assemblage

$$\left\{\frac{P^A_{\{\alpha i\}} W P^A_{\{\alpha i\}}}{\mathrm{Tr}(WP^A_{\{\alpha i\}})}, \mathrm{Tr}(WP^A_{\{\alpha i\}})\right\}_{i:\mathrm{Tr}(WP^A_{\{\alpha i\}})>0} \tag{2}$$

of the states of knowledge present in it, with the corresponding probabilities, or simply the state of knowledge

$$\Sigma_i P^A_{\{\alpha i\}} W P^A_{\{\alpha i\}}. \tag{3}$$

Here the decision arises. The assemblage (2) is *always* different from W unless the quantity A happens to be trivial, but the state of knowledge (3) may be the same as the original state of knowledge W. This possibility will be made use of in the manner of EPR, with the definition

(α_d) *In a state of knowledge W concerning Σ, the value of the quantity A of Σ can be determined without perturbing W*, if

$$W = \Sigma_i P^A_{\{\alpha i\}} W P^A_{\{\alpha i\}} \tag{4}$$

and the set $\{\alpha_i\}_i$ of the eigenvalues of A is non-empty.

If this relation between W and A exists, then in fact part of the anomaly which is related to the quantum-mechanical reductions of states (without acceptance of the result) is eliminated: the weak reduction of states does not alter the existing state of knowledge. From Chapter VI we know that in the orthodox view it is not permissible to say that the assumption (4) eliminates the actual "perturbing" part of the measurement (1), but here EPR take a different view, as follows.

The relation defined by (α_d) can be applied for the desired interpretation of the premise of (EPR) by assuming that, if the state s of the object Σ allows a knowledge W, such that the value of the quantity A of Σ can be determined without perturbing W, *this determination can also be made without perturbing the state s.* Thus we have the following version of (EPR):

(EPR_1) Let s be a state and A a quantity of Σ. If s allows a knowledge W of itself which is related to A by (α_d), then A is real in s.

In this form, the condition of reality does not have the full evidentness which EPR were able to give it. They did so only by introducing a second object coupled to Σ in an appropriate way. From the purely logical aspect, the above formulation provides everything which EPR did, and the lack of evidentness is replaced by greater logical simplicity. We shall therefore start by basing EPR's argument on (EPR_1) and only afterwards give their own formulation.

Variation VI

First, a new concept is to be introduced:

(β_d) Two quantities A and A' of an object Σ are called an *EPR pair* if

(1) there exists a state of knowledge W of Σ such that the value of A and that of A' can both be determined when that state of knowledge is present and without perturbing it: briefly, A and W are related by (α_d), and so are A' and W,

(2) there exists no state of knowledge W of Σ which gives (with certainty) both the value of A and that of A'.

EPR themselves took, as such a pair, in their sixteenth paragraph, a position coordinate and the corresponding momentum component of a particle. This has the considerable advantage of providing a familiar physical picture. But in a von Neumann formulation of quantum mechanics, this is not a possible choice for the purpose in question, since (1) in (β_d) cannot be satisfied for even one quantity having a continuous spectrum, and it is likely to be difficult to achieve any modification of this part of (β_d) (within a von Neumann formulation). We must therefore be content with indicating a less obvious class of EPR pairs. For this purpose we take in the Hilbert space $\mathfrak{H}$ of Σ a direct decomposition

$$\left.\begin{aligned} \mathfrak{H} &= \Sigma_m^{\oplus}\mathfrak{H}_m \oplus \mathfrak{H}_0, \\ 1 < \dim \mathfrak{H}_m &< +\infty, \dim \mathfrak{H}_0 = 0 \text{ or } > 1, \end{aligned}\right\} \tag{1}$$

of which there are obviously many. By decomposition we can clearly take orthonormal bases of $\mathfrak{H}$, $\{\phi_\mu\}_\mu$ and $\{\phi'_\mu\}_\mu$, such that no two ϕ_μ and ϕ'_ν are proportional. Also, we can evidently choose corresponding self-adjoint operators A and A' such that the ϕ_μ and ϕ'_μ are complete systems of *non-degenerate* eigenvectors of A and A'. The quantities belonging to A and A' form an EPR pair if W in (β_d) (1) is chosen so that the $\mathfrak{H}_m$ in (1) are its eigenspaces for eigenvalues greater than zero, and $\mathfrak{H}_0$ its eigenspace for the eigenvalue zero; (β_d)(2) is satisfied, since A and A' have no common eigenvector. Hence

(γ_s) there exist EPR pairs of quantities A and A' of Σ.

The cases just constructed are also cases of quantities which cannot be measured simultaneously with arbitrary accuracy, but can be measured simultaneously with limited accuracy (cf. section IV.2).

Our next objective is to prove

(Δ) From (Q_3) and (EPR_1) it follows that, for every EPR pair of quantities A and A' of Σ, there is a state s of Σ such that A and A' are real in s.

Let A and A' be an EPR pair. From (β_d) (1) there exists a corresponding W that is related both to A and to A' by (α_d). From (Q_3), there exists s corresponding to W, such that W is a knowledge of s. This s therefore satisfies the premise of (EPR_1) as regards both A and A', and it follows from (EPR_1) that A and A' are real in s. This proves (Δ). We now add a further definition:

(γ_d) A state s of Σ is called an *EPR state* if there is an EPR pair of quantities of Σ which are both real in s.

Then we can prove

(θ) From (Q_3) and (EPR_1) it follows that there is at least one EPR state of Σ.

The proof is as follows. According to (γ_s), EPR pairs exist. From (Δ), for every such pair A and A' there is a state s such that A and A' are real in s. From (γ_d), every such s is an EPR state. In addition, we have

(δ_s) If s is an EPR state of Σ, there is no complete knowledge of s.

Let s be an EPR state. From (γ_d) there then exists an EPR pair A and A' such that A and A' are real in s. If W were a complete knowledge of s, (β_s) would show that both A and A' are known in W, but this is impossible by (β_d) (2). Thus (δ_s) is proved; cf. EPR's tenth paragraph. The argument is completed by proving the desired result:

(Λ_1) The assertions (Q_1), (Q_3) and (EPR_1) are incompatible in the theory $\mathfrak{t}$.

From (θ), (Q_3) and (EPR_1), there exists an EPR state of Σ. From (δ_s), there exists no complete knowledge of such a state. From (Q_3) and (EPR_1), the assertion in (Q_1) regarding completeness is then disproved.

VARIATION VII

The meta-theorem (Λ_1) is all that was to be proved, from the purely logical aspect, in relation to the assertion regarding completeness (Q_1): the latter has been proved incompatible with the EPR condition of reality. It has already been mentioned, however, that EPR do not apply this condition in the form (EPR_1) which we have used so far, but in another form, although it is logically equivalent to (EPR_1) according to the theory $\mathfrak{t}$. The difference lies in the wording of the premise of (EPR_1). We obtained this as an interpretation of the premise of our original formulation (EPR), which followed closely EPR's own wording. The purpose of the interpretation was to state, in a manner which should apply to the underlying formulation of quantum mechanics, what is meant by saying that, without in any way perturbing the state s of an object Σ, we can acquire a knowledge of s which allows us to predict with certainty the value of a quantity A of Σ. EPR achieve this by means of a second object. We shall therefore now use I instead of Σ, and II for the newly defined object (EPR's nomenclature is the opposite of this). Analogously to the procedure in Variation V, we first put forward a relation (cf. (α_d)) between a quantity of A of I and a state of knowledge $W^{\otimes}$ of $I \otimes II$:

(δ_d) *The value of the quantity A of I can be determined without perturbing $W_I^{\otimes}$, when the state of knowledge of $I \otimes II$ is $W^{\otimes}$*, if A has a non-empty set $\{\alpha_i\}_{i \in J}$ of eigenvalues and if a measurement $\{Q_k\}_{k \in K}$ of II and one-to-one mapping $f: J \rightarrowtail K$ exist such that

(1) $\mathrm{Tr}(W^{\otimes}(1 \otimes Q_k)) = 0$ for $k \notin f(J)$,

(2) for all $i \in J$, if $\mathrm{Tr}(W^{\otimes}(P^A_{\{\alpha i\}} \otimes 1)) > 0$, then $\mathrm{Tr}(W^{\otimes}(1 \otimes Q_{f(i)})) > 0$ and

$$\mathrm{Tr}\left(\frac{(1 \otimes Q_{f(i)})\, W^{\otimes}(1 \otimes Q_{f(i)})}{\mathrm{Tr}(W^{\otimes}(1 \otimes Q_{f(i)}))}(P^A_{\{\alpha i\}} \otimes 1)\right) = 1,$$

(3) $\mathrm{Tr}(W^{\otimes}\Sigma_{i \in J}(P^A_{\{\alpha i\}} \otimes 1)) = 1$.

In order to clarify the relation between (α_d) and (δ_d), the words "without perturbing $W_I^{\otimes}$" from (α_d) have been placed in the expression being defined. The justification for this will be given shortly. As we know from section VI.2, it is of course impossible for $W^{\otimes}$ itself not to be perturbed.

In the same way as (α_d) was used to obtain the premise of (EPR_1), (δ_d) will be included in the premise of the new wording (EPR_2), assuming that, whenever the state s of I allows a knowledge $W^{\otimes}$ of $I \otimes II$ with $W_I^{\otimes}$ as the knowledge of s, such that when $W^{\otimes}$ is present the value of the quantity A of I can be determined without perturbing $W_I^{\otimes}$, then *this is also done without perturbing the state s*. Thus we have

(EPR_2) Let I and II be two objects; let s be a state of I and A a quantity of I. If there exists a state of knowledge $W^{\otimes}$ of $I \otimes II$ with $W_I^{\otimes}$ as a state of knowledge of s, related to A in accordance with (δ_d), then A is real in s.

Although, as we shall see immediately, this postulate is equivalent to (EPR_1), its premise makes more plausible the idea that, on the assumption given, the state s of the system I is not perturbed by determining the value of A, and therefore the value of A must have been present even before the measurement in question. EPR's justification of this was approximately as follows (their twelfth to fifteenth paragraphs). Let the systems I and II be initially separated and then subject to a temporary interaction which finally gives for the combined system a state of knowledge $W^{\otimes}$ (EPR take a state of maximum knowledge, but this is not important at the moment; cf. Variation VIII), which is related by (δ_d) to the given quantity A of I. Then let the interaction cease. The conditions stated in (δ_d) ensure that a suitable measurement of II will determine the value of A, since (1) and (2) in (δ_d) are essentially equivalent to (β_{31}) and (β_{32}) in section VI.3; the only difference being that now the $P^A_{\{\alpha i\}}$ need not form a complete system, and this is precisely compensated by (3) in (δ_d). It also seems certain (cf. Mathematical Appendix 1 in section VI.3) that $W_I^{\otimes}$ is related to A by (α_d), and this will in fact be proved rigorously in Variation VIII. Thus (δ_d), and hence the premise of (EPR_2), give at least as much as (α_d) or the premise of (EPR_1). In addition, however, EPR comment: "Since at the time of measurement the two systems no longer interact, no real change can take place in the first system in consequence of anything that may be done to the second system" (their fifteenth paragraph, with the words "first" and "second" interchanged). They continue immediately. "This is, of course, merely a statement of what is meant by the absence of an interaction between the two systems". From this we learn that what the authors really intend to offer in support of the idea that the state of I cannot be altered by a measurement of II they themselves would rather have expressed as follows: at the time of the measurement of II, the situation is such that anything done with system II no longer has a real influence on system I; the state of I is unaffected by the measurement of II. In the next section, this treatment will be compared with the orthodox view.

Variation VIII

The question of the relation between the conditions (EPR_2) and (EPR_1) is essentially answered by the following theorem.

(ε_s) Let two objects I and II and a quantity A of I be given. If a state of knowledge $W^{\otimes}$ of $I \otimes II$ is related to A by (δ_d), then A and $W_I^{\otimes}$ are related by (α_d). If conversely a state of knowledge W of I is related to A by (α_d), then there is a state of (maximum) knowledge $W^{\otimes}$ of $I \otimes II$ with $W_I^{\otimes} = W$, related to A by (δ_d), if

$$\dim \mathfrak{H}^{II} \geqslant \dim \mathfrak{H}^{I}.$$

The proof of this theorem (in the purely epistemic formulation of quantum mechanics) is given in Mathematical Appendix 1 to the present section. From (ε_s) there follows at once

(Π) Let the systems I and II be given, together with a state s and a quantity A of I. The premise of (EPR_1) is satisfied if and only if that of (EPR_2) is satisfied (for s and A in each case). Thus (for given I and II) (EPR_1) and (EPR_2) are equivalent.

The proof from (ε_s) is trivial and may be omitted. From (Π) we see that (EPR_2) gives nothing, from the purely logical point of view, that is not given by (EPR_1). The situation in which one would deduce that the quantity A is real in the state s is just given a better physical illustration than would be possible with only a single object.

From (Π) and (Λ_1) we have trivially

(Λ_2) In the theory $\mathfrak{t}$, the assertions (Q_1), (Q_3) and (EPR_2) are incompatible.

Although this can be obtained directly from (Π) and (Λ_1) as a second principal result comparable with (Λ_1), it is useful to glance over the argument which gave (Λ_1) via (EPR_1), as it applies to the direct use of (EPR_2) to obtain (Λ_2). One can easily see that the reasoning in Variation V can be exactly imitated, replacing the state of knowledge W of I which there appeared throughout as in the premise of (EPR_1) by the state of knowledge $W^{\otimes}$ of $I \otimes II$ which occurs in the premise of (EPR_2), if, in addition to the relation between (α_d) and (δ_d) which is provided by (ε_s), we can obtain a corresponding relation involving *two* quantities, A and A'. These two quantities occur in the concept of an EPR pair and are, in particular, vital to the construction of an EPR state in (Δ) and (θ). Omitting obvious relationships, we shall give only the most significant resulting theorem:

(ζ_s) Let the objects I and II be again given, and let $\dim \mathfrak{H}^{II} \geqslant \dim \mathfrak{H}^{I}$. If each of the two quantities A and A' of I is related by (α_d) to a state of knowledge W of I, then there exists a state of (maximum) knowledge $W^{\otimes}$ of $I \otimes II$, with $(W^{\otimes})_I = W$, such that A and $W^{\otimes}$ are related by (δ_d), and so are A' and $W^{\otimes}$.

The proof of (ζ_s) is given in Mathematical Appendix 2 to this section. It has been mentioned in Variation VI that EPR choose for their EPR pair A and A' a position coordinate x_I and the corresponding momentum component p_I of a particle I. For the measurements $\{Q_k\}_{k \in K}$ and $\{Q'_k\}_{k \in K}$ of II which occur implicitly in (ζ_s) through (δ_d), they choose measurements of a position coordinate x_{II} and the corresponding momentum component p_{II} of a particle II. Purely formally, $x_I \otimes \mathbf{1} - \mathbf{1} \otimes x_{II}$ and $p_I \otimes \mathbf{1} + \mathbf{1} \otimes p_{II}$ are interchangeable as operators in $\mathfrak{H}^I \otimes \mathfrak{H}^{II}$. The state of knowledge $W^{\otimes}$ of $I \otimes II$ whose existence is asserted in (ζ_s) is then regarded by EPR as being described by a common eigenvector $\Phi^{\otimes}$ of these two interchangeable operators. This construction is impossible in a von Neumann formulation of quantum mechanics, as is the use of x_I and p_I to form an EPR pair as defined by (β_d).

Variation IX

Lastly, we come to the question, which has not yet been discussed, of the compatibility of the assertion (Q_2) regarding completeness, introduced in Variation II in addition to (Q_1), with EPR's condition of reality. It has already been mentioned that EPR do not make a clear distinction between these two assertions regarding completeness. In fact, the construction of an EPR state in Variation V does not provide any information as to the compatibility question now posed as regards (Q_2). Instead, it seems certain that, in order to arrive at incompatibility, we must bring in a further condition not yet mentioned, which pertains to the relation between the states of a system I and those of a system $I \otimes II$ containing I as a subsystem. The condition is not uniquely determined by our present objective, but a formulation that is especially simple and classically plausible is

(T) Let two systems I and II be given. For every state of knowledge $W^{\otimes}$ of $I \otimes II$, and every state $s^{\otimes}$ of $I \otimes II$ such that $W^{\otimes}$ is a state of knowledge of $s^{\otimes}$, there exists a state s of I such that $W_I^{\otimes}$ is a state of knowledge of s and in which a quantity A of I is real if and only if the quantity $A \otimes 1$ is real in $s^{\otimes}$.

The existence of s for a given state of knowledge $W^{\otimes}$ of $s^{\otimes}$, such that $W_I^{\otimes}$ is a state of knowledge of s, hardly calls for comment. Classically, however, the decisive condition which relates s directly to $s^{\otimes}$ is also almost obvious: $s^{\otimes}$ specifies a reality behaviour for the quantities of I, in particular, and we have only to ask whether this occurs in a way that could also arise from a state of I. EPR evidently regard it as entirely certain that this is so, as in (T). But, as will be shown in section 3, (T) is *not* satisfied in the ontic formulation of quantum mechanics.

The analysis to clarify the relation between (Q_2) and (EPR_1) (or (EPR_2)) will be initiated by the theorem

(η_s) Let I and II be given. For every state of knowledge $W^{\otimes}$ of $I \otimes II$ for which $W_I^{\otimes}$ is not maximum, there exists a quantity A of I which (1) is related to $W_I^{\otimes}$ by (α_d), (2) does not have a value known in $W_I^{\otimes}$.

To prove this, it is sufficient to take an A which is described by the same operator as $W_I^{\otimes}$. Then (1) follows immediately from (α_d), and (2) follows since $W_I^{\otimes}$ is not maximum, by hypothesis. We also have

(Σ) From (EPR_1) (or (EPR_2)) and (T) it follows that, if $W^{\otimes}$ is a state of knowledge of $I \otimes II$ for which $W_I^{\otimes}$ is not maximum, and if $s^{\otimes}$ is a state of $I \otimes II$, then $W^{\otimes}$ is not a state of complete knowledge of $s^{\otimes}$.

If $W^{\otimes}$ is not even a state of knowledge of $s^{\otimes}$, then by (α_s) it is not a state of complete knowledge of $s^{\otimes}$. Let $W^{\otimes}$, therefore, be a state of knowledge of $s^{\otimes}$. From (η_s), there exists a quantity A of I satisfying the conditions (1) and (2) in (η_s). From (T), there exists a state s of I satisfying the condition in (T). Since A and $W_I^{\otimes}$ are related by (α_d) according to (1) in (η_s), and $W_I^{\otimes}$ is a state of knowledge of s, (EPR_1) shows that the quantity A is real in s. Hence, from (T), the quantity $A \otimes 1$ is real in $s^{\otimes}$. From (2) in (η_s), however, the value of A in $W_I^{\otimes}$ is unknown, and so therefore is that of $A \otimes 1$ in $W^{\otimes}$. Hence $W^{\otimes}$ is not a state of complete knowledge of $s^{\otimes}$ (see (β_s)). Lastly, we have

(θ_s) There exist states of maximum knowledge $\Phi^{\otimes}$ of $I \otimes II$ for which $\Phi_I^{\otimes}$ is not maximum.

This is certainly true for every $\Phi^{\otimes}$ that is not a $\otimes$ product in $\mathfrak{H}^I \otimes \mathfrak{H}^{II}$. Thus we reach

(Ω) The assertions (Q_2), (Q_3), (T) and (EPR_1) (or (EPR_2)) are incompatible in the theory $\mathfrak{t}$.

For, according to (θ_s), there is a state of maximum knowledge $\Phi^{\otimes}$ of $I \otimes II$ such that $\Phi_I^{\otimes}$ is not maximum. From (Q_3), $\Phi^{\otimes}$ corresponds to a state $s^{\otimes}$ of $I \otimes II$ such that $\Phi^{\otimes}$ is a knowledge of $s^{\otimes}$. Then, from (EPR_1), (T) and (Σ), $\Phi^{\otimes}$ is not a complete knowledge of $s^{\otimes}$. Thus there exists a state of $I \otimes II$ and a state of knowledge of it which is maximum but not complete, in contradiction with (Q_2).

Mathematical Appendix to section VII.2

1. Proof of (ε_s)

First, let (δ_d) be valid for A and $W^{\otimes}$. Then we have $\{Q_k\}_k$ and f as specified in (δ_d). From condition (2) of (δ_d), as in the Mathematical Appendix to section VI.3, part 4,

$$\mathrm{Tr}(W^{\otimes}(P^A_{\{\alpha i\}} \otimes Q_{f(i)})) = \mathrm{Tr}(W^{\otimes}(1 \otimes Q_{f(i)})) \tag{1}$$

for all $i \in J$. From (3) of (δ_d),

$$\mathrm{Tr}(W^{\otimes}(1-\Sigma_i P^A_{\{\alpha i\}})) = 0, \tag{2}$$

and therefore

$$\mathrm{Tr}(W^{\otimes}(1 \otimes Q_{f(i)})) = \mathrm{Tr}(W^{\otimes}(P^A_{\{\alpha i\}} \otimes Q_{f(i)})) + \sum_{j:\, j \neq i} \mathrm{Tr}(W^{\otimes}(P^A_{\{\alpha j\}} \otimes Q_{f(i)})). \tag{3}$$

From (1) we have

$$\mathrm{Tr}(W^{\otimes}(P^A_{\{\alpha j\}} \otimes Q_{f(i)})) = 0 \tag{4}$$

for $j \neq i$. From (2), we get the expansion

$$W^{\otimes}(1 \otimes Q_{f(i)}) = W^{\otimes}(P^A_{\{\alpha i\}} \otimes Q_{f(i)}) + \sum_{j:\, j \neq i} W^{\otimes}(P^A_{\{\alpha j\}} \otimes Q_{f(i)}), \tag{5}$$

and so, using (4),

$$W^{\otimes}(1 \otimes Q_{f(i)}) = W^{\otimes}(P^A_{\{\alpha i\}} \otimes Q_{f(i)}). \tag{6}$$

This corresponds to (3) in the Mathematical Appendix to section VI.3, part 5. In the same way as there, we reach the further conclusions

$$(1 \otimes Q_{f(i)}) W^{\otimes}(1 \otimes Q_{f(i)}) = (P^A_{\{\alpha i\}} \otimes 1) W^{\otimes}(P^A_{\{\alpha i\}} \otimes 1) \tag{7}$$

(cf. (1) in that Appendix) and

$$\Sigma_{k \in K}(1 \otimes Q_k) W^{\otimes}(1 \otimes Q_k) = \Sigma_{i \in J}(P^A_{\{\alpha i\}} \otimes 1) W^{\otimes}(P^A_{\{\alpha i\}} \otimes 1) \tag{8}$$

(corresponding to (3) in part 1 of that Appendix). The relation (α_d) between A and $W_I^{\otimes}$ which is now to be proved follows in the same way as there.

Conversely, let (α_d) be valid for some A and W in $\mathfrak{H}^I$. Then there exists an orthonormal basis $\{\phi_\mu\}_{\mu \in M}$ of $\mathfrak{H}^I$ such that for W we have

$$\left.\begin{aligned} W &= \Sigma_\mu w_\mu P_{\phi_\mu}, \\ w_\mu \geqslant 0, \quad \Sigma_\mu w_\mu &= 1, \end{aligned}\right\} \tag{9}$$

and for A

$$P^A_{\{\alpha i\}} \phi_\mu = \delta_{ig(\mu)} \phi_\mu, \tag{10}$$

where $g\colon M \rightarrowtail J$ is a suitable mapping of the subscripts. Also, let $\{\psi_\nu\}_{\nu \in N}$ be any orthonormal basis of $\mathfrak{H}^{II}$. Since $\dim \mathfrak{H}^{II} \geqslant \dim \mathfrak{H}^{I}$, there is a one-to-one subscript mapping, $h\colon M \rightarrowtail N$. We then put

$$\Phi^\otimes = \Sigma_{\mu \in M} \sqrt{w_\mu} \cdot \phi_\mu \otimes \psi_{h(\mu)} \tag{11}$$

to describe the desired state of knowledge of $I \otimes II$. Finally, let us take

$$Q_i = \sum_{\mu:\, g(\mu)=i} P_{\psi_{h(\mu)}} \tag{12}$$

and, if this orthogonal system is not yet complete, make it complete in any manner as $\{Q_k\}_{k \in K}$. Then $\{Q_k\}_{k \in K}$ represents the desired measurement of II. Since $J \subseteq K$, f can be taken as an inclusion mapping. This gives all the parameters A, $\Phi^\otimes$, $\{Q_k\}_{k \in K}$ and f which appear in (1)–(3) of (η_d).

We now have to show that these have been correctly chosen. From (11) it easily follows that $(\Phi^\otimes)_I = W$ (see, e.g. Ludwig 1954, §V.2); (1) of (η_d) follows immediately from (11) and the choice of the Q_k whose subscripts are not in J as orthogonal to all the Q_i in (12). To prove (2) of (η_d), we first obtain from (11) and (12)

$$\frac{(1 \otimes Q_i) P_{\Phi^\otimes} (1 \otimes Q_i)}{\mathrm{Tr}(P_{\Phi^\otimes}(1 \otimes Q_i))} = P_{\sum_{\mu:\, g(\mu)=i} \sqrt{w_\mu} \cdot \phi_\mu \otimes \psi_{h(\mu)}}. \tag{13}$$

It is also generally true that

$$\mathrm{Tr}(P_{\sum_{\mu:\, g(\mu)=i} \sqrt{w_\mu} \cdot \phi_\mu \otimes \psi_{h(\mu)}} (P^A_{\{\alpha i\}} \otimes 1)) = \frac{\|(P^A_{\{\alpha i\}} \otimes 1) \sum_{\mu:\, g(\mu)=i} \sqrt{w_\mu} \cdot \phi_\mu \otimes \psi_{h(\mu)}\|^2}{\| \sum_{\mu:\, g(\mu)=i} \sqrt{w_\mu} \cdot \phi_\mu \otimes \psi_{h(\mu)}\|^2}. \tag{14}$$

Calculation of the numerator and denominator by means of (10) easily shows that this probability is equal to unity. Finally, (3) of (η_d) follows since $(\Phi^\otimes)_I = W$, using the assumed relationship of W to A.

2. Proof of (ζ_s)

From the proof of the second part of (ε_s) (part 1 of this Appendix), we know that $\Phi^\otimes$, $\{Q_k\}_{k \in K}$ and f exist for A and W, and similarly $\Phi^{\otimes\prime}$, $\{Q_k'\}_{k \in K'}$ and f' for A' and W, such that $(\Phi^\otimes)_I = (\Phi^{\otimes\prime})_I = W$ and the conditions (1)–(3) of (η_d) are valid for A, $\Phi^\otimes$, $\{Q_k\}_{k \in K}$, f and for A', $\Phi^{\otimes\prime}$, $\{Q_k'\}_{k \in K'}$, f'. The only thing that remains to be proved is therefore that $\Phi^\otimes$ and $\Phi^{\otimes\prime}$ can be chosen so as to be equal. This is proved as follows.

According to the spectral resolution of W

$$\left.\begin{aligned} W &= \Sigma_m \tilde{w}_m P_m, \\ \tilde{w}_m &> 0,\ \tilde{w}_m \neq \tilde{w}_n \text{ for } m \neq n, \end{aligned}\right\} \tag{1}$$

we can divide the basis $\{\phi_\mu\}_\mu$ or $\{\phi_\mu'\}_\mu$ of $\mathfrak{H}^I$ (used to construct $\Phi^\otimes$ or $\Phi^{\otimes\prime}$) into sections such that the vectors in each section form a basis for P_m (see (9) in part 1). The ϕ_μ or ϕ_μ' which pass through $1-\Sigma_m P_m$ can be ignored, since their coefficients in $\Phi^\otimes$ are zero; cf. (11) in part 1. This gives a new assignment of subscripts $\phi_{m,\,\kappa_m}$ or $\phi'_{m,\,\kappa_m}$, and for each m there exists a unitary matrix $u^{(m)}$ such that

$$\phi'_{m,\,\kappa_m} = \Sigma_{\lambda_m} u^{(m)}_{\lambda_m \kappa_m} \phi_{m,\lambda_m}. \tag{2}$$

When it is not desired to make $\Phi^\otimes$ and $\Phi^{\otimes\prime}$ equal, the bases $\{\psi_\nu\}_\nu$ and $\{\psi_\nu'\}_\nu$ in $\mathfrak{H}^{II}$ can be chosen entirely arbitrarily. Here we apply the condition that the choice of $\{\psi_\nu\}_\nu$ is arbitrary, but the other basis is chosen in accordance with the matrices occurring in (2):

$$\psi'_{m,\,\kappa_m} = \Sigma_{\lambda_m}(u^{(m)}_{\kappa_m \lambda_m})^{-1}\psi_{h(m,\,\lambda_m)}, \tag{3}$$

with

$$\Sigma_{\kappa_m} u^{(m)}_{\lambda_m \kappa_m}(u^{(m)}_{\kappa_m \iota_m})^{-1} = \delta_{\lambda_m \iota_m}. \tag{4}$$

Since the $u^{(m)}$ are unitary, the $\psi'_{m,\,\kappa_m}$ are orthogonal and normalized. They can be made complete in any manner if necessary. From (2)–(4) we have

$$\Sigma_{\kappa_m}\phi'_{m,\,\kappa_m} \otimes \psi'_{m,\,\kappa_m} = \Sigma_{\kappa_m}\phi_{m,\,\kappa_m} \otimes \psi_{h(m,\,\kappa_m)} \tag{5}$$

for all m, and so, from (1), the $\Phi^{\otimes}$ and $\Phi^{\otimes\prime}$ given in (11) of part 1 are equal.

VII.3 Comments on the foregoing analysis of the EPR argument

In section 2 our basis has been the purely epistemic formulation of orthodox quantum mechanics (or the purely statistical formulation, which would show no formal difference), from section II.4. A possible extension by inclusion of the concept of the state of an individual object, on the stronger basis (α') and (β') or the weaker basis (α)–(δ), has been considered from the viewpoint of whether this will maintain logical compatibility between certain completeness postulates and certain reality postulates. The meta-theorems (Λ) and (Ω) then showed that certain combinations of the relevant postulates are logically incompatible. This proves that the extension in question is logically impossible if all the postulates concerned are to be maintained, but does not show which combinations of the conditions can be used without involving contradictions. We shall now consider the latter point.

If the orthodox standpoint is adopted, one may hope that the epistemic or statistical extension of quantum mechanics, with states of individual objects, already set out in section II.5, which may now be conversely regarded as an extension of the purely epistemic or purely statistical quantum mechanics of section II.4 by the inclusion of these individual states, will lead to positive results as regards the completeness postulates. This is in fact so. The individual states are described, as in section II.5, by the vectors ϕ (to within a constant factor) in the underlying Hilbert space $\mathfrak{H}$, and the quantity A is said to have the value α in the state ϕ if ϕ is an eigenvector of A corresponding to the eigenvalue α. This determines the position with regard to (α') and (β') in section 2. Then (Q_1)–(Q_3) follow immediately. For (Q_1) we take, in a given state s, the state of knowledge that is described by the same vector $\phi \in \mathfrak{H}$ as s is. Then, if A has the value α in s, $A\phi = \alpha\phi$, and hence $\|P^A_{\{\alpha\}}\,\phi\|^2 = 1$. If A does not have the value α in s, $A\phi \neq \alpha\phi$, and hence $\|P^A_{\{\alpha\}}\,\phi\|^2 < 1$. According to the definition (γ_d) in section 2, ϕ describes a state of complete knowledge of s. To prove (Q_2) we first show that a state of maximum knowledge ϕ which is a knowledge of the state s is described by the same vector as s. For, let ψ be the vector in $\mathfrak{H}$ which describes s; if $\phi \nsim \psi$, we could choose the quantity A that is described by P_ϕ and take the eigenvalue $\alpha = 1$. This choice would give $A\psi \neq \alpha\psi$, and also $\|P^A_{\{\alpha\}}\,\phi\|^2 = 1$, i.e. ϕ would not be a state of knowledge of s (cf. (α'_d) in section 2). Thus we must have $\phi \sim \psi$, i.e. ϕ describes s. We saw previously that the state of knowledge of s provided by ϕ is then a complete knowledge. This proves (Q_2). To prove (Q_3), with a given state of knowledge W, let us take a state s that is described by an eigenvector of W with a positive eigenvalue. If then $\mathrm{Tr}(WP^A_{\{\alpha\}}) = 1$, we have $P^A_{\{\alpha\}}\phi = \phi$, and so $A\phi = \alpha\phi$. But if $\mathrm{Tr}(WP^A_{\{\alpha\}}) = 0$, we have $WP^A_{\{\alpha\}} = 0$ and $P_\phi P^A_{\{\alpha\}} = 0$, so that

$A\phi \neq \alpha\phi$. According to the definition (α_d'), W is then a state of knowledge of the state s described by ϕ.

The chosen (orthodox) extension of the purely epistemic (or purely statistical) quantum mechanics thus satisfies the two completeness postulates (Q_1) and (Q_2) and the postulate (Q_3), which is really self-evident. Is it the only such theory which does so? We shall now show that this is certainly true if a further condition is added which concerns only the concept of the state of an individual object:

(Q_4) If s and s_1 are different states of the object Σ, then there exist a quantity A and a value α such that A has the value α in s but not in s_1.

In the model described above, (Q_4) is satisfied together with (Q_1)–(Q_3), as can be seen immediately. We now proceed in the opposite direction. Let the purely epistemic (or purely statistical) quantum mechanics be extended by the basic concepts (α') and (β') of section 2, and let (Q_1)–(Q_4) be valid in this extended form. We then have to show that there exists a description of states by vectors (to within a constant factor) in $\mathfrak{H}$ such that the quantity A has the value α in the state s if and only if $A\phi = \alpha\phi$ for the representatives. To do so, we consider the relation that P_ϕ is a knowledge of the state s, restricted to states P_ϕ of maximum knowledge ((α_d') in section 2). Since the states of maximum knowledge are described in a one-to-one manner by the vectors in $\mathfrak{H}$ (apart from a constant factor), we get a certain relation between the $\phi \in \mathfrak{H}$ and the states s of the object Σ, which must provide the desired description; for brevity, we denote it by R. If ϕRs, then (Q_2) shows that the state of knowledge P_ϕ, being maximum, is also complete. Then (γ_d') in section 2 shows that a quantity A has the value α in s if and only if $A\phi = \alpha\phi$. We now have only to prove that R provides a genuine description. Let $\phi \in \mathfrak{H}$ be given. According to (Q_3) there exists a state s such that ϕRs. If ϕRs and ϕRs_1, the condition (γ_d') shows that a quantity A has the value α in s if and only if it has the value α in s_1. From (Q_4), $s = s_1$. But if ϕRs and $\phi_1 Rs$, it similarly follows that $A\phi = \alpha\phi$ if and only if $A\phi_1 = \alpha\phi_1$. This immediately gives $\phi \sim \phi_1$. Finally, let a state s be given. According to (Q_1) there exists a state of complete knowledge W of s. For this, (γ_d') in section 2 implies that A has the value α in s if and only if $\mathrm{Tr}(WP^A_{\{\alpha\}}) = 1$. Now let us assume that W is not maximum, and let ϕ be an eigenvector of W having a positive eigenvalue. From $\mathrm{Tr}(WP^A_{\{\alpha\}}) = 1$ it follows that $\|P^A_{\{\alpha\}}\, \phi\|^2 = 1$, while conversely there exist pairs (A, α) such that $\|P^A_{\{\alpha\}}\, \phi\|^2 = 1$ but $\mathrm{Tr}(WP^A_{\{\alpha\}}) < 1$. From (Q_3), there corresponds to ϕ a state s_1 of which P_ϕ is a knowledge, and this is a complete knowledge because of (Q_2) and because P_ϕ is maximum. Thus A has the value α in s_1 if and only if $\|P^A_{\{\alpha\}}\, \phi\|^2 = 1$. The relation between s and s_1 is therefore as follows. If A has the value α in s, then it has the value α in s_1 also; but there are quantities A and values α such that A has the value α in s_1 but not in s. Hence $s \neq s_1$. From (Q_4) it follows that A and α exist such that A has the value α in s but not in s_1. This contradicts the first of the two partial assertions just made concerning s and s_1. We have thus proved

(Γ) An extension of the purely epistemic (or purely statistical) quantum mechanics (section II.4) by including the states of individual objects on the basis of (α') and (β') in section 2 satisfies the postulates (Q_1)–(Q_4) if and only if there exists a description of states by all vectors (apart from a constant factor) in the underlying Hilbert space such that a quantity A has the value α in the state s if and only if the mathematical representatives are such that $A\phi = \alpha\phi$.

Let us now consider the conditions (EPR) (in the form (EPR_1) or (EPR_2)) and (T) of section 2, which are critical from the orthodox standpoint. From (Λ) in section 2, (EPR) is not satisfied in the model just described. Whether (T) is or is not satisfied might perhaps be determined from (Ω) in section 2, but since (EPR) is not satisfied, (Ω) by itself does not tell us anything about (T). A direct analysis, however, easily shows that (T) also is not satisfied in this model. Let $\Phi^{\otimes}$ be a state of $I \otimes II$, having the form $\Sigma_\lambda \phi_\lambda \otimes \chi_\lambda$, with an orthonormal basis $\{\phi_\lambda\}_\lambda$ for $\mathfrak{H}^I$ and all $\chi_\lambda \neq 0$. Let the quantity P_ϕ be chosen for a state ϕ of I; it is real in ϕ and has the value 1. In $\Phi^{\otimes}$, however, $P \otimes 1$ is not real, i.e. as is easily seen, $P_\phi \otimes 1$ does not have $\Phi^{\otimes}$ as an eigenvector. Thus (T) cannot be satisfied. The question now is whether (EPR) or (T) can ever be satisfied without contradiction. This is in fact possible for both, with the following model. We consider, for the set of all quantities A of an object Σ which have at least one eigenvalue, functions s which assign to each quantity A one of its eigenvalues. Thus (α') in section 2 is satisfied. Then, if A is a quantity, α a value and s a state of this type, we say that A has the value α in s if A has at least one eigenvalue and $s(A) = \alpha$. Thus (β') in section 2 is satisfied. In this model we clearly have (K) in section 2: every quantity having at least one eigenvalue is real in every state. (EPR_1) is a general consequence of (K); its premise implies the assumption that the relevant quantity A has an eigenvalue. But (T) also is satisfied. It is easily seen that A, as an operator in $\mathfrak{H}^I$, has an eigenvalue if and only if $A \otimes 1$, as an operator in $\mathfrak{H}^I \otimes \mathfrak{H}^{II}$, has an eigenvalue. The given function $s^{\otimes}$ for $I \otimes II$ therefore induces, by restriction to all quantities of the form $A \otimes 1$, a function s for I. This provides all that is necessary for the proof of (T).

The model thus presented serves only the purpose for which it was intended: to prove that (EPR) and (T) are separately (and, as it turns out, even together) at least *logically* possible conditions. (Q_4) is satisfied in the model. (Q_1) is not satisfied, in the strong sense that no state of knowledge is a complete knowledge of any of the allowed states (cf. Variation IV). The most profitable is the consideration of (Q_2) and (Q_3). The set of states of knowledge W can be divided into two classes, according as W corresponds to a property $P \neq 0$ such that $\text{Tr}(WP) = 0$, or not. A state of knowledge in the first class is not a knowledge of any state. A state of knowledge in the second class is a knowledge of every state. Hence (Q_2) is trivially satisfied, in the sense that there is no maximum knowledge as a knowledge of any state. Moreover, (Q_3) is *not* valid. Thus the model as a whole does not offer much help. Of course this does not imply any conclusion as to the existence of physically significant models which satisfy (EPR) (in the form (EPR_1) or (EPR_2)) and (T). Nothing appears to be known on this point, and EPR at least presented no model for it.

Some information on the last-mentioned topic can be obtained by looking at EPR's undertaking in terms of the problem (discussed in detail in section 1) of the existence of a classical theory of hidden variables for quantum mechanics. This problem is in fact raised (in a very general manner) by EPR, but they place the emphasis quite differently from von Neumann and the authors who follow him. The starting-point is the same: the question whether classical states of a quantum-mechanical object are possible (classical only in the sense, initially, that these states provide definite values for every quantity). Von Neumann and his followers thereafter concentrate their interest firstly on the establishment of relationships in which each individual state must be in correspondence with the structural and in particular the functional connections between quantities (or properties) in a manner such as (β_2)–(ε_2) in section 1, and secondly on the requirement that a sufficient number of classical states are present, relative to the number of probabilistic descriptions of state. EPR, on the other hand, impose no such requirement. Instead, they concentrate on the derivation of a sufficient condition for a quantity to be real in a state. In our analysis of EPR's arguments, the second von Neumann condition was included in (Q_3), which appears in the two decisive theorems (Λ) and (Ω) (section 2). The first von Neumann condition was not included, and for this reason we were able to put forward at the beginning of this section, with the ontic formulation of quantum mechanics, a theory of "hidden parameters" which satisfies (Q_1)–(Q_4) and is indeed the only theory which does so, but it does not satisfy the first von Neumann condition. For example, an object Σ in the state ϕ can (with certainty) have the property $E \cup F$ with $E \perp F$ without having (with certainty) the property E or F; this violates (β_2) in section 1, and is illustrated by the two-hole experiment (sections I.3 and IV.4). As regards the aim of EPR's argument, there may be more reasonable models than the one given above, for (EPR) and (T) in section 2. One point, however, is certain: the relationship between the classical states involved and the probabilistic descriptions of state in quantum mechanics will violate at least one of the conditions which formed the basis of the proofs of impossibility in section 1. The conclusion to be drawn from such a violation can, of course, be meaningfully discussed only when an otherwise acceptable proposal is made.

Let us now proceed to a closer study of the EPR reality condition as defined in section 2. In Variation V this condition was first of all specified in the form (EPR), without saying exactly what is meant by the words "it is possible, without in any way perturbing the state s of an object Σ, to acquire a knowledge of s such that the value of the quantity A of Σ can be predicted with certainty". It was stated that there is a definition complying with the orthodox requirements, and therefore necessarily different from the one chosen by EPR. This orthodox form will now be given and compared with EPR's proposal. Firstly, we have corresponding to (α_d) in section 2 the definition

(α_d) The value of the quantity A of Σ can be determined in the presence of the assemblage $\{W_i, w_i\}_i$ of states of knowledge W_i, without perturbing this assemblage, if A is known in all the W_i.

Corresponding to (EPR_1) in section 2, we have the reality condition

(OR_1) Let s be a state and A a quantity of Σ. If there corresponds to s an assemblage $\{W_i, w_i\}_i$ of states of knowledge, such that at least one W_i is a state of knowledge of s and $\{W_i, w_i\}_i$ is related to A by (α_d), then A is real in s.

As in Variation V, the meaning is that, given the premise of (OR_1), any determination of the value of the quantity A does not perturb the assumed state of s. But now, as we shall show, the premise is a much stronger one.

For we have

(Δ) (OR_1) follows from (EPR_1) and (α') and (β') in section 2.

Let the premise of (OR_1) be valid. From this we shall prove the premise of (EPR_1). By hypothesis, there exists $\{W_i, w_i\}_i$ with the conditions specified in (OR_1). One of these is that some W_i, say W_{i_0}, is a state of knowledge of s. Moreover, every W_i, and in particular W_{i_0}, provides a knowledge of the value of A, i.e. $\mathrm{Tr}(W_i P_\alpha^A) = 1$ with a suitable α. Hence $W_i P_\alpha^A = P_\alpha^A W_i = W$. For an eigenvalue α' of A, other than α, $W_i P_{\alpha'}^A = 0$. Hence $W_i = \Sigma_\alpha P_\alpha^A W_i P_\alpha^A$, summed over the eigenvalues of A. W_{i_0} is the state of knowledge required for the premise of (EPR_1). (Δ) is now proved: (OR_1) follows from (EPR_1), and in particular we have seen that the premise of (EPR_1) follows from that of (OR_1). The converse statements are not true: (EPR_1) does not follow from (OR_1), and the premise of (OR_1) does not follow from that of (EPR_1). This is shown by the model described at the beginning of the present section, for which (OR_1) is valid but not (EPR_1). Let the premise of (OR_1) be valid. Then some W_i, say W_{i_0}, is a knowledge of the given state ϕ. From (α_d') in section 2 it follows that $B\phi = \beta\phi$ if $\mathrm{Tr}(W_{i_0} P_\beta^B) = 1$. Now A is known in every W_i, including W_{i_0}, so that

$\text{Tr}(W_{i_0}P_\alpha^A) = 1$ for a certain α. Thus also $A\phi = \alpha\phi$, i.e. A is real in ϕ. But (EPR$_1$) is not valid; for example, let A be specified by the complete system of eigenvectors $\{\phi_\mu\}_\mu$, and let ϕ be a finite linear combination of some of the ϕ_μ, $\phi = \Sigma_i c_i \phi_{\mu_i}$, but not an eigenvector of A; let a state of knowledge W be chosen so that it too has the ϕ_μ as eigenvectors, and in particular $P = \Sigma_i P_{\phi_{\mu_i}}$ as a spectral operator corresponding to a positive eigenvalue. Then W is a knowledge of ϕ, as can be shown immediately from the definition (α_d') in section 2. Likewise, W is related to A by (α_d). Hence the premise of (EPR$_1$) is valid but not the conclusion. This fulfilment of the premise of (EPR$_1$) is then, of course, also a case where the premise of (OR_1) is not satisfied, since (OR_1) is valid in the model. Thus the premise of (OR_1) is stronger than that of (EPR$_1$).

By means of arguments exactly similar to the above, one can deal with an orthodox condition of reality (OR_2) which is related to (EPR$_2$) in the same way as (OR_1) to (EPR$_1$). We first define, in analogy with (α_d),

(β_d) Let two systems I and II be given. The value of a quantity A of I can be determined in the presence of the assemblage $\{W_r^\otimes, w_r\}_r$ for $I \otimes II$, without perturbing this assemblage, if A has a non-empty set $\{\alpha_i\}_{i \in J}$ of eigenvalues and if a measurement $\{Q_k\}_{k \in K}$ of II and a one-to-one mapping $f: J \rightarrowtail K$ exist such that for every r (1) the conditions (1)–(3) in (δ_d) (section 2) are satisfied and (2) there exists a $k \in K$ such that

$$\text{Tr}((W_r^\otimes)_{II} Q_k) = 1.$$

This expresses the facts that, for every $W_r^\otimes$ as specified in (β_d), the measurement of the quantity A of system I can take place by a measurement of II, and that moreover the result of this measurement is already known in $W_r^\otimes$. Thus any lack of knowledge of the value of A lies in the assumption that only an *assemblage* of states of knowledge is present. Then (EPR$_2$) corresponds to

(OR_2) Let two systems I and II be given. Then, if A is a quantity of I and s a state of I, and if there exists an assemblage $\{W_r^\otimes, w_r\}_r$ for $I \otimes II$ such that at least one $(W_r^\otimes)_I$ is a state of knowledge of s and $\{W_r^\otimes, w_r\}_r$ is related to A by (β_d), then A is real in s.

There is no difficulty in proving

(θ) For given systems I and II, (OR_1) and (OR_2) are equivalent, with (α') and (β') in section 2 as the basis.

Together with the equivalence of (EPR$_1$) and (EPR$_2$) from (Π) in section 2, this gives

(Λ) (OR_2) follows from (EPR$_2$) with (α') and (β') in section 2.

This result can also be derived directly. From the premise of (OR_2) we get the assemblage $\{W_r^\otimes, w_r\}_r$. One of the $W_r^\otimes$ in this assemblage describes s, and every $W_r^\otimes$, including those which describe s, is related to A by (δ_d) in section 2 (this has given (β_d)), and hence we have the premise of (EPR$_2$). The standard model easily shows that (EPR$_2$) does not, conversely, follow from (OR_2); this is because of the condition (2) in (β_d).

Finally, let us once again consider the EPR experiment (cf. Variation VII). At the beginning of the experiment, the two systems I and II are to be set up separately, and one has the maximum knowledge ϕ^0 and ψ^0 respectively of these systems, in terms of the purely epistemic formulation of quantum mechanics; one then has also the maximum knowledge $\Phi^0 = \phi^0 \otimes \psi^0$ of $I \otimes II$. In this situation, there is no difference between the conditions (EPR$_1$) and (OR_1), or (EPR$_2$) and (OR_2), within an extension of the purely epistemically formulated quantum mechanics that satisfies (α') and (β') in section 2. For, let I be assumed in the state s^0, so that ϕ^0 is a knowledge of s^0. Then, if A is a quantity of I which can be shown to be real in s^0 on the basis of (EPR_1) and the *prevailing* state of knowledge, the state of knowledge which exists according to the premise of (EPR_1) must be the given ϕ^0. However, with a state of maximum knowledge, the value of a quantity can be determined without perturbing this state of knowledge only if it is already known, as follows at once from (α_d) in section 2. Then, with $\{\phi^0, 1\}$ as a (trivial) assemblage, we have immediately the premise of (OR_1). Thus A is real in s^0, in the sense used in (OR_1). The converse follows from (Δ). A quantity A of I is therefore shown by (EPR_1), for a state of knowledge ϕ^0 of I, to be real in s^0 if and only if the same applies to (OR_1). This clearly occurs because ϕ^0 is maximum. The same result is obtained for (EPR_2) and (OR_2). If, in accordance with (δ_d) in section 2, the value of A can be determined without perturbing Φ^0, then ϕ^0 is an eigenvector of A, since from (3) in (δ_d) we must have $\text{Tr}(P_{\Phi^0}\, P^A_{\{\alpha i\}} \otimes 1) > 0$ for at least one α_i. Then the result stated follows from the representation $\Phi^0 = \phi^0 \otimes \psi^0$, using (2) in ($\delta_d$). But, if A is known in ϕ^0, the premise of (OR_2) is easily obtained, and we again have the equivalence of (OR_2) and (EPR_2) with respect to Φ^0 (simply because Φ^0 is a product).

Thus, although the initial situation in the EPR experiment is not at all critical as regards the difference between the two conditions of reality, the occurrence of an interaction between I and II causes a radical change: "When two systems, of which we know the states by their respective representatives, enter into temporary physical interaction due to known forces between them, and when after a time of mutual influence the systems separate again, then they can no longer be described in the same way as before, viz. by endowing each of them with a representative of its own. I would not call that *one* but rather *the* characteristic trait of quantum mechanics, the one that enforces its entire departure from classical lines of thought" (Schrödinger 1935b, §1). In fact the interaction between I and II has the effect of converting $\Phi^0 = \phi^0 \otimes \psi^0$ into a new state of knowledge Φ^1 which is *not* a product. The purely epistemic formulation then still automatically provides states of knowledge $(\Phi^1)_I$ of I and $(\Phi^1)_{II}$ of II, but whereas Φ^1 is still maximum, the constituent states of knowledge $(\Phi^1)_I$ and $(\Phi^1)_{II}$ are certainly not so. We may further assume, with EPR, that the two systems are no longer interacting at the time to which the state of knowledge Φ^1 applies. Schrödinger (*ibid.*) describes the resulting configuration as follows: "The best possible knowledge of a *whole* does not necessarily include the best possible knowledge of all its *parts*, even though they may be entirely separated and therefore virtually capable of being 'best possibly known', i.e. of possessing, each of them, a representative of its own. The lack of knowledge is by no means due to the interaction being insufficiently known—at least not in the way that it could possibly be known more completely—it is due to the interaction itself." Still without abandoning the purely epistemic formulation of quantum mechanics, we can also consider an EPR pair of quantities A and A' of I (cf. Variation VI), which is related to the existing situation in that the existing Φ^1 is related to A and A' by (δ_d) in section 2, so that the existing $(\Phi^1)_I$ in particular is unperturbed when the value of A is determined and when the value of A' is determined by a measurement of system II (cf. (α_d) and (ζ_s) in section 2). However, it is not possible to connect an EPR pair in this way with every Φ^1 that is not a product. A sufficient condition, e.g. for the EPR pairs constructed in Variation VI in section 2, is that all positive eigenvalues of $(\Phi^1)_I$ are at least doubly degenerate. Thus we must assume that the interaction which converts Φ^0 into Φ^1 is a suitable one. EPR, who start from quite definite quantities A and A' (position and momentum), express no opinion on this point.

The situation which occurs, in the manner described above, after the end of the interaction is characterized by a sharp divergence between the two conditions of reality, in contrast to the initial situation. In an extension of the epistemic formulation, represented by (α'), (β') and (Q_3) in section 2, the premise of (EPR_2) (and hence of (EPR_1)) can be obtained relative to Φ^1 for the two quantities A and A' with the same state s^1 of I. From (Q_3) there is a state s^1 described by $(\Phi^1)_I$. Any such state satisfies the premise of (EPR_2) as regards A and A', with Φ^1 as the corresponding knowledge of $I \otimes II$. EPR, of course, imagine simply that the system I after the interaction is in a *definite* state s^1 described by $(\Phi^1)_I$ and derived from s^0. If we now assume that (EPR_2) also is valid in the extension concerned, we can deduce that A and A' are real in the state s^1, which is then found to be a state for which there is no complete knowledge permitted by quantum mechanics ((Λ_2) in section 2). On the other hand, the existing situation no longer allows the premise of (OR_2) (and hence (OR_1)) to be obtained. The assemblage of states of knowledge of $I \otimes II$ there required is *de facto* the trivial assemblage $\{\Phi^1, 1\}$. If there existed a state s^1 satisfying the premise of (OR_2) for both A and A', both quantities would have to be already known in $(\Phi^1)_I$. But the way in which A and A' are obtained shows that no one state of knowledge allowed by quantum mechanics can give them simultaneously known values; see (β_d) in section 2. Thus (OR_2) cannot be applied. This of course does not imply that (OR_2) (and (OR_1)) is invalid: (EPR_2) can be used in the extension concerned only if it is valid in that extension, and then, as we know from (Λ), (OR_2) is valid also. But the existing contingent situation here considered does not allow the *use* of (OR_2).

In fact the application of the orthodox condition of reality becomes possible again only when one of the two measurements which lead to determinations of the values of A and A' has actually been carried out on system II. First of all, it is easily seen that these two measurements cannot be made simultaneously, in the sense of orthodox quantum mechanics (cf. section IV.2): a simultaneous measurement would lead via the reduction of states to a knowledge of system I, in particular, which makes both quantities A and A' known simultaneously, and this is not admissible according to the construction of A and A'. We therefore have to decide whether A and A' is to be measured. If, for example, A is measured by $\{Q_k\}_{k \in K}$ on II, the state of knowledge before observing the result of the measurement is represented by the assemblage

$$\left\{\frac{(1 \otimes Q_k)\, P_{\Phi^1}(1 \otimes Q_k)}{\text{Tr}(P_{\Phi^1}(1 \otimes Q_k))},\ \text{Tr}(P_{\Phi^1}(1 \otimes Q_k))\right\}_{k:\ \text{Tr}(P_{\Phi^1}(1 \otimes Q_k)) > 0}.$$

From (1)–(3) in (δ_d) (section 2), the value of A is known in every state of knowledge in this assemblage. Since one of these states of knowledge is the "actual" one that will be forthwith determined by observing the result, and this state of knowledge, according to (Q_3), describes a state s^2, we now have the premise of (OR_2) (or (OR_1)) for s^2 and A together, and can say that A actually has in s^2 a definite value, although this is unknown before the observation. A', on the other hand, is certainly not real in s^2, in the orthodox sense.

The foregoing analysis corresponds to Bohr's attitude to the problem of the EPR paradox (cf. Bohr 1935, 1939, 1948, 1949, and our Chapter I): "From our point of view we now see that the wording of the above-mentioned criterion of physical reality proposed by Einstein, Podolsky and Rosen contains an ambiguity as regards the meaning of the expression 'without in any way disturbing a system' " (Bohr 1935, p. 700, col. 1). Here Bohr refers to EPR's reality condition, as quoted at the beginning of Variation V. This condition was then provisionally formulated as (EPR), and it was pointed out that everything depends on precisely what is meant by "without in any way disturbing a system". The ambiguity in this was elucidated by two explications, one leading to the premise of (EPR_1) (or (EPR_2)) and the other to that of (OR_1) (or (OR_2)). From EPR's standpoint it is impossible to understand why in the second stage distinguished above, i.e. when Φ^1 is present, one cannot say that the quantities A and A' of system I both have definite values, even if these are not known through $(\Phi^1)_I$: for each of these two quantities, the value can be determined from system II by a measurement when that system is no longer interacting with I and therefore the measurement cannot lead to any perturbation of the state of I. This argument corresponds to taking the premise of (EPR_1) or (EPR_2) instead of (EPR). As regards EPR's argument, Bohr (*ibid.*) concedes immediately that system I is not perturbed in the usual sense by the measurement of II: "Of course there is in a case like that just considered no question of a mechanical disturbance of the system under investigation during the last critical stage of the measuring procedure". But he decisively rejects the proposition that for this reason alone the relevant quantities of I can be taken to have real values: "But even at this stage there is essentially the question of *an influence on the very conditions which define the possible types of predictions regarding the future behavior of the system*". In fact the "very conditions" mentioned here by Bohr include, first of all, the correlation which the state of knowledge Φ^1 of the whole system $I \otimes II$ creates between the possible results of a measurement of II and the values of quantities of system I that can be predicted from any such results. In addition to assuming a perturbation-free measurement, EPR in their condition make an essential use of just the possibilities of prediction for system I that arise from this correlation. But a value of a quantity of I, thus unambiguously correlated with a possible result of a measurement of II, is precisely as real or as unreal as that result. For the correlation itself is entirely relative: *if* a certain result is obtained in a measurement of II or a particular value of a quantity is really present, *then* a corresponding property can be predicted with certainty for the correlated value in I. *The problem of reality, therefore, is simply transferred from system I to system II.* For system II, however, no measurement has been made in the second stage, characterized by the state of knowledge Φ^1 which gives rise to the correlation. Here we reach the second part of Bohr's "very conditions". On his view, no phenomenon exists until the measurement of II has been made; the EPR experiment at first meets only the requirement that certain initial conditions (using ϕ^0 and ψ^0) are regarded as having been constructed. It must be supplemented by a further measurement of II, to be regarded as completed, which *definitely* makes possible a prediction relating to I and in particular the testing thereof. The measurements of II which belong to an EPR pair of I would, however, be mutually exclusive. A decision must therefore be made as to which of the two measurements is to be carried out. This treatment using Bohr's standpoint corresponds to the fact that the condition (OR_2) (or (OR_1)), because of its stronger premise, can be applied only after the measurement of II has been made, not before. Of the two conditions derived, EPR use the first and ignore the second. On Bohr's view, both are essential. The first shows that the decision about the reality problem has to be made in system II, and the second makes it: "Since these conditions constitute an inherent element of the description of any phenomenon to which the term 'physical reality' can be properly attached, we see that the argumentation of the mentioned authors does not justify their conclusion that quantum-mechanical description is essentially incomplete" (Bohr, *ibid.*).

The analysis of the EPR experiment shows also, however, how careful one must be in applying the orthodox concept of state. Bohr (1939, p. 21), without fully adopting any such concept, such as that of the ontic formulation of quantum mechanics in section II.5, but nevertheless using the word "state", takes up EPR's argument in reference to the stage described by Φ^1 and the two possibilities of a measurement of A and A', saying first that "since in both cases the object itself has been treated alike, we should . . . apparently be able to assign to one and the same state of the object two well-defined physical attributes", namely a value of A and a value of A'. But then follows the warning that "in fact . . . no well-defined use of the concept of 'state' can be made as referring to the object [*I*] separate from the body [*II*] with which it has been in contact, until the external conditions involved in the definition of this concept are unambiguously fixed by a further suitable control of the auxiliary body [*II*]". This view can be confirmed by means of the ontic form of quantum mechanics. At the beginning of this section it has been shown that this and only this form can extend the purely epistemic form into a theory having the concept of the state of an individual object, for which the completeness conditions (Q_1) and (Q_2) (contradicting EPR's condition) and two further postulates (Q_3) and (Q_4) are satisfied. At the start of the EPR experiment, a state of maximum knowledge ϕ^0 of the system I was assumed. This corresponds unambiguously to a state $\bar{\phi}^0$ of which one can then say with certainty that I is in that state: the procedure was such that a quantity A is real in $\bar{\phi}^0$ if and only if the value of A is known in ϕ^0. Since ϕ^0 is complete, there can thus be no question of unknown properties of realness, from the orthodox standpoint. But what is the situation after the interaction, i.e. when Φ^1 is the relevant state of knowledge? Then $(\Phi^1)_I$ is no longer maximum, and hence not complete even from the orthodox standpoint. Can one now say without further analysis that the system I is in a state, though an unknown state? This would surely bring about an inconsistency: the state of knowledge Φ^1 of $I \otimes II$ is maximum, so that one can indeed say, as one could initially, that $I \otimes II$ is in a definite state $\bar{\Phi}^1$. This state defines the reality conditions, first of all for the whole system: a quantity $A^{\otimes}$ of this system is real in $\bar{\Phi}^1$ if and only if its value is known in Φ^1. This applies, in particular, to quantities of the form $A \otimes 1$, where A is a quantity of system I. Let us assume that system I is in a state $\bar{\phi}^1$. Then every quantity A of I whose operator has ϕ^1 as an eigenvector would (in this situation) be real. For all such A, it would obviously have to be true that $A \otimes 1$ is real in this situation, and hence in $\bar{\Phi}^1$. Then the value of such an $A \otimes 1$ in Φ^1, i.e. the value of A in $(\Phi^1)_I$, would necessarily be known. From this we can very easily deduce that $(\Phi^1)_I = P_{\phi^1}$, in contradiction to the fact that $(\Phi^1)_I$ is not maximum. We must therefore accept that it is impossible, in the situation defined by Φ^1, to refer to the state of system I. Schrödinger expresses this very well (1935a, §10): " 'The whole is in a definite state, but its parts separately are not.' 'How can that be? A system must be in some state or other.' 'No. State is ψ-function, is maximum sum of knowledge. I need not have acquired this knowledge; I may have been too lazy to do so. Then the system is not in any state.' 'All right, but in that case the agnostic exclusion of a question also does not apply and I can say that the partial system is in some state (= ψ-function) but I do not know which.' 'Stop. I'm afraid that won't do. You cannot say "I do not know which": there is maximum knowledge of the whole system.' *The insufficiency of the ψ-function as a model substitute is due entirely to the fact that it is not always available.* When one has it, it can certainly be used to describe the state. But sometimes one does not have it where it is expected. Then one cannot postulate that it is 'actually a particular one, but unknown'; the view adopted prohibits this. It is a sum of knowledge, and knowledge known to none is no knowledge."

References

[The references marked *, †, ‡ are recommended for reading in connection with Chapter VI, section VII.1, and sections VII.2 and 3 respectively.]

ALBERT, A. A. (1934), On a certain algebra of quantum mechanics, *Annals of Mathematics* **35,** 65–73.
* AMAI, S. (1962), On observation in quantum mechanics, *Progress of Theoretical Physics* **28,** 401–402.
* (1963), Theory of measurement in quantum mechanics, *Progress of Theoretical Physics* **30,** 550–562.
* (1964), On the entropy problem in the theory of measurement in quantum mechanics, *Progress of Theoretical Physics* **31,** 931–933.

BECK, G. and NUSSENZVEIG, H. M. (1958), Uncertainty relation and diffraction by a slit, *Nuovo Cimento* [10] **9,** 1068–1076.

‡ BELL, J. S. (1964), On the Einstein Podolsky Rosen paradox, *Physics* **1,** 195–200.

† (1966), On the problem of hidden variables in quantum mechanics, *Reviews of Modern Physics* **38,** 447–452.

BIRKHOFF, G. and VON NEUMANN, J. (1936), The logic of quantum mechanics, *Annals of Mathematics* **37,** 823–43.

BLOKHINTSEV, D. I. (1964), *Quantum Mechanics,* Reidel, Dordrecht. (Originally published in Russian, in 1963.)

† BOCCHIERI, P. and LOINGER, A. (1957), Some comments on the problem of hidden parameters, *Zeitschrift für Physik* **148,** 308–313 [in German].

‡ BOHM, D. (1951), *Quantum Theory,* Prentice-Hall, New York.

(1952), A suggested interpretation of the quantum theory in terms of "hidden" variables, *Physical Review* **85,** 166–179, 180–193.

(1971), On Bohr's views concerning the quantum theory, in: T. Bastin, *Quantum Theory and Beyond,* Cambridge University Press, Cambridge, pp. 33–40.

‡ BOHM, D. and AHARONOV, Y. (1957), Discussion of experimental proof for the paradox of Einstein, Rosen, and Podolsky, *Physical Review* **108,** 1070–1076.

‡ (1960), Further discussion of possible experimental tests for the paradox of Einstein, Podolsky and Rosen, *Nuovo Cimento* [10] **17,** 964–976.

† BOHM, D. and BUB, J. (1966), A proposed solution of the measurement problem in quantum mechanics by a hidden variable theory, *Reviews of Modern Physics* **38,** 453–469.

BOHR, AA. (1963), Preface, in: N. Bohr, *Essays 1958–1962 on Atomic Physics and Human Knowledge,* Interscience, New York.

BOHR, N. (1925), Atomic theory and mechanics, in *Atomic Theory and the Description of Nature* (1934), Cambridge University Press, Cambridge, pp. 25–51 (also published in *Nature* **116,** 845–852).

(1927), The quantum postulate and the recent development of atomic theory, *ibid.* pp. 52–91 (also published in *Nature* **121,** 580–590 (1928)).

(1929a), Introductory survey, *ibid.* pp. 1–24.

(1929b), The quantum of action and the description of nature, *ibid.* pp. 92–101.

(1929c), The atomic theory and the fundamental principles underlying the description of nature, *ibid.* pp. 102–119.

(1935), Can quantum-mechanical description of physical reality be considered complete?, *Physical Review* **48,** 696–702.

(1937), Causality and complementarity, *Philosophy of Science* **4,** 289–298.

(1938), Natural philosophy of human cultures, in *Atomic Physics and Human Knowledge* (1958), Wiley, New York, pp. 23–31 (also published in *Nature* **143,** 268–272 (1939)).

(1939), The causality problem in atomic physics, in *New Theories in Physics,* International Institute of Intellectual Co-operation, Paris, pp. 11–45.

(1948), On the notions of causality and complementarity, *Dialectica* **2,** 312–319.

(1949), Discussion with Einstein on epistemological problems in atomic physics, in *Atomic Physics and Human Knowledge* (1958), Wiley, New York, pp. 32–66 (also published in Schilpp (1949), pp. 199–241).

(1954), Unity of knowledge, *ibid.* pp. 67–82.

(1955), Atoms and human knowledge, *ibid.* pp. 83–93.

(1956), Mathematics and natural philosophy, *Scientific Monthly* **82**, 85–88.
(1957), Physical sciences and the problem of life, in *Atomic Physics and Human Knowledge* (1958), Wiley, New York, pp. 94–101.
(1958), Quantum physics and philosophy—causality and complementarity, in *Essays 1958–1962 on Atomic Physics and Human Knowledge* (1963), Interscience, New York, pp. 1–7.
(1960), The unity of human knowledge, *ibid.* pp. 8–16.
(1961), Reminiscences of the founder of nuclear science and of some developments based on his work, *ibid.* pp. 30–73 (also published in *Proceedings of the Physical Society* **78**, 1083–1115).
(1962a), Light and life revisited, *ibid.* pp. 23–29.
(1962b), The Solvay meetings and the development of quantum physics, *ibid.* pp. 79–100.
BOPP, F. (1961), Statistical mechanics in the perturbation of the state of a physical system by observation, in *W. Heisenberg und die Physik unserer Zeit*, Vieweg, Braunschweig, pp. 128–149 [in German].
BORN, M. (1926), Quantum mechanics of collision processes, *Zeitschrift für Physik* **38**, 803–827 [in German].
(1955a), Is classical mechanics in fact deterministic?, *Physikalische Blätter* **11**, 49–54 [in German; translated in *Physics in my Generation* (1956), Pergamon, London, pp. 164–170].
(1955b), On determinism, *Physikalische Blätter* **11**, 314–315 [in German].
(1955c), Continuity, determinism and reality, *Det Kongelige Danske Videnskabernes Selskab, Matematisk-Fysiske Meddelelser* **30**, No. 2.
(1958), Predictability in classical mechanics, *Zeitschrift für Physik* **153**, 372–388 [in German].
(1959), Predictability in classical mechanics, *Physikalische Blätter* **15**, 342–349 [in German].
(1961), Comments on the statistical interpretation of quantum mechanics, in: F. Bopp, *W. Heisenberg und die Physik unserer Zeit*, Vieweg, Braunschweig, pp. 103–118 [in German].
(1971), *The Born–Einstein Letters*, Macmillan, London. (Originally published in German, in 1969.)
BORN, M., HEISENBERG, W. and JORDAN, P. (1926), On quantum mechanics. II, *Zeitschrift für Physik* **35**, 557–615 [in German].
BORN, M. and HOOTON, D. J. (1955), Statistical dynamics of multiply periodic systems, *Zeitschrift für Physik* **142**, 201–218 [in German].
BORN, M. and JORDAN, P. (1925), On quantum mechanics, *Zeitschrift für Physik* **34**, 858–888 [in German].
(1930), *Elementary Quantum Mechanics*, Springer, Berlin [in German].
‡ BREITENBERGER, E. (1965), On the so-called paradox of Einstein, Podolsky and Rosen, *Nuovo Cimento* [10] **38**, 356–360.
BRILLOUIN, L. (1964), *Scientific Uncertainty and Information*, Academic, New York.
BROGLIE, L. DE (1953), *Will Quantum Physics Remain Indeterministic?*, Gauthier–Villars, Paris [in French].
(1957), *Measurement Theory in Wave Mechanics*, Gauthier–Villars, Paris [in French].
(1960), *Non-Linear Wave Mechanics. A Causal Interpretation*, Elsevier, Amsterdam. (Originally published in French, in 1956.)
BUB, J. (1968a), Hidden variables and the Copenhagen interpretation—a reconciliation, *British Journal for the Philosophy of Science* **19**, 185–210.
* (1968b), The Daneri–Loinger–Prosperi quantum theory of measurement, *Nuovo Cimento* [10] **57B**, 503–520.
BÜCHEL, W. (1965), *Philosophical Problems of Physics*, Herder, Freiburg [in German].
BUNGE, M. (1955), Strife about complementarity, *British Journal for the Philosophy of Science* **6**, 1–12, 141–154.
(1956), Survey of the interpretations of quantum mechanics, *American Journal of Physics* **24**, 272–286.
(1967), *Foundations of Physics*, Springer, Berlin, chapter 5.
CAPASSO, V., FORTUNATO, D. and SELLERI, F. (1970), Von Neumann's theorem and hidden-variable models, *Rivista del Nuovo Cimento* [1] **2**, 149–199.
‡ CLAUSER, J. F., HORNE, M. A., SHIMONY, A. and HOLT, R. A. (1969), Proposed experiment to test local hidden-variable theories, *Physical Review Letters* **23**, 880–884.
CONDON, E. U. (1929), Remarks on uncertainty principles, *Science* **69**, 573–574.
‡ COOPER, J. L. B. (1950), The paradox of separated systems in quantum theory, *Proceedings of the Cambridge Philosophical Society* **46**, 620–625.
* COOPER, L. N. and VAN VECHTEN, D. (1969), On the interpretation of measurement within the quantum theory, *American Journal of Physics* **37**, 1212–1220.
‡ COSTA DE BEAUREGARD, O. (1965), The "paradox" of the Einstein and Schrödinger relations and the time-thickness of the quantum transition, *Dialectica* **19**, 280–289 [in French].
* DANERI, A., LOINGER, A. and PROSPERI, G. M. (1962), Quantum theory of measurement and ergodicity conditions, *Nuclear Physics* **33**, 297–319.
* (1966), Further remarks on the relations between statistical mechanics and quantum theory of measurement, *Nuovo Cimento* [10] **44B**, 119–128.
DE *Broglie*, L. (q.v.)
D'*Espagnat*, B. (q.v.)
DIRAC, P. A. M. (1925), The fundamental equations of quantum mechanics, *Proceedings of the Royal Society* **A109**, 642–653.

(1926a), Quantum mechanics and a preliminary investigation of the hydrogen atom, *Proceedings of the Royal Society* **A110,** 561–579.
(1926b), On quantum algebra, *Proceedings of the Cambridge Philosophical Society* **23,** 412–418.
(1930), *The Principles of Quantum Mechanics*, Clarendon Press, Oxford.
DIXMIER, J. (1957), *Operator Algebras in Hilbert Space*, Gauthier–Villars, Paris [in French].
DOMBROWSKI, H. D. (1969), On simultaneous measurements of incompatible observables, *Archive for Rational Mechanics and Analysis* **35,** 178–210.
DOMBROWSKI, H. D. and HORNEFFER, K. (1964), The concept of a physical system from the mathematical viewpoint, *Nachrichten der Akademie der Wissenschaften in Göttingen, Mathematisch-Physikalische Klasse*, 67–100 [in German].
DUHEM, P. (1906), *The Object and Structure of Physical Theories*, Rivière, Paris [in French].
EINSTEIN, A. (1936), Physics and reality, *Journal of the Franklin Institute* **221,** 349–382.
(1948), Quantum mechanics and reality, *Dialectica* **2,** 320–324 [in German].
(1949), Reply to criticisms, in: P. A. Schilpp, *Albert Einstein: Philosopher–Scientist*, Library of Living Philosophers, Evanston, pp. 665–688.
(1953a), Introductory remarks on fundamental concepts, in: *Louis de Broglie: Physicist and Philosopher*, Michel, Paris, pp. 4–15 [in German].
(1953b), Elementary considerations on the interpretation of the foundations of quantum mechanics, in: *Scientific Papers Presented to Max Born*, Oliver & Boyd, Edinburgh, pp. 33–40 [in German].
EINSTEIN, A., PODOLSKY, B. and ROSEN, N. (1935), Can quantum-mechanical description of physical reality be considered complete?, *Physical Review* **47,** 777–780.
ESPAGNAT, B. D' (1965), *Ideas of Modern Physics: Interpretations of Quantum Mechanics and Measurement*, Hermann, Paris [in French].
* EVERETT, H. III (1957), "Relative state" formulation of quantum mechanics, *Reviews of Modern Physics* **29,** 454–462.
EXNER, F. (1919), *Lectures on the Physical Principles of the Sciences*, Deuticke, Vienna [in German].
FÉNYES, I. (1952), A probability-theory justification and interpretation of quantum mechanics, *Zeitschrift für Physik* **132,** 81–106 [in German].
† FEYERABEND, P. K. (1956), A note on the von Neumann proof, *Zeitschrift für Physik* **145,** 421–423 [in German].
* (1957), On the quantum theory of measurement, in: S. Körner, *Observation and Interpretation*, Butterworths, London, pp. 121–130.
(1958), Complementarity, *Aristotelian Society Supplement* **32,** 75–104.
(1961), Niels Bohr's interpretation of the quantum theory, in: H. Feigl and G. Maxwell, *Current Issues in the Philosophy of Science*, Holt, Rinehart & Winston, New York, pp. 371–390, 398–400.
‡ (1962), Problems of microphysics, in: R. G. Colodny, *Frontiers of Science and Philosophy*, Allen & Unwin, London, pp. 189–283.
(1968), On a recent critique of complementarity. Part I, *Philosophy of Science* **35,** 309–331.
(1969), On a recent critique of complementarity. Part II, *Philosophy of Science* **36,** 82–105.
FEYNMAN, R. P. and HIBBS, A. R. (1965), *Quantum Mechanics and Path Integrals*, McGraw-Hill, New York.
FRANK, P. G. (1957), *Philosophy of Science*, Prentice-Hall, Englewood Cliffs.
FREISTADT, H. (1957), The causal formulation of quantum mechanics of particles (the theory of de Broglie, Bohm and Takabayasi), *Nuovo Cimento* [10] **5,** Supplement, 1–70.
‡ FURRY, W. H. (1936a), Note on the quantum-mechanical theory of measurement, *Physical Review* **49,** 393–399.
‡ (1936b), Remarks on measurement in quantum theory, *Physical Review* **49,** 476.
GLEASON, A. M. (1957), Measures on the closed subspaces of a Hilbert space, *Journal of Mathematics and Mechanics* **6,** 885–893.
* GREEN, H. S. (1958), Observation in quantum mechanics, *Nuovo Cimento* [10] **9,** 880–889.
GRÜNBAUM, A. (1957), Complementarity in quantum physics and its philosophical generalization, *Journal of Philosophy* **54,** 713–727.
GUDDER, S. P. (1968a), Hidden variables in quantum mechanics reconsidered, *Reviews of Modern Physics* **40,** 229–231.
(1968b), Dispersion-free states and the exclusion of hidden variables, *Proceedings of the American Mathematical Society* **19,** 319–324.
(1969), On the quantum logic approach to quantum mechanics, *Communications in Mathematical Physics* **12,** 1–15.
(1970a), On hidden-variable theories, *Journal of Mathematical Physics* **11,** 431–436.
(1970b), Projective representations of quantum logics, *International Journal of Theoretical Physics* **3,** 99–108.
GUDDER, S. P. and BOYCE, S. (1970), A comparison of the Mackey and Segal models for quantum mechanics, *International Journal of Theoretical Physics* **3,** 7–21.
GUENIN, M. (1966), Axiomatic foundations of quantum theories, *Journal of Mathematical Physics* **7,** 271–282.

GUNSON, J. (1967), On the algebraic structure of quantum mechanics, *Communications in Mathematical Physics* **6**, 262–285.
* HAAKE, F. and WEIDLICH, W. (1968), A model for the measuring process in quantum theory, *Zeitschrift für Physik* **213**, 451–465.
HALMOS, P. R. (1950), *Measure Theory*, van Nostrand, New York.
HANSON, N. R. (1959a), Copenhagen interpretation of quantum theory, *American Journal of Physics* **27**, 1–15.
(1959b), Five cautions for the Copenhagen interpretation's critics, *Philosophy of Science* **26**, 325–357.
HEISENBERG, W. (1925), The quantum-theory reinterpretation of kinematic and mechanical relationships, *Zeitschrift für Physik* **33**, 879–893 [in German].
(1927), The intuitive content of quantum-theory kinematics and mechanics, *Zeitschrift für Physik* **43**, 172–198 [in German].
(1930), *The Physical Principles of the Quantum Theory*, University of Chicago Press, Chicago (German edition: *Die physikalischen Prinzipien der Quantentheorie*, Hirzel, Leipzig).
(1951), Fifty years of the quantum theory, *Die Naturwissenschaften* **38**, 49–55 [in German].
(1955), The development of the interpretation of the quantum theory, in: W. Pauli, *Niels Bohr and the Development of Physics*, Pergamon Press, Oxford, pp. 12–29.
(1959), *Physics and Philosophy*, Allen & Unwin, London.
(1961), Planck's discovery and the philosophical problems of atomic physics, in: *On Modern Physics*, Orion Press, London, pp. 3–19.
(1967), Quantum theory and its interpretation, in: S. Rozental, *Niels Bohr*, North-Holland, Amsterdam, pp. 94–108.
(1969), *The Part and the Whole*, Piper, Munich [in German].
HILBERT, D., VON NEUMANN, J. and NORDHEIM, L. (1928), The principles of quantum mechanics, *Mathematische Annalen* **98**, 1–30 [in German].
HUND, F. (1954), *Matter as Field*, Springer, Berlin [in German].
‡ INGLIS, D. R. (1961), Completeness of quantum mechanics and charge-conjugation correlations of theta particles, *Reviews of Modern Physics* **33**, 1–7.
JAMMER, M. (1966), *The Conceptual Development of Quantum Mechanics*, McGraw-Hill, New York.
JÁNOSSY, L. (1952), The physical aspects of the wave-particle problem, *Acta Physica Academiae Scientiarum Hungaricae* **1**, 423–467.
JÁNOSSY, L. and NÁRAY, Zs. (1957), The interference phenomena of light at very low intensities, *Acta Physica Academiae Scientiarum Hungaricae* **7**, 403–424.
JÁNOSSY, L. and ZIEGLER, M. (1963), The hydrodynamical model of wave mechanics. I, II, III, *Acta Physica Academiae Scientiarum Hungaricae* **16**, 37–48; (1964) *ibid.* 345–53; (1966) *ibid.* **20**, 233–251.
* JAUCH, J. M. (1964), The problem of measurement in quantum mechanics, *Helvetica Physica Acta* **37**, 293–316.
(1968), *Foundations of Quantum Mechanics*, Addison-Wesley, Reading (Mass.).
JAUCH, J. M. and PIRON, C. (1963), Can hidden variables be excluded in quantum mechanics?, *Helvetica Physica Acta* **36**, 827–837.
(1969), On the structure of quantal proposition systems, *Helvetica Physica Acta* **42**, 842–848.
* JAUCH, J. M., WIGNER, E. P. and YANASE, M. M. (1967), Some comments concerning measurements in quantum mechanics, *Nuovo Cimento* [10] **48B**, 144–151.
JEANS, J. H. (1942), *Physics and Philosophy*, Cambridge University Press, Cambridge.
JÖNSSON, C. (1961), Electron interferences at a series of artificially produced fine slits, *Zeitschrift für Physik* **161**, 454–474 [in German].
JORDAN, P., VON NEUMANN, J. and WIGNER, E. (1934), On an algebraic generalization of the quantum mechanical formalism, *Annals of Mathematics* **35**, 29–64.
KAKUTANI, S. and MACKEY, G. W. (1946), Ring and lattice characterizations of complex Hilbert space, *Bulletin of the American Mathematical Society* **52**, 727–733.
KAMBER, F. (1964), The structure of the calculus of assertions in a physical theory, *Nachrichten der Akademie der Wissenschaften in Göttingen, Mathematisch-Physikalische Klasse*, 103–124 [in German].
(1965), Two-valued probability functions in orthocomplementary lattices, *Mathematische Annalen* **158**, 158–196 [in German].
‡ KEMBLE, E. C. (1935), The correlation of wave functions with the states of physical systems, *Physical Review* **47**, 973–974.
(1937), *The Fundamental Principles of Quantum Mechanics*, McGraw-Hill, New York.
(1938), Operational reasoning, reality, and quantum mechanics, *Journal of the Franklin Institute* **225**, 263–275.
KOCHEN, S. and SPECKER, E. P. (1967), The problem of hidden variables in quantum mechanics, *Journal of Mathematics and Mechanics* **17**, 59–87.
† KOMAR, A. (1962), Indeterminate character of the reduction of the wave packet in quantum theory, *Physical Review* **126**, 365–369.

KOOPMAN, B. O. and VON NEUMANN, J. (1932), Dynamical systems of continuous spectra, *Proceedings of the National Academy of Sciences* **18**, 255–263.
* KRIPS, H. P. (1969a), Fundamentals of measurement theory, *Nuovo Cimento* [10] **60B**, 278–290.
* (1969b), Axioms of measurement theory, *Nuovo Cimento* [10] **61B**, 12–24.
‡ (1969c), Two paradoxes in quantum mechanics, *Philosophy of Science* **36**, 145–152.
LANDÉ, A. (1965), *New Foundations of Quantum Mechanics*, Cambridge University Press, Cambridge.
LAUE, M. VON (1955), Is classical mechanics in fact deterministic?, *Physikalische Blätter* **11**, 269–270 [in German].
* LOINGER, A. (1968), Comments on a recent paper concerning the quantum theory of measurement, *Nuclear Physics* **A108**, 245–249.
LONDON, F. and BAUER, E. (1939), *The Theory of Observation in Quantum Mechanics*, Hermann, Paris [in French].
LÜDERS, G. (1951), The change of state caused by the process of measurement, *Annalen der Physik* [6] **8**, 322–328 [in German].
* LUDWIG, G. (1953), The process of measurement, *Zeitschrift für Physik* **135**, 483–511 [in German].
(1954), *The Principles of Quantum Mechanics*, Springer, Berlin [in German].
* (1958), The ergodic theorem and the concept of macroscopic observables. I. II, *Zeitschrift für Physik* **150**, 346–374; **152**, 98–115 [in German].
* (1961), Solved and unsolved problems of the measurement process in quantum mechanics, in: F. Bopp, *W. Heisenberg und die Physik unserer Zeit*, Vieweg, Braunschweig, pp. 150–181 [in German].
(1970), *Interpretation of the Concept of a Physical Theory and Axiomatic Basis of the Hilbert-Space Structure of Quantum Mechanics in Terms of the Principal Theorems of Measurement*, Springer, Berlin [in German].
MACKEY, G. W. (1963), *The Mathematical Foundations of Quantum Mechanics*, Benjamin, New York.
MARGENAU, H. (1931), Causality and modern physics, *Monist* **41**, 1–36.
(1932), Probability and causality in quantum physics, *Monist* **42**, 161–188.
‡ (1936), Quantum-mechanical description, *Physical Review* **49**, 240–242.
(1937), Critical points in modern physical theory, *Philosophy of Science* **4**, 337–370.
(1949), Reality in quantum mechanics, *Philosophy of Science* **16**, 287–302.
(1950), *The Nature of Physical Reality*, McGraw-Hill, New York.
(1958), Philosophical problems concerning the meaning of measurement in physics, *Philosophy of Science* **25**, 32–33.
(1963a), Measurements in quantum mechanics, *Annals of Physics* **23**, 469–485.
(1963b), Measurements and quantum states, *Philosophy of Science* **30**, 1–16, 138–157.
‡ MARGENAU, H. and WIGNER, E. P. (1962), Comments on Professor Putnam's comments, *Philosophy of Science* **29**, 292–293.
‡ (1964), Reply to Professor Putnam, *Philosophy of Science* **31**, 7–9.
MARLOW, A. R. (1965), Unified Dirac–Von Neumann formulation of quantum mechanics. I. Mathematical theory, *Journal of Mathematical Physics* **6**, 919–927.
MESSIAH, A. (1961), *Quantum Mechanics*, Vol. I, North-Holland, Amsterdam.
MEYER-ABICH, K. M. (1965), *Correspondence, Individuality and Complementarity*, Steiner Verlag, Wiesbaden [in German].
MISES, R. VON (1930), Causal and statistical law in physics, *Erkenntnis* **1**, 189–210 [in German].
MISRA, B. (1967), When can hidden variables be excluded in quantum mechanics? *Nuovo Cimento* [10] **47**A, 841–859.
* MOULD, R. A. (1962), Quantum theory of measurement, *Annals of Physics* **17**, 404–417.
NAGEL, E. (1961), *The Structure of Science*, Harcourt, New York.
NEUMANN, J. VON (1927), Thermodynamics of quantum-mechanical ensembles, *Nachrichten von der Gesellschaft der Wissenschaften zu Göttingen, Mathematisch-Physikalische Klasse*, 273–291 [in German].
(1929), Proof of the ergodic theorem and of the H theorem in the new mechanics, *Zeitschrift für Physik* **57**, 30–70 [in German].
(1932), The operator method in classical mechanics, *Annals of Mathematics* **33**, 587–642 [in German].
(1936), On an algebraic generalization of the quantum mechanical formalism. Part I, *Matematicheskiĭ Sbornik* **1**, 415–484.
(1937), Continuous geometries with a transition probability, *unpublished manuscript*, reviewed by I. Halperin in: A. H. Taub, *J. von Neumann. Collected Works* (1961), Pergamon Press, Oxford, vol. IV, pp. 191–194.
(1955), *Mathematical Foundations of Quantum Mechanics*, Princeton University Press, Princeton. (Originally published in German, in 1932.)
OCHS, W. (1970), Can quantum states be represented as ensembles of dispersion-free states? I, *Zeitschrift für Naturforschung* **25a**, 1546–1555 [in German].
(1971), Can quantum theory be presented as a classical ensemble theory? *Zeitschrift für Naturforschung* **26a**, 1740–1753.

(1972a), On the foundation of quantal proposition systems, *Zeitschrift für Naturforschung* **27a**, 893–900.
(1972b), On the covering law in quantal proposition systems, *Communications in Mathematical Physics* **25**, 245–252.
(1972c), On Gudder's hidden-variable theorems, *Nuovo Cimento* [11] **10B**, 172–184.
PARK, J. L. (1968), Quantum theoretical concepts of measurement, *Philosophy of Science* **35**, 205–231, 389–411.
PARK, J. L. and MARGENAU, H. (1968), Simultaneous measurability in quantum theory, *International Journal of Theoretical Physics* **1**, 211–283.
PAULI, W. (1933), The general principles of wave mechanics, *Handbuch der Physik* (2nd edition) **24/1**, 83–272, Springer, Berlin [in German].
(1953), Comments on the problem of hidden parameters in quantum mechanics and on the pilot-wave theory, in: *Louis de Broglie: Physicist and Philosopher*, Michel, Paris, pp. 33–42 [in French].
(1954), Probability and physics, *Dialectica* **8**, 112–124 [in German].
* PEARLE, P. (1967), Alternative to the orthodox interpretation of quantum theory, *American Journal of Physics* **35**, 742–753.
* PERES, A. and ROSEN, N. (1964a), Measurement of a quantum ensemble by a classical apparatus, *Annals of Physics* **29**, 366–377.
* (1964b), Macroscopic bodies in quantum theory, *Physical Review* **135B**, 1486–1488.
‡ PERES, A. and SINGER, P. (1960), On possible experimental tests for the paradox of Einstein, Podolsky and Rosen, *Nuovo Cimento* [10] **15**, 907–915.
PETERSEN, A. (1963), The philosophy of Niels Bohr, *Bulletin of the Atomic Scientists* **19** (7), 8–14.
(1968), *Quantum Physics and the Philosophical Tradition*, MIT Press, Cambridge (Mass.).
PIRON, C. (1964), Quantum axiomatics, *Helvetica Physica Acta* **37**, 439–468 [in French].
POLIKAROW, A. (1962), The interpretation of quantum mechanics, *Wissenschaftliche Zeitschrift der Humboldt-Universität zu Berlin, Mathematisch-Naturwissenschaftliche Reihe*, 1–24 [in German].
POPPER, K. R. (1959), *The Logic of Scientific Discovery*, Hutchinson, London, chapter IX. (Originally published in German, in 1935.)
(1967), Quantum mechanics without "the observer", in: M. Bunge, *Quantum Theory and Reality*, Springer, Berlin, pp. 7–44.
PRUGOVEČKI, E. (1967), On a theory of measurement of incompatible observables in quantum mechanics, *Canadian Journal of Physics* **45**, 2173–2219.
‡ PUTNAM, H. (1961), Comments on the paper of David Sharp, *Philosophy of Science* **28**, 234–237.
‡ (1964), A reply to Margenau and Wigner, *Philosophy of Science* **31**, 1–6.
REICHENBACH, H. (1956), *The Direction of Time*, University of California Press, Berkeley.
ROSENFELD, L. (1953), Strife about complementarity, *Science Progress* **41**, 393–410.
(1963), Niels Bohr's contribution to epistemology, *Physics Today* **16** (10), 47–54.
* (1965), The measuring process in quantum mechanics, *Progress of Theoretical Physics*, Supplement, Extra Number, 222–231.
‡ RUARK, A. E. (1935), Is the quantum-mechanical description of physical reality complete?, *Physical Review* **48**, 466–467.
‡ SACHS, M. (1968), On pair annihilation and the Einstein–Podolsky–Rosen paradox, *International Journal of Theoretical Physics* **1**, 387–407.
SCHEIBE, E. (1964), *Contingent Propositions in Physics*, Athenäum, Frankfurt [in German].
(1967), Bibliography on fundamental problems in quantum mechanics, *Philosophia Naturalis* **10**, 249–290 [in German].
SCHILPP, P. A. (1949), *Albert Einstein: Philosopher–Scientist*, Library of Living Philosophers, Evanston.
SCHRÖDINGER, E. (1926), The relationship between the quantum mechanics of Heisenberg, Born and Jordan and my own, *Annalen der Physik* [4] **79**, 734–756 [in German].
(1935a), The present situation in quantum mechanics, *Die Naturwissenschaften* **23**, 807–812, 823–828, 844–849 [in German].
‡ (1935b), Discussion of probability relations between separated systems, *Proceedings of the Cambridge Philosophical Society* **31**, 555–563.
‡ (1936), Probability relations between separated systems, *Proceedings of the Cambridge Philosophical Society* **32**, 446–452.
† SCHULZ, G. (1959), Critique of von Neumann's disproof of causality in quantum mechanics, *Annalen der Physik* [7] **3**, 94–104 [in German].
SCOTT, W. T. (1968), The consequences of measurement in quantum mechanics. I, II. *Annals of Physics* **46**, 577–592; **47**, 489–515.
SEGAL, I. E. (1947), Postulates for general quantum mechanics, *Annals of Mathematics* **48**, 930–948.
‡ SHARP, D. H. (1961), The Einstein–Podolsky–Rosen paradox re-examined, *Philosophy of Science* **28**, 225–233.
SHE, C. Y. and HEFFNER, H. (1966), Simultaneous measurement of noncommuting observables, *Physical Review* **152**, 1103–1110.

SHERMAN, S. (1956), On Segal's postulates for general quantum mechanics, *Annals of Mathematics* **64,** 593–601.
SHIMONY, A. (1963), Role of the observer in quantum theory, *American Journal of Physics* **31,** 755–773.
SHPOL'SKII, E. V. (1969), *Atomic Physics*, Iliffe, London. (Originally published in Russian, in 1951.)
SLATER, J. C. (1929), Physical meaning of wave mechanics, *Journal of the Franklin Institute* **207,** 449–455.
SOLVAY CONFERENCE (1928), *Electrons and Photons*, Institut International de Physique Solvay and Gauthier–Villars, Paris.
SPECKER, E. (1960), The logic of non-simultaneously-decidable propositions, *Dialectica* **14,** 239–246 [in German].
STEGMÜLLER, W. (1969), *Scientific Explanation and Justification*, Springer, Berlin [in German], Chapter VII.
SÜSSMANN, G. (1957), An analysis of measurement, in: S. Körner, *Observation and Interpretation*, Butterworths, London, pp. 131–136.
(1958), The measurement process, *Abhandlungen der Bayerischen Akademie der Wissenschaften, Mathematisch-Naturwissenschaftliche Klasse*, Neue Folge, No. 88 [in German].
(1963), *Introduction to Quantum Mechanics*, Bibliographisches Institut, Mannheim [in German].
TAKABAYASI, T. (1952), On the formulation of quantum mechanics associated with classical pictures, *Progress of Theoretical Physics* **8,** 143–182.
(1953), Remarks on the formulation of quantum mechanics with classical pictures and on relations between linear scalar fields and hydrodynamical fields, *Progress of Theoretical Physics* **9,** 187–222.
TOLMAN, R. C. (1938), *The Principles of Statistical Mechanics*, Oxford University Press, Oxford.
TOMONAGA, S.-I. (1966), *Quantum Mechanics*, Vol. II, North-Holland, Amsterdam.
VARADARAJAN, V. S. (1962), Probability in physics and a theorem on simultaneous observability, *Communications on Pure and Applied Mathematics* **15,** 189–217.
(1968), *Geometry of Quantum Theory*, van Nostrand, Princeton.
VIGIER, J.-P. (1956), *The Structure of Micro-Objects in the Causal Interpretation of the Quantum Theory*, Gauthier–Villars, Paris.
VON *Laue*, M. (q.v.)
VON *Mises*, R. (q.v.)
VON *Neumann*, J. (q.v.)
VON *Weizsäcker*, C. F. (q.v.)
* WAKITA, H. (1960), Measurement in quantum mechanics. I-III, *Progress of Theoretical Physics* **23,** 32–40; (1962) **27,** 139–144, 1156–1164.
WEIDLICH, W. (1960), The interpretation of quantum mechanics, *Zeitschrift für Naturforschung* **15a,** 651–654 [in German].
* (1967), Problems of the quantum theory of measurement, *Zeitschrift für Physik* **205,** 199–220 (1967).
* WEIDLICH, W. and HAAKE, F. (1969), On the quantum statistical theory of the measuring process, *Journal of the Physical Society of Japan* **26,** Supplement, 231–232.
WEIZEL, W. (1953), Derivation of the quantum theory from a classical, causally determined model. I, II, *Zeitschrift für Physik* **134,** 264–285; **135,** 270–273 [in German].
(1954), Derivation of the quantum-mechanical wave equation of the multi-particle system from a classical model, *Zeitschrift für Physik* **136,** 582–604 [in German].
WEIZSÄCKER, C. F. VON (1941), The interpretation of quantum mechanics, *Zeitschrift für Physik* **118,** 489–509 [in German].
(1955), Complementarity and logic, *Die Naturwissenschaften* **42,** 521–529, 545–555 [in German].
(1961), The unity of physics, in: F. Bopp, *W. Heisenberg und die Physik unserer Zeit*, Vieweg, Braunschweig, pp. 23–46 [in German].
(1971), The Copenhagen interpretation, in: T. Bastin, *Quantum Theory and Beyond*, Cambridge University Press, Cambridge, pp. 25–31.
WEYL, H. (1931), *The Theory of Groups and Quantum Mechanics*, Dover, New York. (Originally published in German, in 1928.)
WIENER, N., SIEGEL, A., RANKIN, B. and MARTIN, W. T. (1966), *Differential Space, Quantum Systems and Prediction*, MIT Press, Cambridge (Mass.), chapter 5.
* WIGNER, E. P. (1963), The problem of measurement, *American Journal of Physics* **31,** 6–15.
‡ WOLFE, H. C. (1936), Quantum mechanics and physical reality, *Physical Review* **49,** 274.
ZIERLER, N. (1961), Axioms for non-relativistic quantum mechanics, *Pacific Journal of Mathematics* **11,** 1151–1169.
† ZINNES, I. I. (1958), Hidden variables in quantum mechanics, *American Journal of Physics* **26,** 1–4.

Index

Index